Electronic Gadgets for the Evil Genius

Electronic Gadgets for the Evil Genius

BOB IANNINI

McGraw-Hill
New York   Chicago   San Francisco   Lisbon
London   Madrid   Mexico City   Milan   New Delhi
San Juan   Seoul   Singapore   Sydney   Toronto

The McGraw·Hill Companies

Library of Congress Cataloging-in-Publication Data

Iannini, Bob.
  Electronic gadgets for the evil genius / Bob Iannini.
      p.   cm.
  Includes index.
  ISBN 0-07-142609-4
    1. Electronics—Amateurs' manuals. 2. Electronic apparatus and appliances—Design and
construction. I. Title.
  TK9965.I25   2004
  621.381—dc22   2004049070

12 13 14 15 QPD/QPD   0 1 0 9 8 7

ISBN 978-0-07-142609-1
MHID 0-07-142609-4

*The sponsoring editor for this book was Judy Bass and the production supervisor
was Sherri Souffrance. It was set in Times Ten by MacAllister Publishing Services, LLC.
The art director for the cover was Anthony Landi.*

Printed and bound by Quebecor/Dubuque.

This book is printed on recycled, acid-free paper containing a minimum of 50 percent recycled,
de-inked fiber.

McGraw-Hill books are available at special quantity discounts to use as premiums and sales promo-
tions, or for use in corporate training programs. For more information, please write to the Director
of Special Sales, McGraw-Hill Professional, Two Penn Plaza, New York, NY 10121-2298. Or contact
your local bookstore.

Contents

Contents

Acknowledgments

I wish to express thanks to the employees of Information Unlimited and Scientific Systems Research Laboratories for making these projects possible. Their contributions range from many helpful ideas to actual prototype assembly. Special thanks go out to department heads Rick Upham, general manager in charge of the lab and shop and layout designer; Sheryl Upham, order processing and control; Joyce Krar, accounting and administration; Walter Koschen, advertising and system administrator; Chris Upham, electrical assembly department; Al (Big Al) Watts, fabrication department; Sharon Gordon, outside assembly; and all the technicians, assemblers, and general helpers at our facilities in New Hampshire, Florida, and Hong Kong that have made these endeavors possible.

Also not to be forgotten is my wife Lucy, who has contributed so much with her support and understanding of my absence due to long hours in front of the computer necessary for preparing this manuscript.

I wish to acknowledge the contributions, including the front cover material, of Durlin Cox of Resonance Research in Baraboo, Wisconsin. Resonance Research is a supplier of large, museum-quality, electrical display devices.

My gratitude goes to Tim Ventura of American Gravitics and his contributions to the lifter craft technology found in Chapter 1 "Antigravity Project."

Finally, I would like to thank Zarir Sholapura of Zeonics and his contributions to high-voltage and high-energy research.

Special Notes

General Information

All projects in this book have been built and tested in our laboratories at Information Unlimited/Scientific Systems. The gadgets perform as described. Builders having difficulties may contact our technical help department.

Part Sources

Most parts are readily available through electronic supply houses, electrical supply houses, and hardware/builder supply outlets. Certain specialized and proprietary parts such as transformers, capacitors, *printed circuit boards* (PCB), tooled fabrication, and optical parts are manufactured in our shops. They are indicated with database numbers on the parts list and must be obtained by email or through our web site at www.amazing1.com. Without these special items, many of these projects would not be possible. Advanced builders may attempt substitution with the risk of reduced performance or failure.

Safety

Several projects are noted as being dangerous electrically, kinetically, and optically. Such projects must not be attempted unless the builder is experienced in his or her related field. Safety is stressed with all necessary compliances required for the trade.

Labeling is shown as required if the finished product is being offered to the trade. Labels are available and should be used if your project is to be exposed or demonstrated to the public.

Unfortunately, many scientific devices must use dangerous parameters to properly produce the desired target effects.

Printed Circuit Boards

Some projects are shown using PCB. They may also be built on vector or circuit boards. Attempt to follow the layout as shown when transposing component locations and wire routing.

Project Origins

Some of the devices shown are fall-out technology acquired from our project contracts as performed for the government and other related agencies. Others are the result of our own research efforts. We are constantly engaged in research for the trade and in many interesting research and development programs.

Other Available Options

Most of the projects are available as partial or complete kits and may be laboratory assembled upon request.

Our offshore assembly operation can competitively supply all types of specialty transformers, coils, and power supplies in volume for the trade.

Our engineering services specialize in high-voltage power supplies, shock and *electromagnetic pulse* (EMP) pulsers, Marx impulse generators, and all related components.

Contact Us

Information Unlimited
Phone Number: 603-673-6493
Fax Number: 603-672-5406
Web site: www.amazing1.com
Email: riannini@metro2000.net

Introduction

On a typical day, electronics enable us to turn on the TV or the computer, use the cellular phone, hear the "beeps and blips" of electronic toys and games, have a medical checkup, listen to the news on space exploration and world conflicts, even drive our car.

Electronic Gadgets for the Evil Genius presents a "hands-on" approach allowing the electronic enthusiast to construct many devices that are not as well known. These "action" projects demonstrate exciting and useful concepts of this diversified field.

Tesla Coils

The spectacular and highly visible effect produced by the Tesla coil has amazed and fascinated people for years. These high-frequency, high-voltage devices possess qualities unlike conventional electricity. Tesla output energy defies most insulation materials; transmits energy without wires; produces heat, light, and noise; and harmlessly passes through human tissue, causing virtually no sensation or shocking effect in the person.

Considerable research, money, and effort have been dedicated to the actual construction of similar, large Tesla coil-type devices capable of producing 200-foot lightning bolts and powering lights 25 miles away. Nikola Tesla was the originator of this research, along with countless other contributions to the electrical sciences. As more progress was made in these fields, it was soon realized that he indeed was responsible for many advances in the development of energy production.

Tesla is finally being credited for his work and is taking his rightful place as a truly great electrical genius in this field. His main theory of wireless energy transmission, however, is still much in question, and many dedicated groups are hopeful in obtaining some breakthrough that will resurrect it. The Tesla coil is the basis for much of this interesting research and still amazes all who come in contact with this visual and audible effect.

Plasma Devices

Electrical plasma can be anything from a simple electric arc from a welder to a complex entity of closed-circuit, toroidal flowing currents such as ball lightning. A nuclear explosion produces a plasma; the sun and a simple fire are forms of plasma. Plasma guns may be the weapons of the future, and plasma propulsion may power spacecraft to near the speed of light. Plasma confinement may be a key to fusion energy, and rotating plasma fields may hold the secrets to levitation devices such as hover boards. Gas displays using energized plasma can also provide spectacular visual effects.

Lasers

Lasers first appeared in the early sixties as a crystalline rod of aluminum oxide (ruby) placed within a helical flash tube. A high-powered flash of light caused a stimulated emission of a powerful light pulse within the ruby rod, capable of punching holes in the hardest of metals. Since then, the laser (an acronym for *light amplification by stimulated emission of radiation*) has been a part of our lives from laser printers and recorders to complicated eye surgery and mega-power-directed beams of energy protecting us from a potential missile attack. To the experimenter, the laser possesses an almost magical property due to its ability to transport energy over a distance.

Lasers are classified in relation to their power output and are closely regulated and labeled:

- *Class 2* lasers can produce up to 1 milliwatt of optical power. Some popular applications include alignment, intrusion alarms, and point-to-point communications. Even though this class of lasers has the lowest power, a Class 2 laser pointing in your direction can appear as the brightest object at distances over 10 miles.

- *Class 3a* lasers can produce up to 5 milliwatts. Some applications include laser pointers, gun sights, disc readers, holography, small light shows, and other visual effects.

- *Class 3b* lasers can produce up to 500 milliwatts and are used for printing, disk burning, range finding, target designation, and night vision illumination to name a few applications.

- *Class 4* lasers have unlimited power output. They are the workhorses of the group, with the capability to work with many materials from the hardest of metals to simple wood engraving. Class 4 laser surgery now provides a precision never before thought possible, and eye surgery for cataract removal is now sworn by all who have it done. High-power light shows similar to those at Walt Disney World use Class 4 optical lasers. Bluish-green lasers easily penetrate seawater because of compatible colors allowing covert point-to-point communications. By penetrating seawater for submarine communications, the concentration of energy into a micro-sized area opens up high-temperature and fusion research. Projecting energy over great distances can power powering interplanetary spacecraft. The integration of multiple micro-sized plasma explosions may provide the magical levitating vehicles often depicted as spacecraft. Also, weapons that use directed beam energies into the megajoules in battles against aircraft, ground vehicles, and other difficult targets are now possible. Antipersonal weapons designed to neutralize and disable personel use energies into the kilojoules using timed, pulsed laser diodes have kill ranges well in excess of one mile and are backpackable. These systems are lightweight and require complex optical conditioning.

Many high-powered lasers use carbon dioxide as the laser's medium. These devices are efficient, transmit through air, and are easy to build. They can be made to generate continuous power output into the tens of thousands of watts. An experimenter that can easily build this type of laser that will burn and cut many materials Lasers can be are *pulsed,* obtaining enormous peak powers into the gigawatts. This power is not to be confused with energy, as the power pulses last for fractions of a second, whereas energy must be integrated over a 1-second period. A true measure of pulsed laser energy is by its output in joules.

Research in the field of lasers still remains very fertile with many new and exciting developments still yet to come.

Ultrasonics

The field of ultrasonics remains a relative gray area with few available hobbyist-level projects. Ultrasonic energy is produced by a piezoelectric or magnetostrictive transducer powered by a signal generator. Ultrasonics can be used for cleaning where a solvent transmits these vibrations dislodging unwanted materials and dirt. Plastic materials can be welded or cut by rapid vibrations, causing frictional heating. Acoustical ultrasonics is often used for discouraging animals against intruding into a certain area. It is also used for range finding and can be an excellent intrusion detection device.

High-sound-pressure, acoustical energy is very inefficient, owing to the physics of energy transfer between two surfaces of dissimilar densities. Standing waves impede this energy coupling and make it difficult to obtain high-decibel output. Energy transfer between two surfaces is optimized when both materials have like densities, which is why sound travels better through water. Air is many times less dense than a liquid and its lack of density therefore offers a greater challenge to the researcher in overcoming the problem of successful energy transfer. Nevertheless, sonic transducers are effectively used with horns and other means to vibrate as much air as possible.

The effects of high-sound-pressure sonic energy can provide an excellent low-liability deterrent to unauthorized intrusion. A vertical wall of pain can be generated, causing nausea, dizziness, and extreme paranoia, which usually discourages the intrusion. No after-effects are produced once out of the field. However, sound pressure levels exceeding 140 decibels per minute can be harmful and should be avoided.

Listening to low-level ultrasonic sounds can be interesting to a nature enthusiast. Many insects and mammals emit sounds well out of human hearing range. Many man-made devices, such as rotating machinery, generate high-frequency sound and enable the detection of leaking air, water, or leaking electricity in the form of corona usually indicating a potential fault. Thus, directional ultrasonic listening can be a valuable tool.

Electrokinetics

The properties of magnets have long fascinated man since the discovery of lodestones by the ancient Greeks centuries ago. Even in today's advanced technology, the ability to attract and repel magnetized objects still remains a mystery. Magnetism, in spite of its mysterious properties, is perhaps the most important force known to man. Without the knowledge of how to use magnets, everyday motors, transformers, communications, and most forms of transportation would be next to impossible. Electricity generation would not exist.

The Star Wars defense initiative has opened up many doors to the potential use of this technology. Electrokinetically launching objects at hypervelocities much like the effect of a meteor shower will create a destructive barrier to incoming ballistic missiles. The propelling of radioactive waste and many other materials safely into outer space, the potential levitation of terrestrial vehicles, and of course projectile-type weapons are other applications.

Even though magnetic properties do not give way to variances, they do manifest properties in different forms. Motors use magnets to produce rotating mechanical energy from electricity and, of course, the opposite where generators use magnets and rotating

energy to produce electrical power. Transformers take advantage of changing magnetic fields as a function of time, whereas relays and solenoids produce linear motion.

Electrokinetic accelerators utilize magnetic forces where a conducting and movable armature is placed between two parallel conducting rails. A force is now produced in the armature as a result of the interactive magnetic fields occurring around the armature and the current-carrying rails. Remember a current-carrying conductor produces a proportional magnetic field, which is basic high school physics.

Those who are familiar with vectors will recall the Lorentz JXB forces where a force is produced between a current-carrying conductor (the armature) and the magnetic B field produced in the rails. Acceleration of the armature occurs over the entire length of the rails and can reach unheard of velocities compared to chemical combustion.

Even though pulsed magnetics is not new, little information exists to provide a "hands-on" approach for the hobbyist or experimenter. Positive interest exists in this field for using this technology as a viable potential for the previously mentioned applications, as well as the nonevasive use of shockwaves in breaking up kidney and gallstones. Even a form of "magnetic destructors" is intended for use in robot wars and contests.

We therefore feel a "how-to" book demonstrating projects for the serious electronic and mechanical experimenter, as well as for technical interest groups, will prove to be popular. The projects here will fall within the realms of both the amateur and the serious experimenter. All the projects will contain a briefing of mathematical theory for those interested, along with a simple explanation of the operation. Individual chapters will have headings suggesting the required competence and experience of the reader, as well as any hazards that may be encountered. All the construction projects will also contain a full bill of materials with sources necessary to complete the device as described.

About the Author

Bob Iannini runs Information Unlimited, a firm dedicated to the experimenter and technology enthusiast. Founded in 1974, the company holds many patents, ranging from weapons advances to children's toys. Mr. Iannini's 1983 *Build Your Own Laser, Phaser, Ion Ray Gun & Other Working Space-Age Projects*, now out of print, remains a popular source for electronics hobbyists.

Part One

Chapter One

Antigravity Project

An antigravity project provides a means of levitating an object purely by electrical forces (see Figure 1-1). Motors, fans, jets, or magnets are not used. A propulsive thrust is created by the reactionary forces of an ion wind. This phenomenon is an excellent means for future transportation once a few engineering problems are solved, and a vehicle could operate in an almost frictionless environment.

Construction requires minimal electronic experience in building the electronics, as well as patience with a steady hand in constructing the craft. The project is presented into two sections, the ion power supply and the craft. Expect to spend between $25 to $50 for parts, noting many are available through Information Unlimited (www.amazing1.com). The complete parts list is at the end of the chapter as Table 1-1.

Theory of Operation

The following equations show motion obtained via the reactionary effect of a volume of air accelerated by electric charges. A thin, positively charged emitter wire is located in a charge that is in proximity to a smooth, attracting surface. Air particles are now charged in proximity to this thin emitter wire and are attracted to the negative space charge around the smooth surface. It appears that maximum thrust (or effect) requires moving as much air mass as fast as possible in a given amount of time, expressed as the following:

```
Thrust = mv/t   m = mass of air
v = velocity   t = time
```

The power input to produce this movement is related to $(\frac{1}{2}/mv^2)/t$ energy in joules.

If we now define system efficiency as the ratio of "power out" to "power in," we obtain

```
Eff = mv/t x 1/2 mv²/t = 2/v
```

Efficiency now becomes inversely proportional to the velocity of the air and therefore suggests the utilization of large masses or volumes at low velocities to be efficient. This is not to say that the effect on the maximum lifting capability follows these same guidelines.

It is known that air molecules and ions are elastic on impact at low velocities. High velocities have a tendency to cause molecular disassociation with accompanying secondary ionization.

This secondary ionization will cause a net decrease in the reactionary effect or thrust due to a reversal of direction of the now oppositely charged particles. The objective now becomes to move as much mass as possible at a low velocity or energy where the maximum amount of elastic collision takes place with a

Figure 1-1   *Antigravity lifter in flight*

minimum of destructive molecular disassociation and secondary ionization.

Your lifter requires a high DC voltage at a relatively low current. The driver power supply schematic is shown in Figure 1-2. It generates 30,000 volts at a load current of 1 *milliampere* (ma). This is usually sufficient to power lifters up to 36 inches per side when properly constructed using up to $1\frac{1}{2}$-inch emitter wire spacing. It easily powers the 8-inch unit shown in these instructions. Even though the current is low, improper contact can result in a harmless yet painful shock.

The output voltage of the driver is obtained using a Cockcroft Walton voltage multiplier with four to five stages of multiplication. Note this method of obtaining high voltages was used in the first atom smasher ushering in the nuclear age! This multiplier section

requires a high-voltage/frequency source for input. Input is supplied by a transformer (T1) producing 6 to 8 *kilovolts* (Kv) at approximately 30 *kilohertz* (kHz). You will note that this transformer is a proprietary design owned by Information Unlimited. The part is small and lightweight for the power produced.

The primary part of T1 is current driven through an inductor (L1) and is switched at the desired frequency by a *field-effect transistor* (FET) switch (Q1). The capacitor (C6) is resonated with the primary of T1 and zero voltage switches when the frequency is properly adjusted. (This mode of operation is very similar to class E operation.) The timing of the drive pulses to Q1 is therefore critical to obtain optimum operation.

The drive pulses are generated by a 555 timer circuit (I1) connected as an astable multivibrator with a

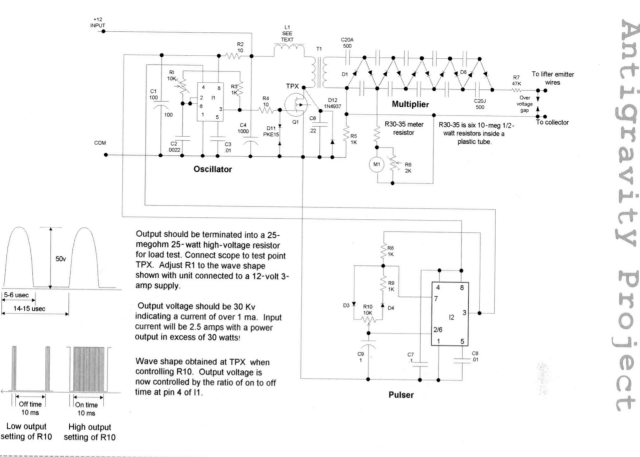

Output should be terminated into a 25-megohm 25-watt high-voltage resistor for load test. Connect scope to test point TPX. Adjust R1 to the wave shape shown with unit connected to a 12-volt 3-amp supply.

Output voltage should be 30 Kv indicating a current of over 1 ma. Input current will be 2.5 amps with a power output in excess of 30 watts!

Wave shape obtained at TPX when controlling R10. Output voltage is now controlled by the ratio of on to off time at pin 4 of I1.

Low output setting of R10   High output setting of R10

Figure 1-2   *Antigravity schematic*

---

repetition rate determined by the setting of the trim-pot (R1) and a fixed-value timing capacitor (C2).

I1 is now turned on and off by a second timer (I2). This timer operates at a fixed frequency of 100 Hz but has an adjustable "duty cycle" (the ratio of the on to off time) determined by the setting of the control pot (R10). I1 is now gated on and off with this controlled pulse providing an adjustment of output power.

An over-voltage protection spark gap is placed across the output and is easily set to break down at 20 to 30 Kv. This usually is sufficient for lifters having a 2- to 3-centimeter separation between the emitter wire and the collector skirts. Even though the output is short circuit protected against continuous overload, constant hard discharging of the output can cause damage and must be limited. A pulse current resistor (R7) helps to protect the circuit from these potential catastrophic current spikes.

Power input is controlled by a switch (S1) that is part of the control pot (R10). The actual power can

be a small battery capable of supplying up to 3 amps or a 12-volt, 3- to 5-amp converter for 115 uses. Power switch S1 must be added to the GRA1 series or you must use other external means to power control.

Construction Steps

This section discusses the construction of the electronic, ion-generating power supply and the lifter craft. The ion generator is built using a *printed circuit board* (PCB) that is individually available or you may use the more challenging *perforated circuit board*. The perforated board approach is more complicated, as the component leads must be routed and used as the conductive metal traces. It is suggested that you follow the figures in this section closely and mark the actual holes with a pen before inserting the parts. Start from a corner and proceed from left to right. Note that the perforated board is the preferred approach for science projects, as the system looks more homemade. The PCB only requires that you identify the particular part and insert it into the respective marked holes. Soldering is then greatly simplified.

Board Assembly Steps

To assemble the board, follow these steps:

1. Lay out and identify all the parts and pieces. Verify the separate resistors with the parts list, as they have a color code to determine their value. Colors are indicated on the parts list.

2. Fabricate a piece of .1-inch grid perforated board to a size of 7.2 × 2.6 inches. Locate and drill holes as shown in Figure 1-3. An optional PCB is available from Information Unlimited.

3. Fabricate the metal heat sink for Q1 from a piece of .063-inch aluminum at 1.5 × .75 inches, as shown in Figure 1-4.

4. Assemble L1 as shown in Figure 1-4.

5. If you are building from a perforated board, insert components starting in the lower left-hand corner, as shown in Figure 1-5 and 1-6. Pay attention to the polarity of the capacitors with polarity signs and all semiconductors. Route leads of components as shown and solder as you go, cutting away unused wires. Use certain leads as the wire runs or use pieces of the included #22 bus wire. Follow the dashed lines on the assembly drawing as these indicate connection runs on the underside of the assembly board. The heavy dashed lines indicate the use of thicker #20 bus wire, as this is a high-current discharge path and common ground connections. See Figures 1-7a and 1-7b for an expanded view.

6. Attach the external leads as shown. Figure 1-6 shows the construction of the safety spark gap made from pieces of #20 bus wire. This prevents high voltages from damaging circuit components when using light or no load connections. The circuit is not designed to operate with continual discharging and indicates a fault or too light of a load if it continually fires. See Figures 1-7a and 1-7b for an expanded view.

7. Double-check the accuracy of the wiring and the quality of the solder joints. Avoid wire

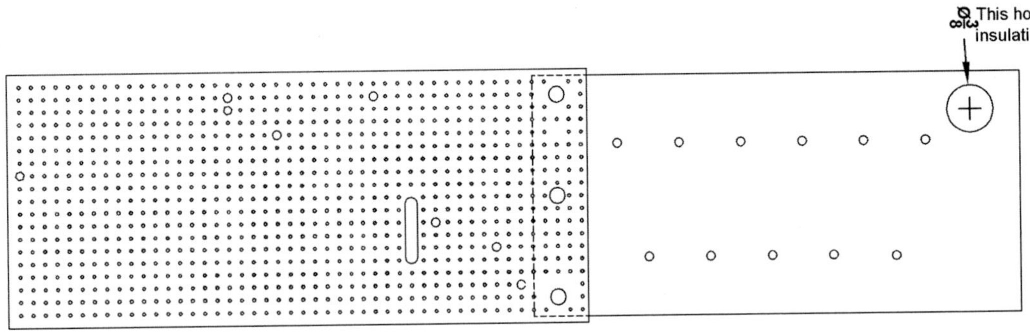

This hole is for plastic insulating guide tube.

The assembly board is in two sections attached together by two outer 6-32 nylon screws and nuts. The middle hole is used to fasten the entire assembly to the base of the enclosure.

The circuit section is 4.8" x 2.9" .1 x .1 perforated board. The high-voltage Plexiglas section is 3.8 x 2.9" .063 thickness. Drill eight .063" holes in the perforated section and eleven in the Plexiglas section located as shown.

Drill the three .125" holes in both sections for attaching together.

Drill and drag the .125" slot as shown. This cutout and the enlarged holes are for mounting transformer T1.

Using the optionally available printed circuit board will still require fabrication of the Plexiglas board.

Hole diameters are not critical.

Always use the lower left-hand corner of perf board for position reference.

Figure 1-3   *Driver board fabrication*

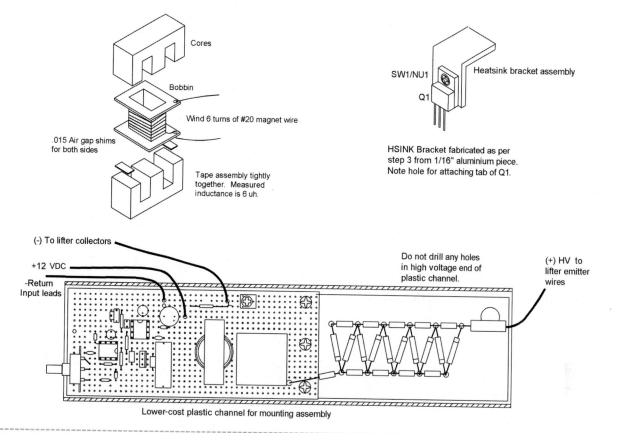

Figure 1-4 *L1 current feed inductor and heat sink bracket*

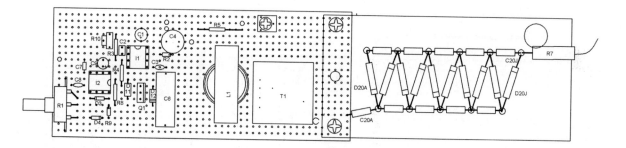

Note polarity of C1, C4, C9, D3, D4, D12, and D20A-D20J.

Note position of I1, I2, Q1.

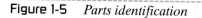

Figure 1-5 *Parts identification*

bridges, shorts, and close proximity to other circuit components. If a wire bridge is necessary, sleeve some insulation onto the lead to avoid any potential shorts.

8. Fabricate a channel from a piece of 1/16-inch plastic material. Add it to the assembly and secure it at its corners using silicon rubber adhesive. You may also enclose it in a suitable plastic box as shown in Figures 1-8 and 1-9. Figure 1-10 shows the simplified channel enclosure that does not include the meter M1.

See Figures 1-7a and 1-7b for enlarged views of this figure.

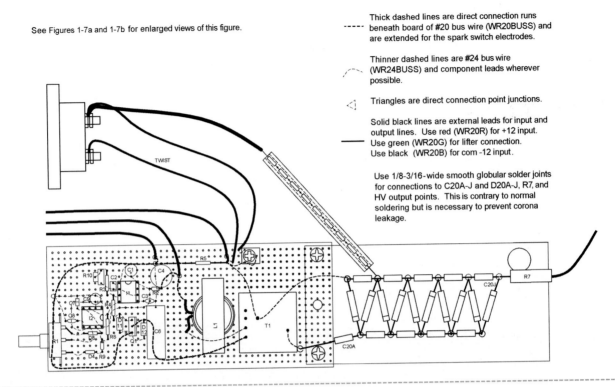

---- Thick dashed lines are direct connection runs beneath board of #20 bus wire (WR20BUSS) and are extended for the spark switch electrodes.

Thinner dashed lines are #24 bus wire (WR24BUSS) and component leads wherever possible.

Triangles are direct connection point junctions.

Solid black lines are external leads for input and output lines. Use red (WR20R) for +12 input. Use green (WR20G) for lifter connection. Use black (WR20B) for com -12 input.

Use 1/8-3/16-wide smooth globular solder joints for connections to C20A-J and D20A-J, R7, and HV output points. This is contrary to normal soldering but is necessary to prevent corona leakage.

Figure 1-6 *Wiring connections and external leads*

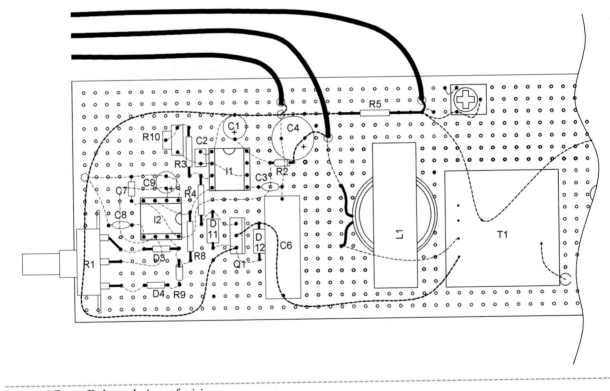

Figure 1-7a *Enlarged view of wiring*

Chapter One

Cut and paste 7a and 7b together.

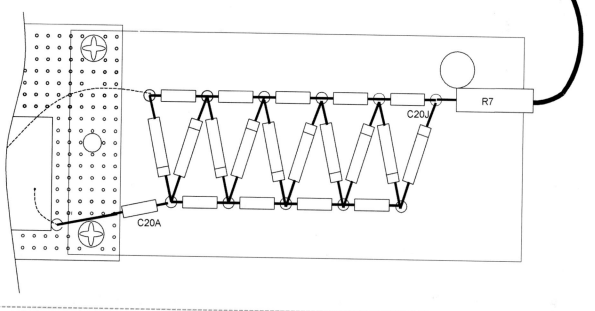

Figure 1-7b *Enlarged view of wiring*

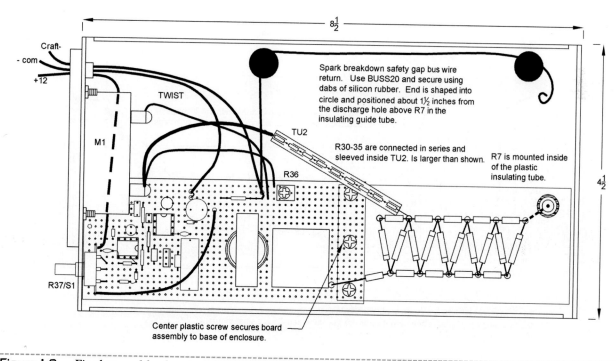

Spark breakdown safety gap bus wire return. Use BUSS20 and secure using dabs of silicon rubber. End is shaped into circle and positioned about 1½ inches from the discharge hole above R7 in the insulating guide tube.

R30-35 are connected in series and sleeved inside TU2. Is larger than shown.

R7 is mounted inside of the plastic insulating tube.

Center plastic screw secures board assembly to base of enclosure.

Figure 1-8 *Final assembly showing metered enclosure*

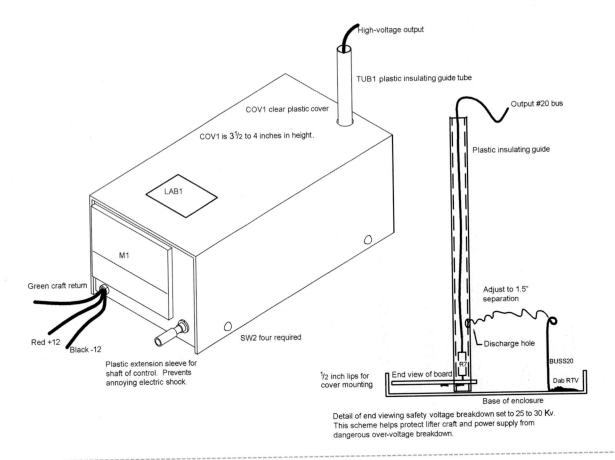

High-voltage output

TUB1 plastic insulating guide tube

COV1 clear plastic cover

COV1 is 3½ to 4 inches in height.

Output #20 bus

Plastic insulating guide

LAB1

M1

Adjust to 1.5"
separation

Green craft return

Discharge hole

Red +12

Black -12

BUSS20

SW2 four required

Dab RTV

Plastic extension sleeve for
shaft of control. Prevents
annoying electric shock.

½ inch lips for
cover mounting

End view of board

R7

Base of enclosure

Detail of end viewing safety voltage breakdown set to 25 to 30 Kv.
This scheme helps protect lifter craft and power supply from
dangerous over-voltage breakdown.

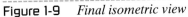

Figure 1-9   *Final isometric view*

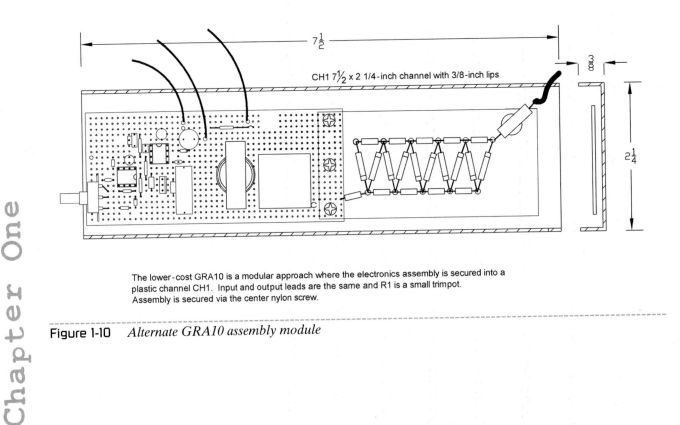

7½

CH1 7½ x 2 1/4-inch channel with 3/8-inch lips

3/8

2¼

The lower-cost GRA10 is a modular approach where the electronics assembly is secured into a
plastic channel CH1. Input and output leads are the same and R1 is a small trimpot.
Assembly is secured via the center nylon screw.

Figure 1-10   *Alternate GRA10 assembly module*

Chapter One

Testing Steps

The following steps are for testing out what you've built:

1. Preset trimpot R1 to midrange and R10 to full *clockwise* (CW). Set the spark gap to 1 to 1 ¼ inches, as shown in Figure 1-9.

2. Obtain a 25-megohm, 20-watt high-voltage resistor. You can make one by connecting 25 1-meg, 1-watt resistors in a series and sleeving them into a plastic tube. Seal the ends with silicon rubber.

3. Obtain a 12-volt, DC, 3-amp power converter or a 12-volt, 4-amp rechargeable battery.

4. Connect up the input to the power converter and output to the 25-meg load resistor. Connect an oscilliscope set to read 100 volts and a sweep time of 5 usecs to drain pin of Q1 for viewing the signal wave shape as R1 is adjusted.

5. Apply power and quickly adjust R1 to the wave shape shown in Figure 1-2. The spark gap may fire intermittently and should be respaced just to the point of triggering. This is usually between 25 to 30 Kv.

6. Rotate R10 *counterclockwise* (CCW) and note the input current smoothly dropping almost to zero. This control varies the ratio of off to on time and nicely controls the system current to the lifter craft emitters.

This "off and on" switching provides a varying, realistic throbbing sound as the craft lifts and produces more thrust.

Note: If you have access to a high-voltage probe meter, such as a B&K HV44, it will be possible to measure the direct output, noting 20 to 30 Kv across the 25-meg load resistor. This equates out to over 30 watts. You will see a smooth change in output as R10 is varied.

Note: The spark gap spacing adjustment and the spacing between the craft ion emitter wires and collectors are strongly dependent on one another. Fine-tune the system by setting the gap on the threshold

of firing before the craft emitter wire starts to break down. Do not exceed 1½ inches (38 mm).

The primary objective of the protective spark gap is to prevent damage to the craft as well as to the power supply circuit. Never allow a continuous breakdown to occur, as damage to the circuit may result.

You may now proceed to the craft assembly section. Figure 1-11 shows the assembly of a suggested launching pad.

The following information provides a step-by-step description of the methods and procedures involved with building a prototype electrokinetic propulsion device that is easily powered by the device shown in these plans. If properly constructed, this device will generate enough force to levitate itself from a resting surface.

1. Obtain the required materials:

   - 2 mm by 6 mm balsa wood strips
   - 30-gauge enameled copper magnet wire for high-voltage power leads
   - 42- to 44-gauge stainless steel wire (for lifter corona wire on the parts list)
   - Aluminum foil
   - One tube of superglue or a hot-glue gun
   - Sewing thread
   - One hobby knife
   - One Scotch brand tape roll

2. Cut the balsa support struts (see Figure 1-12). First, cut the balsa strips in half to create 2-millimeter by 3-millimeter strips. Cut these into two sets: one set of three struts 20 centimeters in length, and a second set of two struts 11 centimeters in length. Bevel the edges of each of the 20-centimeter struts to allow them to be glued later at an angle to the 11-centimeter struts. The beveling should be about 30 degrees in slope, and remember to bevel both ends on the same side of the balsa face.

3. Assemble the balsa struts (see Figure 1-13). Mark each of the 11-centimeter struts at the top (to help you remember which end is up)

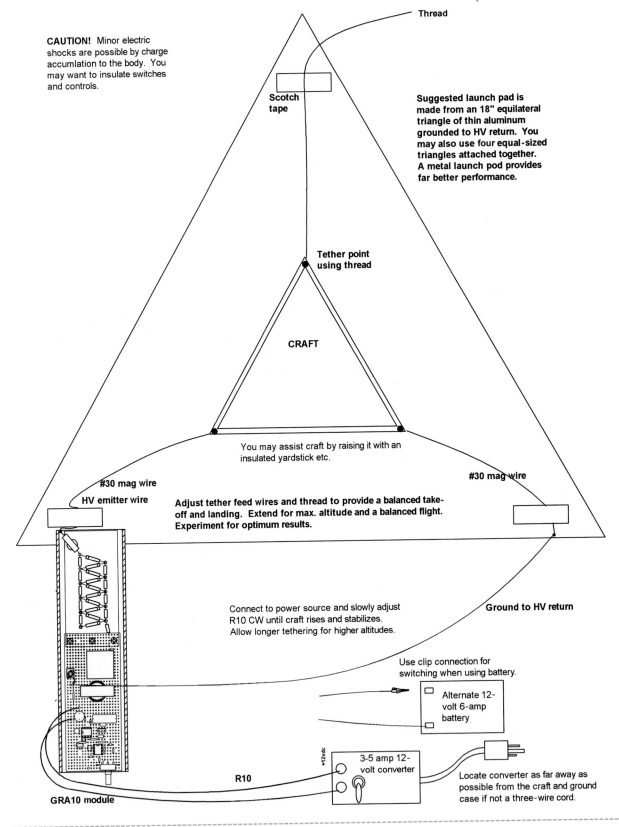

CAUTION! Minor electric shocks are possible by charge accumulation to the body. You may want to insulate switches and controls.

Thread

Scotch tape

Suggested launch pad is made from an 18" equilateral triangle of thin aluminum grounded to HV return. You may also use four equal-sized triangles attached together. A metal launch pod provides far better performance.

Tether point using thread

CRAFT

You may assist craft by raising it with an insulated yardstick etc.

#30 mag wire

#30 mag wire

HV emitter wire

Adjust tether feed wires and thread to provide a balanced take-off and landing. Extend for max. altitude and a balanced flight. Experiment for optimum results.

Ground to HV return

Connect to power source and slowly adjust R10 CW until craft rises and stabilizes. Allow longer tethering for higher altitudes.

Use clip connection for switching when using battery.

Alternate 12-volt 6-amp battery

+12vdc

3-5 amp 12-volt converter

R10

GRA10 module

Locate converter as far away as possible from the craft and ground case if not a three-wire cord.

Figure 1-11   *Final setup showing launching pad*

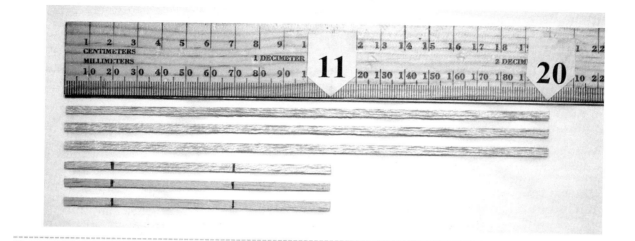

Figure 1-12   *Balsa wood struts*

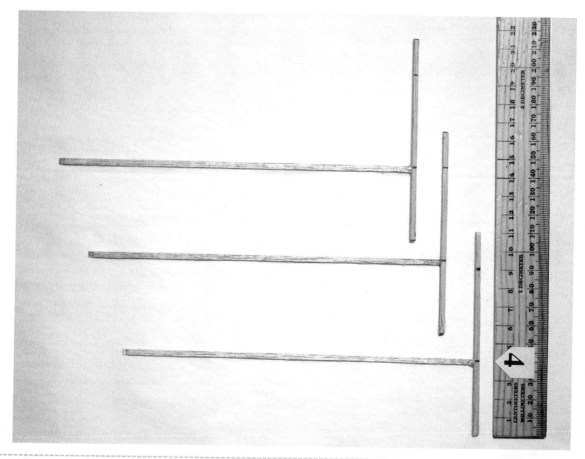

Figure 1-13   *Assembly of balsa wood struts*

and again at a mark 4 centimeters from the bottom. Sparingly use superglue to attach each of the three vertical 11-centimeter struts to a 20-centimeter horizontal strut as shown in the figure. In the figure, the beveled ends of the 20-centimeter struts have been glued at

right angles at the 4-centimeter mark on the vertical struts.

4.   Complete the chassis assembly (see Figure 1-14). Similar to the previous step, glue together the three pieces of the lifter frame using

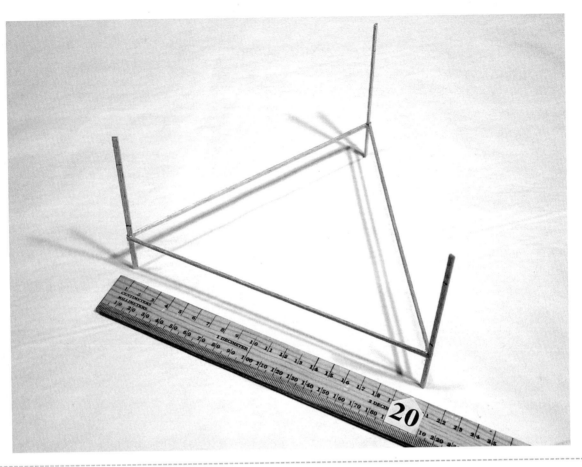

Figure 1-14 *Assembled chassis frame*

superglue. Glue the unconnected ends of the 20-centimeter struts to the other side of the 4-centimeter mark you created on the vertical strut. Ensure that the ends of the lifter line up as shown in the figure.

5. Cut an aluminum foil strip (see Figure 1-15). Cut a strip of aluminum foil 5 centimeters wide and approximately 1 meter in length. This foil strip will be used to surround the bottom part of the lifter.

6. Fold the foil around the chassis (see Figure 1-16). Put glue on the 20-centimeter strip and hold it on the foil until it sets. Notice in the figure how the foil is even with the bottom of the vertical struts. If done correctly, you should have an extra 1 centimeter of foil above the 20-centimeter balsa horizontal strut. Roll the lifter chassis down the length of the foil, gluing each side of the chassis as you go. You must have an extra 1-centimeter lip above the horizontal struts to reduce ion leakage.

7. Fold down the foil edges (see Figure 1-17). Cut the corners around the top of the 1-centimeter lip above the horizontal struts and fold the foil over the top of the strut for each of the three lifter sides. Use a piece of Scotch tape cut in half lengthwise to hold the folded corners as close to the inside of the foil as possible and reduce leakage.

8. Attach the ground wire to the foil (see Figure 1-18). Poke a small hole through the foil skirt and run the ground wire through it, as shown in the figure. The hole should be behind the strut so that the wire is supported by it. Make sure to strip the ground wire of its enamel coating before you connect it. The ground wire must have a section of bare copper to contact the foil in order to work. Give yourself about two extra feet of wire off the lifter to connect it to your power supply's ground.

9. Attach the corona wire (see Figure 1-19). Approximately 3 centimeters up from the top of the foil, or about 2 centimeters from the

Figure 1-15   *Cutting of alumunum foil strip*

Figure 1-16   *Gluing of alumunum strip to frame*

Figure 1-17   *Folding over the aluminum foil edges*

Figure 1-18   *Connection of the ground feed wire*

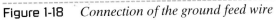

Figure 1-19   *Attaching and connecting of the corona ion emitter wire*

top of the vertical support struts, run a length of 40-gauge, 2.8-millimter corona wire around all three vertical struts and connect it to the 30-gauge power supply wire to be connected to the power supply. Make sure to loop the wire around each of the vertical struts at least once to ensure that they stay in place, and when you come back to the first vertical strut, tie the wire off so that the corona wire runs around the entire frame of the lifter.

If you have correctly followed the steps, you should have a lifter prototype identical to the one shown in Figure 1-20. Use the Testing Guide document to assist you with testing the lifter, and use the Troubleshooting document if you encounter problems while testing.

Notes

Attaching the corona wire can be difficult to hold in place on the vertical struts, especially because the wire is so thin. One method is to use a tiny dab of hot glue to keep the wire in place at each post, thereby freeing your hands to wrap the wire around the next post.

Another way to hold the wire in place is to wrap it around the 30-gauge power-lead wire and then wrap the tip of the 30-gauge wire around the vertical strut. Using this method you can attach both wires to each other and the vertical post at the same time. Ensure that the tip of the 30-gauge wire is completely stripped of enamel before wrapping the corona wire around it.

Tethering Setup

To complete the necessary tethering (see Figure 1-21), follow these steps:

1. Attach tethers to the lifter. Using common household thread, attach three tie-down threads to the lifter, one at each corner of the lifter tied in a double knot around the vertical balsa strut at the point where the strut meets the foil skirt wrapped around the lifter. Cut

Figure 1-20 *Completed lifter assembly*

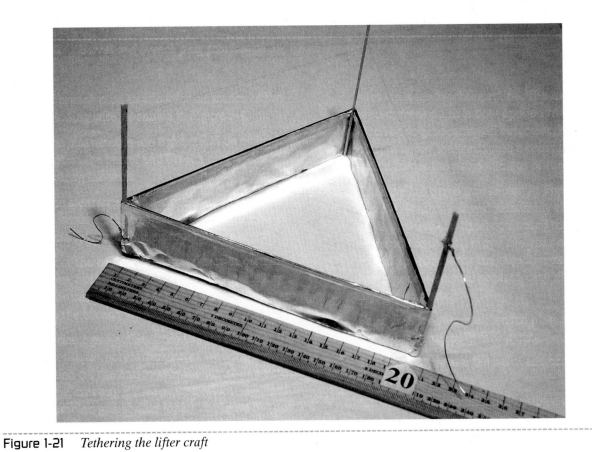

Figure 1-21 *Tethering the lifter craft*

the thread lengths at approximately 1 foot in length each.

2. Attach tethers to the test surface. Use Scotch tape to attach the lengths of thread to the test surface. Place the thread under the tape with approximately 8 inches of thread extending from the tape in order to provide enough slack for the lifter to take off. When you have done this, you should be able to raise the lifter approximately 8 inches in height from the test surface and have all of the thread become taut at the same height. Note in Figure 1-21 how the three tie-downs (colored pieces of tape) are attached to the thread and the test surface is very close to the actual lifter itself. Slack can be adjusted so that the lifter remains stable once it has taken off from the test surface.

Lifter Troubleshooting Guide

This section provides in-depth information on troubleshooting a prototype lifter in the event that it does not work correctly during testing.

Problem 1: The Lifter Is Not Moving, and I Don't Hear a Hissing Noise

If you don't hear a hissing noise, this means the lifter is not receiving high-voltage power from the high-voltage supply. Try these suggestions:

- Unplug the high-voltage supply and wait for the high-voltage charge to dissipate.

- Check the ground connection to the lifter and make sure you have good contact to both the foil skirt and the high-voltage supply's ground wire.

- Make sure the crimped end of your corona wire is hooked snugly around the electrode that comes out of the high-voltage supply.

- Ensure that the enamel coating has been removed from the ends of the ground wire. You can do this by scraping the enamel off with the edge of a hobby knife.

Problem 2: The Lifter Is Not Moving, and the High-Voltage Supply Seems Completely Dead

Have you tried checking the power supply to the high-voltage supply?

- Check the high-voltage supply's on/off switch.

- Check the 12-volt supply that provides power to the high-voltage supply.

- Check the on/off switch on the power strip if you are using one.

Problem 3: The Lifter Is Not Moving, and I Do Hear a Hissing Noise

The hissing noise (and slight breeze) means that the lifter is getting high-voltage power from the high-voltage supply. One of the following suggestions might eliminate the problem:

- Unplug the high-voltage supply and wait for the high-voltage charge to dissipate.

- Check the ground-wire connectivity to make sure it has a connection back to the high-voltage supply.

- Try sliding the entire corona wire down the vertical balsa struts a little until it is closer to the foil skirt. This reduces the distance between the two capacitive elements and increases the lifting power.

Problem 4: The Lifter Is Moving a Little but Is Not Taking Off

The lifter either does not have enough power to lift off or the high-voltage capacitance on the foil skirt is causing an attraction to the test surface. Try one of the following:

- Leave the high-voltage supply plugged in and the power turned on.

- Try gently blowing on the lifter until it moves. This may jostle the lifter off the capacitive spot that it sits on and it may suddenly lift off without warning.

- Try using a long nonconductive piece of plastic to gently lift one edge of the lifter up from the testing surface.

- Check the high-voltage and ground wires to ensure that they are off the testing surface. They may also cause static cling to the testing surface.

Problem 5: The Lifter Is Moving but Has a Lot of Electrical Arcing

The electrical arcing on your lifter is preventing enough capacitance to build up to allow a sustainable lift.

- Unplug the high-voltage supply and wait for the high-voltage charge to dissipate.

- Try sliding the entire corona wire up the vertical balsa struts a little (less than 1 centimeter at a time) until it is farther away from the foil skirt. This increases the distance between the two capacitive elements and it reduces lifting power, but it should also reduce electrical arcing and make liftoff possible.

- Ensure that the foil skirt has been folded over the top of the balsa horizontal strut and has been taped with a piece of Scotch tape on the other side. If the foil has not been folded over

the top of the horizontal balsa strut, it will not have proper surface area capacitance.

- Ensure that no sharp edges are present that would lead to ion leakage on the foil skirt of the lifter. Also ensure that no sharp edges are sticking off the end of the corona wire as they will reduce capacitance.

Problem 6: The Lifter Is Lifting off, but when Arcing Occurs It Suddenly Drops or Loses Height

Sudden drops in lift height may accompany large, electrical arcs between the foil and the corona wire due to a rapid reduction in capacitance in the corona wire and foil skirt of the lifter, which in turn reduces thrust. See Problem 5 for details on troubleshooting this.

Problem 7: The Lifter Is Lifting off and "Bouncing" the Tether, but Thrust Is Not Stable

If the lifter is bouncing at the end of the tether or rapidly swaying back and forth, it is due to irregularities in thrust versus positioning.

- Unplug the high-voltage supply and wait for the high-voltage charge to dissipate.

- Check each of the tethering threads and make sure they are all the same length when fully extended.

- Check each of the tethering threads to make sure they are a reasonable length (8 to 14 inches).

- Check the high-voltage and ground wires to make sure they aren't snagging on something during liftoff. You may try taping the high-voltage and ground wires to a nonconductive surface about 10 inches from the top of the testing surface. (Note: Make sure these wires *do not* cross or else they will short out your power supply.)

- Make sure all three corners of the lifter are about the same weight. Note that the corners with wires attached will weigh a little more but should not be noticeably heavier.

Problem 8: Arcing Is Occurring Between the Ground Wire and the HV Wire

Electrical arcing between the ground wire and the high-voltage wire occurs because the wires are too close. This does not include the arcing that occurs on the lifter itself, which is covered under Problem 5.

- *Immediately* unplug the high-voltage supply and wait for the high-voltage charge to dissipate.

- Move the ground and high-voltage wires farther apart from each other to prevent future arcing.

Note that electrical arcs between these two wires can cause permanent damage to your high-voltage supply power source.

Problem 9: The Lifter Lifts off and Immediately Shorts the HV and Ground Wires

Electrical shorts between the high-voltage and ground wires can occur when the position of these wires changes as the lifter lifts off from the test surface.

- *Immediately* unplug the high-voltage supply and wait for the high-voltage charge to dissipate.

- Reposition the high-voltage and ground wires so that they are less likely to touch and cause an electrical short during liftoff. Taping the wires in place about 10 inches from the testing surface can also reduce the movement of these wires during liftoff.

Problem 10: The Lifter Lifts off and Immediately Pulls to One Side

The lifter may pull to one side when it lifts off, meaning that instead of moving up in a reasonably straight manner from the testing surface, it takes off and flies in a particular direction. This occurs because of weight or thrust instability and is usually due to weight on the corners of the lifter due to incorrectly positioned high-voltage and ground wires.

- Unplug the high-voltage supply and wait for the high-voltage charge to dissipate.

- Try shortening or lengthening the tethering threads to position the lifter where it should be at maximum height.

- Try shortening all three tethering threads to a reasonable length by reducing them approximately 2 inches.

- Ensure that the high-voltage and ground wires are suspended from the testing surface so that they do not add excess weight to the corners of the lifter.

- Ensure that the aluminum foil skirt is reasonably straight and not crumpled around the entire frame of the lifter and that it is cut to the same height all the way around the frame.

- Ensure that the corona wire maintains the same distance from the foil skirt around the entire frame of the lifter and is pulled reasonably tight between vertical balsa struts.

Problem 11: The Corona Wire Flutters During Liftoff and Causes Thrust Problems

The corona wire may flutter during liftoff due to electrostatic forces.

- Unplug the high-voltage supply and wait for the high-voltage charge to dissipate.

- Try straightening the corona wire by hand until it is lined up reasonably straight above the foil skirt.

- Try removing the corona wire from the lifter and rewrapping it with a new piece of wire stretched firmly between posts. (Remember that the posts are balsa and cannot take much stress.)

Problem 12: The Foil Skirt Flutters During Liftoff and Causes Thrust Problems

A reasonable amount of aluminum skirt flutter may occur during testing due to electrostatic forces; however, an excessive amount may cause thrust problems.

- Unplug the high-voltage supply and wait for the high-voltage charge to dissipate.

- Try straightening the corona wire by hand until it is lined up reasonably straight above the foil skirt.

- Try using small dabs of superglue to hold the bottom of the skirt in place on the bottom of the vertical balsa support struts. Be aware that weight is critical and superglue may add too much weight if it is used excessively.

Problem 13: The Lifter Hisses but Just Will Not Lift Off

If you've gone through this entire troubleshooting document and still not found out what the problem is, it could be that your lifter is just too heavy. Some ways to correct weight problems include the following:

- Shaving excess balsa from the frame

- Trimming aluminum foil off the lifter (trim even strips around the bottom to remove the foil)

- Supporting the high-voltage and ground wires off the test surface to reduce wire weight

- Ensuring that you used the correct gauge of stainless steel corona wire

- Shortening the overall length of the high-voltage and ground wires to reduce electrical resistance

- Ensuring that you've used balsa wood and not basswood to construct the lifter (basswood looks almost identical but has a smoother consistency and more almond-like color and texture)

- Ensuring that you haven't used heavy-duty aluminum foil for the foil skirt

Author's Note: General Troubleshooting Guidelines

Here are some extra thoughts on troubleshooting that don't really fit well in any of the previous categories but may assist you during testing:

- First, there needs to be a capacitance between the corona wire and the foil skirt. If sharp edges are present or if these elements are just too close, then too much charge transfer will occur and capacitance will be lost. This is the main reason why the foil is folded over the top of the horizontal balsa support struts: to reduce sharp edges and increase capacitance. This is also why several centimeters of distance occur between the corona wire and the foil skirt. This also reduces the charge transfer and increases capacitance.

- Second, the charge transfer between the corona wire and the foil skirt needs to occur for the Biefeld-Brown effect to work properly. If the corona wire is too heavily shielded due to the wire's enamel coating, the charge transfer will not occur. Similarly, if the distance between the corona wire and the foil skirt is too great, the charge transfer will not occur. Please note that the charge transfer seems to be required only when the device is tested in an atmosphere, because the lifter design has been successfully tested in a vacuum environment.

- Make sure your lifter weighs as little as possible. The required weight appears to be about 2.6 grams maximum, although that is not set in stone. It depends on construction, leakage current, and so on. The prototypes I've constructed weighed so little that I could barely tell when I was holding them; the wires seemed to outweigh the prototype itself.

- The lifter prototype requires time and a little finesse to make it work. I spent nearly two weeks building prototypes before I built my first working model. It only takes me about 20 minutes to build a working model at this point.

The lifter has a tendency to cause attraction to the test surface. This is static cling from the capacitive charge on the foil skirt to the testing surface. You can reduce this by putting something underneath the lifter (I use a plastic straw) to prop it up from the surface during testing.

Table 1-1 Complete parts list

Ref. #	Qty.	Description	DB #
R1		10K trimpot vertical	
R2, 4	2	10-ohm, ¼-watt resistor (br-blk-blk)	
R3, 5, 8, 9	4	1K, ¼-watt resistor (br-blk-red)	
R7	1	47K, 1-watt resistor (yel-pur-or)	
R10		10K trimpot vertical	
C1		100 mfd/25-volt vertical electro-radial leads	
C2		.0022 *microfared* (mfd)/50-volt greenie plastic cap (222)	
C3, 8	2	.01 mfd/50-volt disk (103)	
C4		1000 mfd/25-volt vertical electrolytic capacitor	
C20A–20J	10	500 pfd, 10 Kv ceramic disk cap	#500P/10KV
C6		.22 mfd/250-volt metalized polypropylene	
C9		1 mfd/25-volt vertical electro cap	
C7		.1 mfd/50-volt cap INFO#VG22	
D20A–20J	10	16 Kv, 5 ma avalanche diodes	#VG16
D3, 4	2	IN914 silicon diodes	
D11	1	PKE15 15-volt transient suppressor	
D12		1N4937 fast-switching 1 Kv diode	
Q1		IRF540 MOSFET transistor TO220	
I1, 2	2	555 DIP timer	
T1		Mini-switching transformer, 7 Kv 10 ma.	#IU28K089
L1 #IU6UH	2	6 Uh inductor; see text on assembly	
PBOARD		5- × 2.8- × .1-inch grid perforated board. Fab to size per Figure 1-3.	
PCGRA		Optional PCB	#PCGRA
WR20R	12 inch	#20 vinyl red wire for positive input	
WR20B	12 inch	#20 vinyl black wire for negative input	
WR20G	12 inch	#20 vinyl green wire for output ground to craft return	
WR20KV	4 inch	20 Kv silicon high-voltage wire for output	
WR20 BUSS	18 inch	#20-inch bus wire for spark gap and heavy leads (see Figure 1-8)	
WR24 BUSS	12 inch	#24 bus wire for light leads (see Figure 1-7a, b).	
WR30 MAG	36 inch	#30 magnet wire for tethering the craft	
M1		50-uamp meter shown with 3-inch face	
R30–35	6	10 meg, 1-watt resistor (br-bk-bl)	
R36		2 kHz trimpot for meter adjust	
R37/S1		10K pot and switch	
TOP1		Top 8 ½ - × 4 ½ - × 3 ½ - × ¹/₃₂ -inch black or clear plastic	
COV1		Bottom 8 ½ - × 4 ½ - × 3 ½ - × ¹/₃₂ -inch black or clear plastic	
TUB1		Insulating tube 5- × ³/₈ -inch-thick wall rigid plastic tubing	
TUB2		4- × ³/₈ -inch Tygon sleeving for R30–35	
BU1		³/₈ -inch clamp bushing	
SW2		#6 × ³/₈ sheet metal screws	
HSINK		1.5- × 1-inch .063 AL plate fabbed as per Figure 1-4	
SW1	1	6-32½ Phillips screw	
NUT	4	6-32 kep nut	
SW3	3	6-32 × ³/₈ -inch nylon screws	

Table 1-1 Complete parts list (continued)

Ref. #	Qty.	Description	DB #
WIRE2 MIL10	10 inch	2-millimeter stainless wire for ion emitters	#WIRE2MIL10
1MEG	20	1-meg, 1-watt resistors	#1MEG
12DC/3		115 Vac to 12 Vdc/3-amp converter	#12DC/3
BAT12		12-volt, 4-amp hour rechargeable battery	#BAT12
BC12K		Battery charger kit for above BAT12 (parts available though www.amazing1.com)	#BC12K
M1		50-uamp meter shown with 3-inch face	M1
R30–35	6	10-meg, 1-watt resistor (br-bk-bl)	R30–35
R36		2 kHz trimpot for meter adjust	R36
R37/S1		10K pot and switch	R37/S1

Chapter Two
Low-Power Electrokinetic Gun

This chapter covers a project intended for high school-level experimenters. Supervision is suggested when working in the classroom. The unit operates from a low-voltage source of 12 *volts direct current* (VDC) or a battery (see Figure 2-1). The circuit produces 72 joules of capacitive energy storage at 600 volts. This is labeled as a dangerous electrical device if the protective cover is removed. The unit must not be aimed or pointed in the direction of personal or breakable objects. Projectiles can reach a reasonably high velocity.

This is an advanced intermediate-level project requiring basic electronic skills. Expect to spend $35 to $45. All parts are readily available with specialized parts through Information Unlimited (www.amazing1.com) and are listed in Table 2-1 at the end of the chapter.

Basic Theory

Your electrokinetic accelerator demonstrates two methods of electrokinetic acceleration.

First and Preferred Method

A flat, pancake-shaped accelerator coil is structured to match the dimensions of a circular aluminum ring that serves as the launch vehicle. This closed ring now becomes a shorted secondary winding of a pulse transformer that is as closely coupled to the primary windings as possible. A current pulse is produced in the secondary that is the aluminum ring induced by the current pulse in the primary. The result is opposing magnetic fields that cause a mechanical repulsion pulse propelling the nonmagnetic aluminum ring. If the ring were magnetic, it would be attracted to the coil and completely counteract the repulsive force. It is the current flowing in the aluminum ring that causes the repulsive magnetic field (Lenz's law).

Figure 2-1   *Low-power electro-kinetic gun*

The design of the accelerator coil must minimize the leakage inductance of the coil assembly and limit the impedance in the primary discharge circuit. A disadvantage of this type of kinetic system is that the acceleration event occurs over a short distance interval. Therefore, the resulting peak forces must be very high to achieve a high velocity. The repulsive forces will vary as the inverse of the square of the separation distance between the accelerator coil and the moving projectile.

Second Method

This method uses a solenoid coil with a pitch equal to its diameter. This coil uses a magnetic pulse that attracts a small ball or rod-shaped projectile made from a magnetic material. The ball is accelerated into and along the coil axis, exiting with a velocity. The initial positioning of the ball is critical to obtain the maximum exit velocity. The inner diameter of the coil should only be large enough to allow unobstructed movement of the projectile ball. Again, magnetic leakage and pulse width plays an important part in achieving optimum results.

Circuit Description

A single-ended inverter circuit is shown in Figure 2-2 consisting of a self-oscillating transistor (Q1) that resonantly switches the primary of the voltage step-up transformer (T1) and tuning capacitor (C3). The input power is 12 VDC at 1 amp and is controlled by the charging switch (S1).

The base drive to Q1 is provided by the *feedback winding* (FB) of the T1 that must be properly phased. The bias for Q1 is provided by the resistor (R2). The capacitor (C2) speeds up the switching by providing a low-impedance path for the feedback signal. The resistor (R1) provides the starting bias for turning on Q1. The capacitor (C1) bypasses the switching frequency to ground by providing a low-impedance path. A charging choke (L1) is necessary to limit the switching current during the initial charging cycle of the storage capacitors.

The high-voltage/frequency output of the secondary winding of T1 is rectified by the diode (D1) charging storage capacitors (C5, C6) through the low-impedance winding of the accelerator coil (L2). Resistors (R7, R8) help to balance the voltage charge across the two individual capacitors. The charge across these capacitors is applied to L2 by the switching function of the SCR switch. This energy exchange generates the magnetic pulse necessary to launch the projectile. The SCR switch is controlled by the trigger switch (S2) applying a voltage to the gate of the SCR when ready to fire the device.

Board Assembly Steps

To assemble the board, follow these steps:

1. Lay out and identify all parts and pieces. Verify them with the parts list, and separate the resistors as they have a color code to determine their value. Colors are noted on the parts list.

2. Create a piece of .1-inch × .1-inch grid perforated board at 5 3/5 × 3 3/8 inches. Locate and cut out sections for pins of T1 as shown in Figure 2-3.

3. If you are building from a perforated board, insert components starting in the lower left-hand corner, as shown in Figure 2-3. Pay attention to the polarity of capacitors with polarity signs and all semiconductors. Route the leads of the components as shown and solder as you go, cutting away unused wires. Attempt to use certain leads as the wire runs. Follow the dashed lines on the assembly drawing as these indicate connection runs on the underside of the assembly board. The heavy dashed lines indicate the use of thicker #18 bus wire, as this is a high-current discharge path.

   Note that the Q1 transistor must be mounted so that it mounts flush to its mounting surface at a right angle. This step is important for proper heat sinking and mechanical stability.

4. Secure large storage capacitors, C5 and C6, to the board using silicon rubber cement (*room temperature vulcanizing* [RTV]) or tape them in place, as shown in Figure 2-3.

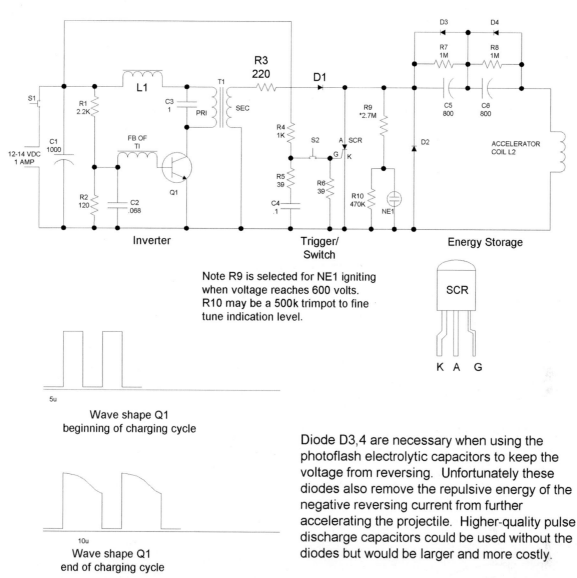

Figure 2-2   *Circuit schematic*

Note R9 is selected for NE1 igniting
when voltage reaches 600 volts.
R10 may be a 500k trimpot to fine
tune indication level.

Wave shape Q1
beginning of charging cycle

Wave shape Q1
end of charging cycle

Diode D3,4 are necessary when using the
photoflash electrolytic capacitors to keep the
voltage from reversing. Unfortunately these
diodes also remove the repulsive energy of the
negative reversing current from further
accelerating the projectile. Higher-quality pulse
discharge capacitors could be used without the
diodes but would be larger and more costly.

5. Wire in SCR1 using short pieces of bus wires
    for extensions through the holes of the board.

6. Attach the following leads, as shown in
    Figure 2-4:

    Three 3-inch leads of #20 (WR20) connected to
    S1 and S2

    Two 5-inch leads of #20 (WR20) for connected
    to NE1

    One 3-inch jump between D2 and + of C5

7. Double-check the accuracy of the wiring and
    the quality of the solder joints. Avoid wire
    bridges, shorts, and close proximity to other
    circuit components. If a wire bridge is neces-

sary, sleeve some insulation onto the lead to
avoid any potential shorts.

Fabrication

To create the fabrication, follow these steps:

1. Fabricate the chassis as shown in Figure 2-5.
    Be sure to make a trial fit and check the sizes
    of the panel components before actually mak-
    ing holes. Kits will usually contain a predrilled
    piece that will mate to the included parts.

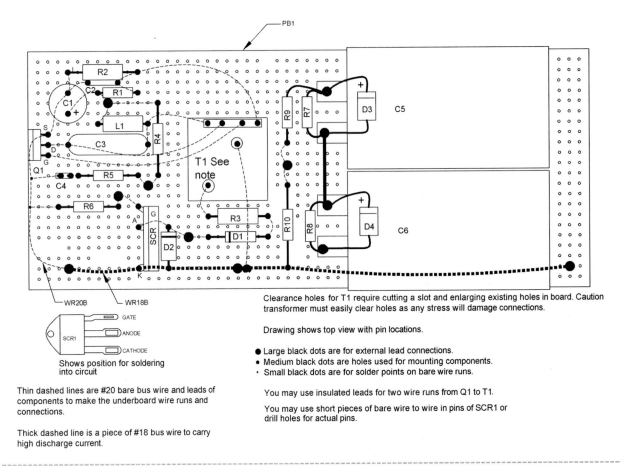

Clearance holes for T1 require cutting a slot and enlarging existing holes in board. Caution transformer must easily clear holes as any stress will damage connections.

Drawing shows top view with pin locations.

● Large black dots are for external lead connections.
● Medium black dots are holes used for mounting components.
· Small black dots are for solder points on bare wire runs.

You may use insulated leads for two wire runs from Q1 to T1.

You may use short pieces of bare wire to wire in pins of SCR1 or drill holes for actual pins.

Shows position for soldering into circuit

Thin dashed lines are #20 bare bus wire and leads of components to make the underboard wire runs and connections.

Thick dashed line is a piece of #18 bus wire to carry high discharge current.

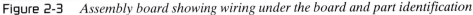

Figure 2-3 *Assembly board showing wiring under the board and part identification*

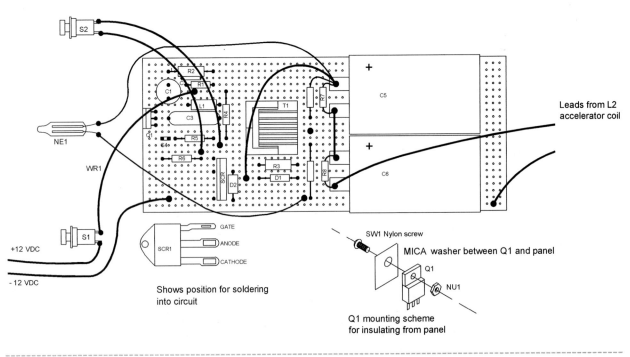

Figure 2-4 *Assembly board showing external wiring*

Fabricate from .063 aluminum as shown. Trial fit and verify diameters of components for mounting before fabbing in the holes.

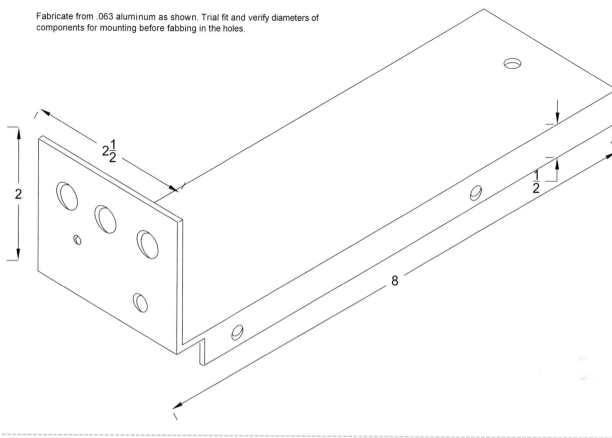

Figure 2-5 *Chassis fabrication*

2.   Fabricate a BK1 bracket, a MAND1 mandrel, and a FLY1 flyway, as shown in Figure 2-6.

3.   Add the cover as shown in Figure 2-7. Note to mate the clearance holes for screws with holes in the chassis section for the (SW4) #6 sheet metal screws.

4.   Wind the L1 accelerator coil, as shown in Figure 2-6. Sleeve in a 1/4-inch section of MAND1 mandrel into the FLY1. Drill small holes for the start and finish of the winding into the side of the bobbin as shown. The measured coil inductance should read 90 to 100 microhenries without the projectile, dropping down to 50 to 60 with the projectile in place. Connect in the coil leads as shown in Figure 2-4.

Note that the objective is to get the projectile body as close to the windings as physically possible. This provides maximum magnetic coupling, and consequently optimum performance.

5.   The final assembly is shown in Figure 2-8. Note the proper mounting of Q1 using a mica washer, shown also in Figure 2-3.

Electrical Pretest

To conduct a pretest, follow these steps:

1.   Connect a 1000-voltmeter across +C5 and − of C6. If you have a scope, you can check the wave shape on the collector of Q1 and verify it, as in Figure 2-2.

Slide the PROJ1 projectile onto the mandrel and point the unit away from people or breakable objects. For maximum velocity, the projectile must be evenly positioned onto the accelerator coil.

Connect 12 volts to the input and push switch S1. Note the voltmeter reading increases in value as long as the switch is held. A current meter on the 12-volt supply will indicate around an ampere of current and will vary through the charging cycle.

Allow the charging to reach 600 volts. Push switch S2 simultaneously while still holding S1

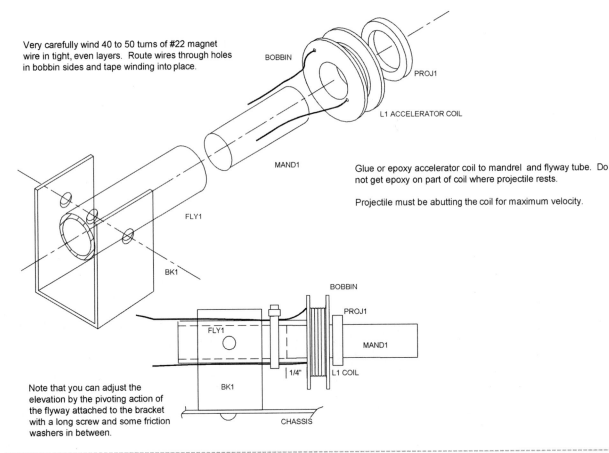

Very carefully wind 40 to 50 turns of #22 magnet wire in tight, even layers. Route wires through holes in bobbin sides and tape winding into place.

BOBBIN

PROJ1

L1 ACCELERATOR COIL

MAND1

Glue or epoxy accelerator coil to mandrel and flyway tube. Do not get epoxy on part of coil where projectile rests.

Projectile must be abutting the coil for maximum velocity.

FLY1

BK1

BOBBIN

PROJ1

FLY1

MAND1

1/4" L1 COIL

BK1

Note that you can adjust the elevation by the pivoting action of the flyway attached to the bracket with a long screw and some friction washers in between.

CHASSIS

Figure 2-6   *Bracket and flyway fabrication*

and note the projectile shooting off the mandrel.

The indicator lamp NE1 should light when the intended charge voltage level is reached. This is a maximum of 700 volts and should not be exceeded. The resistor divider, consisting of R9 and R10, sets this point. You may have to vary these values to get the desired indication at the charge voltage level you want. Without this lamp, you must time the charging cycle and coordinate it with an external meter to obtain the charging time or always use the meter.

Note that the velocity will increase after the first several shots. This is due to the capacitors polarizing their electrolyte.

Obtaining More Velocity (Under No Condition Point or Shoot Toward a Person)

Increase the mandrel diameter to $^{23}/_{32}$ and wind 40 turns of #22 magnet wire in six to eight even wound layers on your .75-inch diameter$\times$.312-wide nylon bobbin #BOBTHER. The measured inductance should be 120 to 150 microhenries. The objective here is to get the body of the projectile as close to all the copper windings as physically possible. This will reduce the reluctance and increase coupling. Further velocity increase can be obtained by removing several turns on the accelerator coil until the current reversing diodes start to blow. Our laboratory unit got down to 50 microhenries, obtaining a very dangerous velocity.

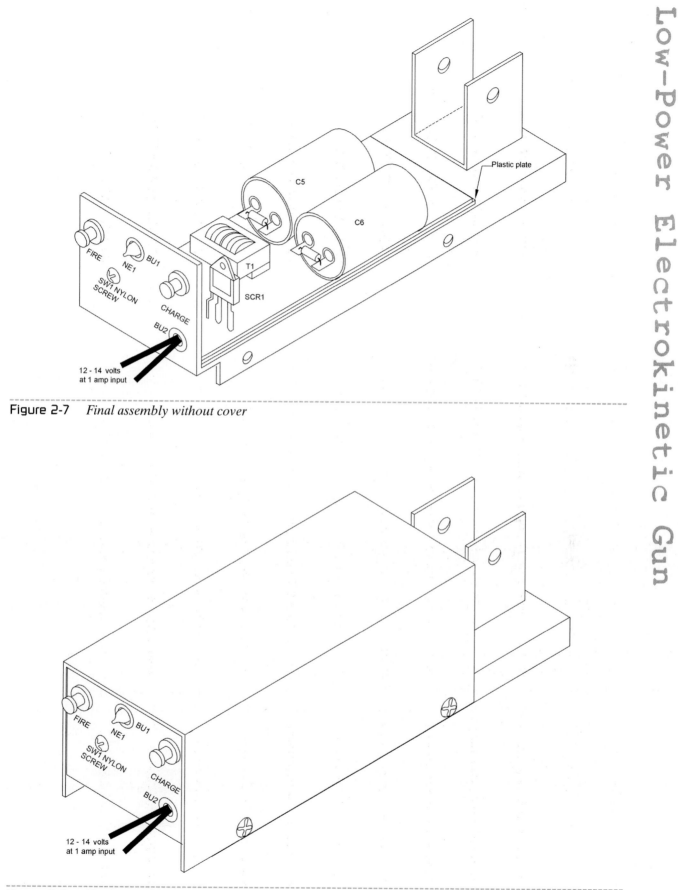

Figure 2-7   *Final assembly without cover*

Figure 2-8   *Final assembly*

Table 2-1 Parts list for the electrokinetic gun

Low-Power Kinetic Gun

Ref #	Qty.	Description
R1		2.2K, $^1/_4$-watt [red-red-red]
R2		120K 1-watt [br-red-br]
R3		220K $^1/_4$-watt [red-red-br]
R4		1K, $^1/_4$-watt [br-blk-red]
R5, 6	2	39K $^1/_4$-watt [or-wh-blk]
R7, 8	2	1 m, $^1/_4$-watt [br-blk-grn]
R9		750K, $^1/_4$-watt [pur-grn-yel]
R10		100K, $^1/_4$-watt [br-blk-yel]
C1		1000 m/25-volt vertical elect. capacitor
C2		.068 m/50-volt plastic capacitor
C3		1 m/250-volt *metal polypropylene capacitor* (mpp)
C4		.1 m/50-volt plastic capacitor
C5, 6	2	800 m/330-volt photo flash capacitor info database #800M/360
L1		3.6 Uh, 1.5-amp inductor [or-bl-blk] 3.6U
T1		Inverter transformer info database #28K074
S1, 2	2	Pushbutton switches SPST
D1		6 Kv, 10 ma high-voltage fast-recovery rectifier info database #VG
D2, 3, 4	3	11 Kv, 1-amp diode 1n5408
SCR1		80-amp scr info database #8070W
NE1		Neon indicator lamp
Q1		Mje3055 NPN TO 220 transistor
PB1		.1 grid perforated circuit board cut to 2.5 × 6 inches
WR18B	6"	#18 bus wire for discharge path (see Figure 2-3)
WR20B	12"	#20 bus wire for long leads on bottom of board

Ref #	Qty.	Description
WR20B	48"	#20 vinyl hookup wire for hookup of external parts
WR22M	10'	#22 magnet wire for winding accelerator coil
SW1		6-32 × $^1/_2$-inch nylon screw for mounting Q1
SW2		6-32 × $^1/_2$-inch screw for mounting bracket
SW3		6-32 × 1.5-inch screw for mounting flyway to bracket (see Figure 2-6)
SW4	4	#6 × $^1/_4$-inch screws for attaching cover to chassis
NUT		6-32 hex nuts
MICA		Mica insulation washer between Q1 and chassis (see Figure 2-4)
CHASSIS		Fabricate from $^1/_{16}$-inch aluminum, as shown in Figure 2-5
COVER		Fabricate from thin plastic or aluminum, as shown in Figure 2-8
BK1		Fabricate bracket (see Figure 2-6) from a 1- × 4- × $^1/_{16}$-inch aluminum or plastic
FLYWAY		2 × $^{11}/_{16}$-inch innerdiameter (ID) × $^1/_{16}$ wall plastic tube
MAND1		$^{11}/_{16}$- × 2-inch cold roll; must sleeve into flyway
PROJ		1- outerdiameter (OD) × $^3/_4$-inch ID $^1/_8$" wall alumunum tubing cut into $^1/_4$-inch length rings
PLATE		2.5- × 6-inch plastic plate to insulate board from chassis
ADAPTR		12-volt, .8-amp adapter for bench use
BHOLD8		8 AA cell vertical battery holder to fit in handle for gun configuration
ENDPLATE	2	Fabricate from .035 G10 or stiff strong plastic
BOBBIN		.75-inch ID × 1.5-inch OD × .312 width nylon bobbin BOBTHER

Chapter Three

High-Energy Pulser

This high-energy pulser is intended to supply the pulsed electrical energy in order to generate a powerful magnetic pulse capable of accelerating objects, constricting cans, shaping metals, exploding and vaporizing materials, accelerating small projectiles to high velocities, powering small rail guns, and other functions where conditioned magnetic energy is required (see Figure 3-1).

Construction and operation is intended for those with experience handling high-voltage, high-energy systems. This is for the advanced builder and requires assembly skills. Expect to spend $200 to $500 mainly for the necessary storage capacitors with most parts being available through Information Unlimited (www.amazing1.com). A parts list can be found at the end of the chapter in Table 3-1.

This device will be referenced for use with our mass accelerator, can crusher, wire exploder, and thermo plasma generator projects in subsequent chapters.

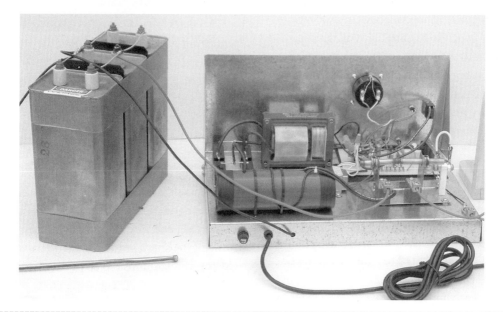

Figure 3-1   *High-energy pulser*

System Description

The described system is a valuable lab tool for the experimenter and researcher dealing with electrokinetic weapons, high-power magnets, *electromagnetic pulses* (EMPs), exploding wires/metal vaporizing, plasma pyrotechnics, high-pulse-powered lasers, electromagnetic launchers, gravity propulsion, and other similar research.

The device provides constant current charging to an external capacitive energy bank up to 5,000 joules with the system. The voltage is adjustable from 500 to 5,000 volts and is preset by a front panel control. Once set, the voltage will remain at this value until the firing period and then automatically recharges to the preset value as long as the safety charge button is activated. A meter monitors the voltage selected. A built-in, triggered spark switch using 3/8 -inch (9.5 mm), pure tungsten electrodes switches the stored bank energy into the target load within a period determined by the load inductance and resistance. A high peak-powered pulse approximately equal to dE/dt is now produced. (Power is the derivative of energy E.)

The system is intended for use as the energy-charging source and switch for many of our related projects. These projects include our mass accelerator driver as an example of a device capable of converting magnetic energy into kinetic energy. Such projects are shown in live action video at www.amazing1.com. Go to our High Voltage page and scroll down to the MASS10. The other available support projects are listed and will be supplemented with more as we complete various research programs, obtain necessary releases, prepare the data on *computer-aided design* (CAD), and so on.

Circuit Theory of Operation

Figure 3-2 shows that a high-voltage, current-limited transformer (T1) is connected to a full bridge rectifier (D1-4) and charges an external storage capacitor (C) through a surge limit resistor (R18). The external storage capacitor is connected between the discharge ground and a spark electrode (G1). The target load is now connected between the discharge ground and a spark switch electrode (G2). You will note that the load (R and L) takes on a complex value, usually highly inductive (not in all cases), with a small series resistance resulting from the inductor wire. Spark switch electrodes G1 and G2 are spaced to approximately 1.2 to 1.5 times the breakdown distance of the current operating voltage used.

A third trigger electrode (TE1) discharges a fast, low-energy, high-voltage pulse into G2, creating a spike of voltage ionizing the gap between G1 and G2, now causing the stored energy in the external storage capacitor to discharge into the load impedance.

The charging voltage of the external storage capacitors is sampled by the resistor divider network (R17) that also provides the series dropping value for the meter (M1). The charging voltage is programmed by control (R8) connected in series with R17. This control sets the trip level of comparator (I1) biasing the relay driver transistor (Q1) where it turns off, deenergizing the primary winding of T1 through relay (RE1) contacts. Once R8 is set to a selected value, it automatically maintains this preset voltage level on the external storage capacitors. A safety pushbutton switch (S3) provides a manual hold for charging the external capacitor.

The *light-emitting diode* (LED) LA1 indicates that the main power is on. LED LA2 indicates the charge has reached its programmed value.

The spark trigger circuit is a conventional *capacitor discharge* (CD) system where the energy on the capacitor (C6) is dumped into the primary of the pulse transformer (T2). This train of positive, high-voltage pulses generated on the secondary winding of T2 is integrated onto capacitors C8 and C9 through isolation diodes D12 and D13. These high-voltage DC pulses cause ionization between the gaps by the discharging action of trigger electrode TE1. The input to this circuit is a voltage doubler consisting of capacitors C4 and C5 and diodes D8 and D9. The "fire" switch (S1) energizes this circuit, causing immediate firing of the spark switch. A *silicon control rectifier* (SCR) switch dumps the charge on C6 and is gate-triggered by a DIAC triggering diode biased into breakdown by the setting of trimpot R14 and timing capacitor C7.

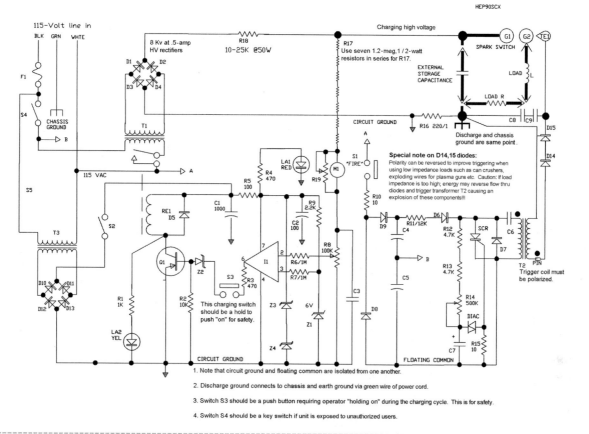

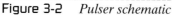

Figure 3-2 *Pulser schematic*

A 12-volt transformer (T3) powers the control circuit, including the RE1 relay. Without these 12 volts, the system cannot be energized without manually actuating RE1. Rectifier diodes D10 through 13 rectify the 12 *volts-alternating current* (VAC) that integrates onto the filter capacitor (C1). Resistor R1 decouples the 12 volts for regulating via zener diodes (Z3, Z4), which are necessary for stable operation of the comparator circuit. The main power for the charging energy is from the 115 VAC lines, fused by F1 and controlled by switch S4.

Special Note

Our lab at Information Unlimited has energy storage facilities consisting of 10 racks of 32 *microfarad* (MFD), 4500-volt oil-filled cans wired in parallel for a total capacity of 1,600 mfd or approximately 13,000 joules at 4,000 volts per rack. All racks connected in parallel produce 130,000 joules. It is paramount at

these energy levels to properly wire and assemble the complete system with adequate spacing and conductor or wire sizes in order to support the multi-megawatt pulses obtained. Explosion shields around the storage racks are used to protect operating personal.us.

The charging time per rack is approximately 10 minutes. All 10 racks would be impractical with this charger, as it would take almost 2 hours. We use a 10,000-volt, 1-amp, current-controlled system that charges 130,000 joules in 1 minute. This 10 Kv, 1-amp high-powered charger is available on special request.

Construction Steps

In this section, it is assumed you are familiar with the use of basic shop tools and have had intermediate assembly experience.

The pulser is built on a metal chassis that is 10 × 17 × 1.5 inches of 22-gauge galvanized steel. It uses a

6,500-volt *root means squared* (RMS), 20-milliampere current-limited transformer. The layout shown should be followed as closely as possible. A higher-capacity transformer can be used with an obvious change in layout size. A suggestion is to connect in parallel up to four of the previous transformers, obtaining 80 milliamperes of charging current. A front panel is used for mounting a voltage meter and all controls. A key switch substituted for S4 is recommended when being used around unauthorized personnel.

The construction steps are as follows:

1. If you purchased a kit, lay out and identify all parts and pieces.

2. Cut a piece of .1-inch grid perforated circuit board 6.25 × 4.25 inches.

3. Insert the components as shown in Figure 3-3 and solder using the leads of the components where necessary for the connection runs. The

dashed line shows the wiring routing and the connections underneath the board. Avoid bare wire bridges and potential shorts. Also avoid cold solder joints as these will surely be a problem. Solder joints should be shiny and smooth, but not globby.

4. Wire the following to the assembly board as shown in Figure 3-3:

To the chassis ground, 8 inches of #18 vinyl wire

To TE1, 4 inches of 20 Kv high-voltage wire

To R18 resistor, 8 inches of #18 vinyl wire

To anodes of D3 and D4, 12 inches of #18 vinyl wire (circuit ground)

To T3 12VAC (2), 8 inches of #22 vinyl wire

To M1 meter (2), 8 inches of #22 vinyl wire

5. Verify all wiring; components; and the orientation of all the diodes, semiconductors, and

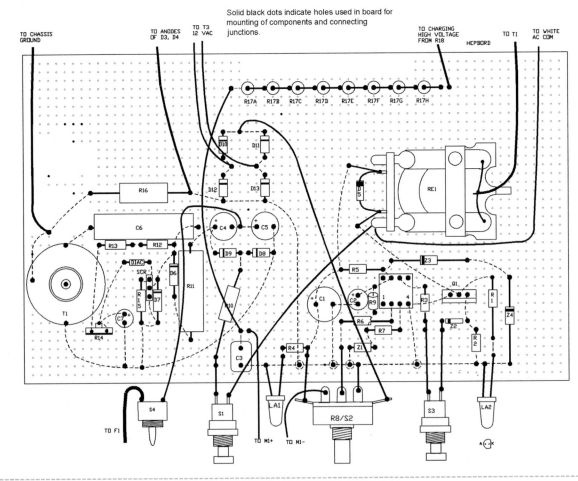

Figure 3-3 *Pulser assembly board*

electrolytic capacitors C1, C2, C4, C5, and C7. Check the quality of soldering, potential shorts, and the cold solder joints. Solder joints should be shiny and smooth, but not globby. Double-check before powering up.

Note that a factory-assembled and tested control board model #HEPBORDE is available on request.

6. The subassembly of the spark switch is as follows (see Figure 3-4):

   a. Fabricate BASE1 from a 4 3/4 - × 2-inch piece of 20-gauge galvanized steel. Bend a 1/2 -inch lip as shown.

   b. Fabricate two BRKT1 brackets from a 2 1/2- × 1 1/4 -inch piece of 20-gauge galvanized steel. Bend a 3/4 -inch lip as shown.

   c. Fabricate two BLK1 blocks at 3/4 × 1 × 1 1/4 inches from *polyvinyl chloride* (PVC) or similar material. They must have good insulating properties.

   d. Fabricate a BLK2 block at 3/8 × 3/8 × 3 inches from virgin Teflon material. It must support the high-voltage trigger pulse.

   e. Carefully solder the COL1 collars to the BRK1 brackets. Rig up a fixture to guarantee coincidental alignment of the tungsten electrodes when completed. You will have to use a propane torch or a super-hot iron for this step.

   f. File off the sharp points on eight 8- × 3/8 - inch sheet metal screws. This is necessary to prevent the PVC material from breaking down due to corona produced by high voltage at sharp points.

   g. Trial-position the parts, locating and drilling the necessary holes for assembling together. Follow the figure for proper locations.

   h. Attach large LUG1 block lugs to either the two sides or lip section of brackets BRKT1. Note the contact must be positive as current pulses are in the kiloamps.

   i. Temporarily preset the main gap to 1/16 of an inch and the trigger gap to 1/8 of an inch.

The spark switch is the heart of the system and is where the stored energy in the capacitors over the charging period is quickly released into the target load as a very high powered pulse. It is important that all connections be able to support the high discharge currents and voltage potentials.

The design shown here is for the HEP90 and is capable of switching up to 3000 joules (with a properly conditioned pulse) being usually sufficient to effectively experiment with MASS DRIVING, CAN CRUSHING, EXPLODING WIRES, MAGNETIZATION, AND OTHER RELATED PROJECTS.

A higher-energy switch capable of switching 20,000 joules is available on special request. Both switches use a high-voltage trigger pulse that depends on a relatively high impedance feed line to the target load for support of the trigger voltage pulse. This is usually not a problem with moderately inductive targets but can be if this value is low. It can be solved by placing several ferrite cores or high u toroids in these lines. The cores produce a high reactance to the trigger pulse but saturate at the main discharge current.

Construction of the spark switch must take into consideration medium mechanical forces resulting from high-peak magnetic fields. This is especially important in the higher-power approach and will require parallel wiring and terminating to reduce inductance and resistance. A sturdy base is always required being securely mounted. A protective shield must be placed over the assembly to protect the operator from potential shrapnel should a fault occur.

NOTES: For reliable triggering, the main gap shoud be set depending on charge voltage used.

The trigger gap should be set to not less than .25" from its contact point on the bracket. If triggering is erradic it is suggested to experiment with this setting.

Important Relationships
Ipk = Ech*SQRT(C/L)
Ipkt = pi/2*SQRT(L*C)

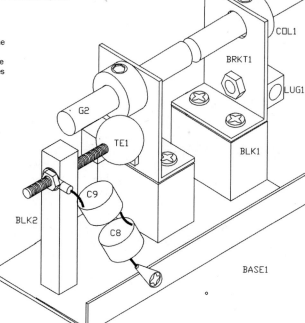

Figure 3-4   *Spark gap switch and ignitor*

Final Assembly

The steps for final assembly are as follows:

1. Fabricate the chassis and panel as shown in Figure 3-5. It is a good idea to fabricate the large hole necessary for the panel meter before doing any forming. The meter we use requires a 4-inch hole. The other smaller holes can be located and drilled after the panel and chassis are attached together.

2. Trial-fit the control board and locate the necessary clearance holes for the controls, indicators, and so on. Note the piece of insulating material between the assembly and the chassis; see the PLATE1 section in Figure 3-6. This can be secured with a couple dabs of *room temperature vulcanizing* (RTV) adhesive com-monly known as silicon rubber. Drill the respective holes as you go, verifying proper location and clearances.

3. Trial-fit the other parts as shown in Figure 3-6 and drill all the necessary mounting and clearance holes. Note the fuse holder FH1/FS1 and the line cord bushing BU2. They are located on the underside of the chassis and are shown as dashed lines.

4. Allow for the proper clearance of high-voltage-carrying components such as the transformer output leads, high-voltage diodes, and resistor R18. Note the high-voltage diodes are mounted to a plastic plate via two-sided tape or RTV adhesive.

5. Secure the control board in place via front panel controls. Secure the rear of the assembly board with several dabs of RTV adhesive when proper operation is verified.

Fabricate front panel from a piece of 21 x 8.5" 22-gauge gal. steel. Fold 2" sides as shown for bracing to chassis section. Fabricate mounting holes for meter.

Fabricate chassis section from a piece of 22 x 15" 22-gauge gal. steel. Fold four 2" panels with $\frac{1}{2}$" lips. Overall size should be 10 x 17 x 2 with $\frac{1}{2}$" lips around bottom.

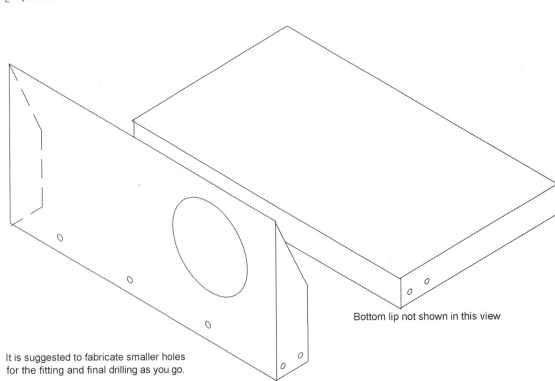

Bottom lip not shown in this view

It is suggested to fabricate smaller holes for the fitting and final drilling as you go.

Figure 3-5 *Chassis fabrication*

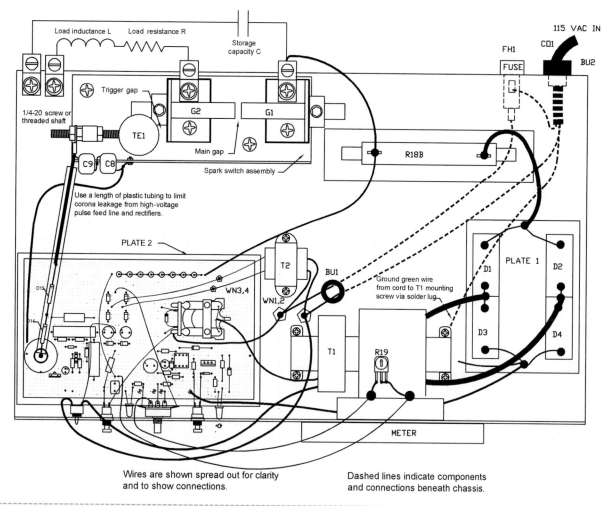

Load inductance L   Load resistance R

Storage capacity C

115 VAC IN

FH1   CO1

FUSE   BU2

Trigger gap

1/4-20 screw or threaded shaft

TE1

G2   G1

Main gap

R18B

Spark switch assembly

C9 C8

Use a length of plastic tubing to limit corona leakage from high-voltage pulse feed line and rectifiers.

PLATE 2

T2

BU1

Ground green wire from cord to T1 mounting screw via solder lug

PLATE 1

D1   D2

WN3,4

WN1,2

D15

T1

R19

D3   D4

METER

Wires are shown spread out for clarity and to show connections.

Dashed lines indicate components and connections beneath chassis.

Figure 3-6   *Final assembly layout*

6. Interconnect all wires and connections. Note the use of wire nuts for attaching leads of T1 and T2.

Electrical Pretest

To conduct an electrical pretest, follow these steps:

1. Short out the transformer output leads using a good-quality clip lead.

2. Connect a 60-watt ballast lamp across the fuse holder as a ballast for initial tests. (Remove the fuse!)

3. Preset the safety toggle switch S4 off or down, rotate the control pot R8/S2 off, adjust the trimpots R14 and R19 to midrange, and plug the power cord CO1 into 115 VAC.

4. Rotate R8/S2 until it snaps on and note LA1 and LA2 coming on.

5. Push the charge button S3, noting that relay RE1 is activating and LA2 is extinguishing as long as S3 is held.

6. Turn S4 on or up and push S3, noting that the ballast lamp from step 2 glows to approximately half brilliance.

7. Push fire button S1 and note a spark occuring between trigger electrode TE1 and the main gaps G1 and G2. Note the trimpot is preset to midrange, but may be adjusted *clockwise* (CW) to increase the discharge rate (see Figure 3-7).

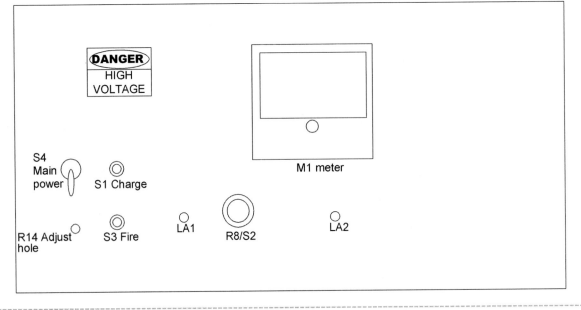

Figure 3-7 *Front panel and controls*

Main Test

To conduct the main test, follow these steps:

1. Unplug the power cord and turn off S2 and S4.

2. Connect a single 30 mfd/4 Kv capacitor or an approximately valued a 5K, 50-watt resistor for C and R, as shown in Figure 3-6.

3. Remove the ballast across the fuse holder and insert a 2-amp slow-blow fuse.

4. Preset the trigger gap to $\frac{1}{8}$ of an inch and the main gap to $\frac{1}{16}$ of an inch.

5. Connect up an accurate voltmeter across the external capacitor.

6. Plug the unit in and turn on S2 and S4. Push S3 and note the external capacitor charging to less than 1 Kv before RE1 deactivates. Note that LA2 is normally on and goes out only during the charging cycle. When the preset charge is reached, it comes back on, indicating the system is charged.

7. Rotate R8/S2 about 30 degrees CW and note that the voltage reaches a higher value before deactivating.

8. Now push fire button S1 and note a momentary heavy arc between the main gap occuring as the charge energy is dumped into the external load resistor.

9. Charge to 2,500 volts as measured by the external meter connected across the capacitor. Adjust R19 across the panel meter so it reads 2.5 on the 5 scale and at midrange. Make a mark on the front panel for a reference setting of 2,500 volts. The panel meter will now provide the charge voltage with a good degree of accuracy dependent on the external meter accuracy. Repeat step 8, noting a heavy arc discharge. Repeat the charge and fire cycles at different voltage settings to become familiar with the controls.

This completes the test and calibration steps. Further operations will require additional support equipment dependent on the particular project you are experimenting with.

Helpful Mathematical Relationships with This Equipment

Energy (joules) $= .5 \times$ farads $\times$ volts^2
System storage energy

Energy (joules) = Q (coulombs) × volts = I (amps) × t (secs) × volts
System storage energy

Charge (coulombs) = C (capacitance) × volts = I (amps) × t (secs)
Charge quantity

Spark gap resistance (ohms) = 2.03×10^{-3} × (spark length inches)/(Q)
Gap resistance

Use $.8 \times 10^{-3}$ for spark length in centimeters.

Peak current (Ipk) = volts × $(C/L)^{.5}$
Ideal current rise in an LC system. Use 75 percent when using oil-filled caps and lower values for photoflash and electrolytics dependent on intrinsic ESR.

Time to (Ipk) = pi/2 x $(LC)^{.5}$
Time to reach peak current value at $1/4$ cycle.

Magnetic flux (B) = (N × I × u)/(A × Le)

B = Webers/meter^2

N × I = amp-turns

u = $4 \times pi \times 10^{-7}$

A = Area of coil face in meters^2

Le = Length between poles in meters

M = Mass in kg

Force (newtons) = B × A × I

Acceleration (meters/t^2) = F/M

Velocity (meters/sec) = $1/2$ at^2

t = from above **time** of current rise

Table 3-1 High-power pulser parts list

Ref. #	Qty	Description	Db #
R1	1	1K, $1/4$ -watt resistor	br-bl-red
R2	1	10K, $1/4$ -watt resistor	br-bl-or
R3, 4	2	470K, $1/4$ -watt resistor	yel-pur-br
R5	1	100-ohm resistor	br-bl-br
R10, 15	2	10-ohm, $1/4$ - or $1/2$-watt resistor	br-bl-bl
R6, 7	2	1 m, $1/4$ -watt resistor	br-bl-gr
R8/S2	1	100K pot/switch	

Ref. #	Qty	Description	Db #
R9	1	2.2K, $1/4$ -watt resistor	red-red-red
R11	1	12K, 3-watt metal oxide resistor	
R12, 13	2	4.7K, $1/4$ -watt resistor	yel-pur-red
R14	1	500K trimpot	
R16	1	220 m/1-watt resistor	red-red-br
R17	7	1.2 m, $1/2$ -watt resistor in series	br-red-gr
C1	1	1000 m/25-volt vertical electrolytic cap.	
C2	1	100 m/25-volt vertical electrolytic cap.	
C3	1	.47 m/50-volt plastic cap.	
C4, 5	2	4.7 m/160-volt electrolytic cap.	
C6	1	3.9 m/350-volt polyester cap. info db# 3.9M	
C7	1	4.7 m/25-volts electrolytic cap.	
T2	1	25 Kv pulse info db# CD25B	
LA1	1	Red LED	
LA2	1	Yel LED	
D5, 6, 7, 8, 9	5	1n4007 1 Kv diode	
D10, 11, 12, 13	2	1n4001 100-volt diode	
Z1, 3, 4	3	6-volt 1-watt zener 1n4735a	
Z2	1	4.3-volt zener 1n5229	
LENS-HOLD	2	LED panel-mount lens holder	
PBOARD	1	6.25 × 4.24-inch PC or perforated board	
S1, 3	2	No pushbutton switch	
S4	1	SPST 115-volt/3-amp toggle switch or key switch	
RE1	1	12 vdc *Double-Pole, Double-Throw* (DPDT) relay info db# RELAY12	
Q1	1	NPN TIP 31 to 220 transistor	
I1	1	741 DIP OP AMP	
DIAC	1	Trigger DIAC	
SCR	1	Mcr 106-8 SCR info db# 8070W	
*HEP-BORDE		Assembled control board module info db# HEPBORDE	
WR18	3	One 4-foot piece of #18 red, gr, bl, vinyl, or plastic wire	

Table 3-1 Continued

Ref. #	Qty	Description	Db #
WR12		One 8-foot piece of #12 red, blk, vinyl, or plastic wire	
COL1	2	3/8-inch steel collars with set screws	
#G1, 2	2	3/8- × 2-inch pure tungsten rods	
BALL78	1	7/8-inch brass ball with 1/4-20 hole	
BRKT1	2	Bracket fab from a 2 1/2 - × 1 1/4 -inch piece of 20-gauge galvanized steel	
BLK1	2	Block fab from a 3/4 - × 1- × 1 1/4 -inch piece of PVC plastic	
BLK2	1	Block fab from 3/8 × 2/8 × 2 1/4 piece of Teflon	
BASE1	1	Base fab from a piece of 4 3/4 × 2 1/2 of 20-gauge galvanized steel	
TE1	1	1/4 -20 × 3-inch threaded rod fabrication with point	
LUG1	4	Heavy-duty screw lugs with 1/4 -inch hole	
LUG2	3	Large solder lug with 1/4 -inch hole	
NU1/420	3	1/4 -20 hex nuts	
SW6	8	#6 blunt sheet metal screws	
SW612	9	6-32 × 1/2 -inch screws with nuts and washers	
C8, 9	2	1000 *picofarads* (PFDs) 15 Kv ceramic capacitors info db# .001/15 KV	
*HEPGAP		Assembled switch module as shown in Figure 3-4	
CHASSIS	1	Fabricated as shown in Figure 3-5	

Ref. #	Qty	Description	Db #
FRONT PANEL	1	Fabricated as shown in Figure 3-5	
PLATE 1	1	Fabricated as shown in Figure 3-6	
PLATE 2	1	Fabricated as shown in Figure 3-6	
FH1	1	Panel-mounted fuse holder	
FS1	1	2-amp fuse	
CO1	1	3-wire #18 line cord	
BU1	1	1/2 -inch plastic bushing	
BU2	1	Clamp bushing for line cord	
WN1	3	Medium-wire nuts	
WN2	2	Small-wire nuts	
METER	1	50 to 100 ua PANEL METER info db# METER50S	
T1	1	6500-volt/20 ma current-limited xfmr info db# 6 kv/20	
T3	1	12-volt/100-milliampere stepdown xfmr info db# 12/.1	
*D1, 2, 3, 4	4	8 Kv, 1-amp block rectifiers info db# H407	
D14, 15	2	30 Kv, 5-milliampere, .1 us RECTIFIERS db# VG30	
*R18	1	10 to 20K, 50-watt resistor info db# 10K/50	
R19	1	2K trimpot	
SW838	8	#8 × 3/8 sheet metal screw	

Chapter Four

Mass Accelerator

The mass accelerator device accelerates a mass shown as a 100-gram aluminum ring to ballistic velocities (see Figure 4-1). It provides an excellent outdoor action demonstration of electrokinetics requiring qualified adult supervision. A video clip can be seen at www.amazing1.com, showing the project vaporizing an object on impact. The project requires experience in the handling of high-voltage, high-energy devices.

Construction of the device requires a working *high-energy pulser,* as described in Chapter 3, "High-Energy Pulser," along with the assembly of the launcher and mass projectiles. It is shown using

energy-storage capacitors currently available on the surplus market. These values can be altered within limits to allow use of the available capacitors that you may have on hand.

Theory of Operation

A circular, nonmagnetic conductive aluminum ring (a projectile) is placed on a flat, pancake-shaped coil. A high-energy storage capacitor is discharged into this coil, producing a high-current pulse. This pulse is coupled into the projectile, inducing an opposing magnetic field that provides a moment of force to accelerate the projectile.

Setup and Operation of the Mass Accelerator

1. Verify a properly operating high-energy pulser system, as described in Chapter 3.

2. Assemble the capacitor bank and safety discharge probe, as shown in Figure 4-2. This probe is connected directly across the storage capacitors and is used to verify that they are fully discharged before going into the circuit

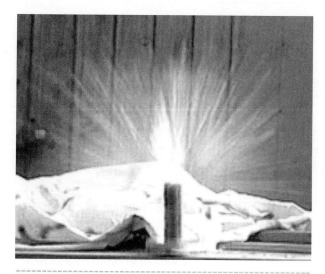

Figure 4-1 *Mass accelerator*

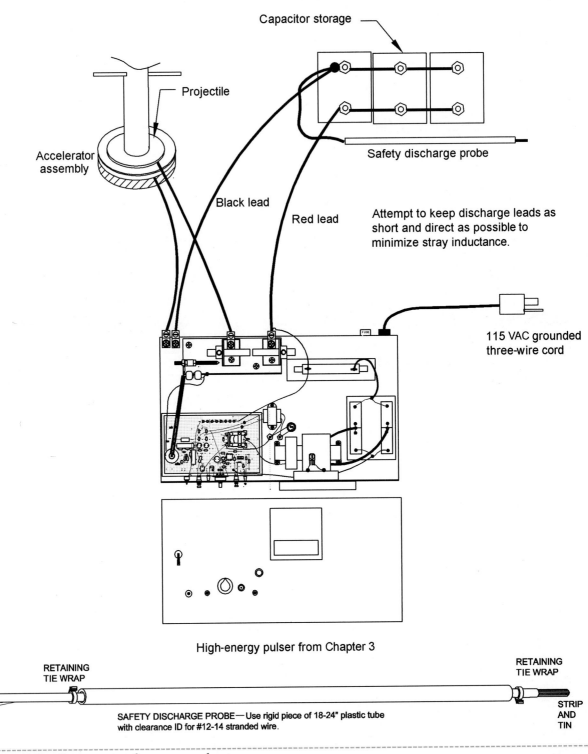

Capacitor storage

Projectile

Accelerator
assembly

Black lead

Red lead

Safety discharge probe

Attempt to keep discharge leads as
short and direct as possible to
minimize stray inductance.

FUSE

115 VAC grounded
three-wire cord

High-energy pulser from Chapter 3

RETAINING
TIE WRAP

RETAINING
TIE WRAP

STRIP
AND
TIN

SAFETY DISCHARGE PROBE—Use rigid piece of 18-24" plastic tube
with clearance ID for #12-14 stranded wire.

Figure 4-2   *System setup for mass accelerator*

to make adjustments. Use #12 wire for all discharge connections. The capacitors shown are
32 *microfarad* (MFD) at 4.5 *kilovolts* (Kv) and
were used in our lab. Other obtainable values
may be used, but we cannot guarantee full
performance.

3. Assemble the accelerator as shown. Figures 4-
    3 and 4-4 require more fabricating but will tol-
    erate energies up to 2,000 joules. Figure 4-5 is
    simpler but cannot exceed 1,000 joules with-
    out breaking up.

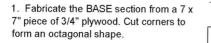

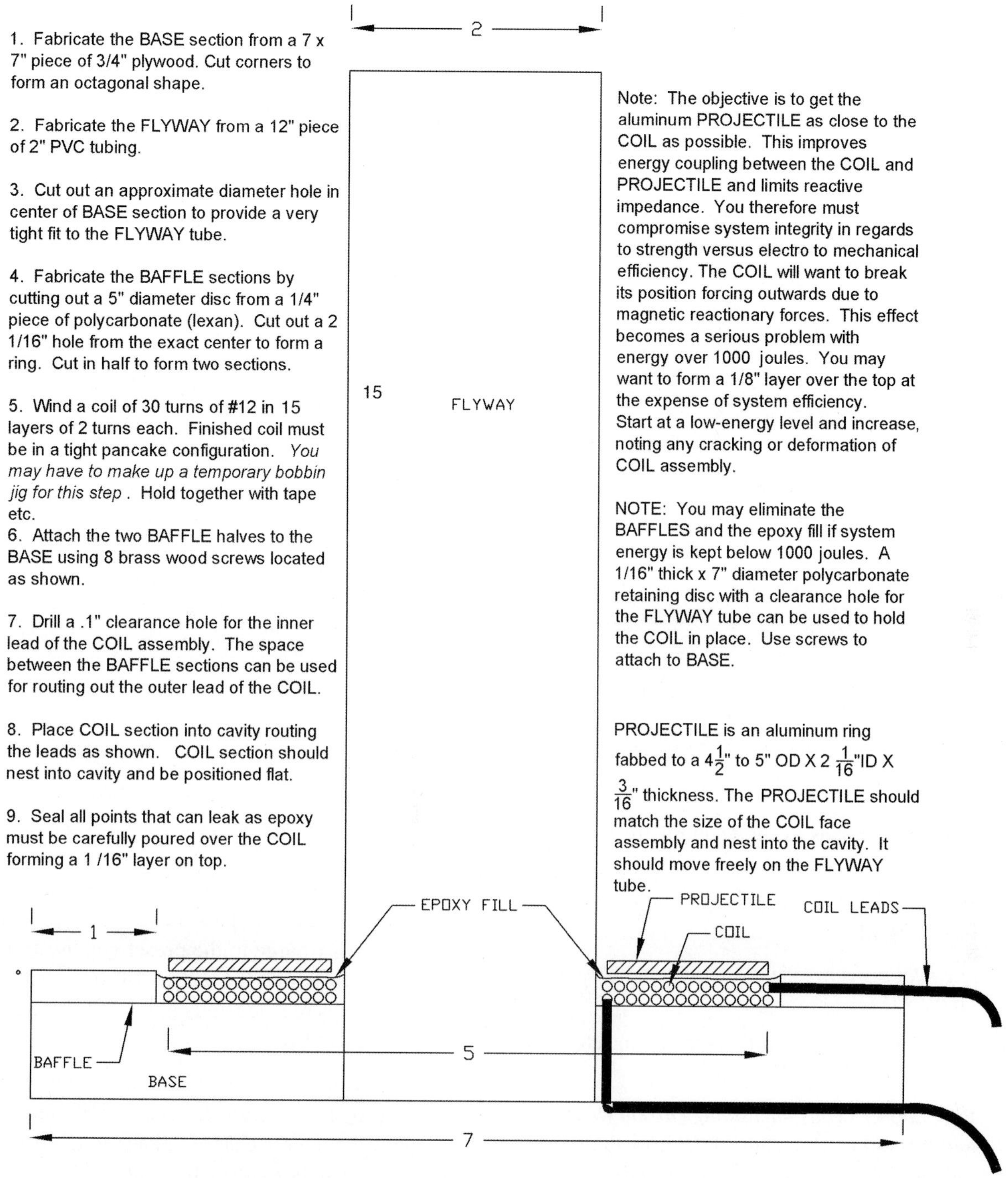

1. Fabricate the BASE section from a 7 x 7" piece of 3/4" plywood. Cut corners to form an octagonal shape.

2. Fabricate the FLYWAY from a 12" piece of 2" PVC tubing.

3. Cut out an approximate diameter hole in center of BASE section to provide a very tight fit to the FLYWAY tube.

4. Fabricate the BAFFLE sections by cutting out a 5" diameter disc from a 1/4" piece of polycarbonate (lexan). Cut out a 2 1/16" hole from the exact center to form a ring. Cut in half to form two sections.

5. Wind a coil of 30 turns of #12 in 15 layers of 2 turns each. Finished coil must be in a tight pancake configuration. *You may have to make up a temporary bobbin jig for this step* . Hold together with tape etc.

6. Attach the two BAFFLE halves to the BASE using 8 brass wood screws located as shown.

7. Drill a .1" clearance hole for the inner lead of the COIL assembly. The space between the BAFFLE sections can be used for routing out the outer lead of the COIL.

8. Place COIL section into cavity routing the leads as shown. COIL section should nest into cavity and be positioned flat.

9. Seal all points that can leak as epoxy must be carefully poured over the COIL forming a 1 /16" layer on top.

15   FLYWAY

Note: The objective is to get the aluminum PROJECTILE as close to the COIL as possible. This improves energy coupling between the COIL and PROJECTILE and limits reactive impedance. You therefore must compromise system integrity in regards to strength versus electro to mechanical efficiency. The COIL will want to break its position forcing outwards due to magnetic reactionary forces. This effect becomes a serious problem with energy over 1000 joules. You may want to form a 1/8" layer over the top at the expense of system efficiency. Start at a low-energy level and increase, noting any cracking or deformation of COIL assembly.

NOTE: You may eliminate the BAFFLES and the epoxy fill if system energy is kept below 1000 joules. A 1/16" thick x 7" diameter polycarbonate retaining disc with a clearance hole for the FLYWAY tube can be used to hold the COIL in place. Use screws to attach to BASE.

PROJECTILE is an aluminum ring fabbed to a $4\frac{1}{2}$" to 5" OD X $2\frac{1}{16}$"ID X $\frac{3}{16}$" thickness. The PROJECTILE should match the size of the COIL face assembly and nest into the cavity. It should move freely on the FLYWAY tube.

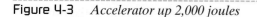

EPOXY FILL — PROJECTILE   COIL LEADS
COIL
BAFFLE
BASE
5
7

Figure 4-3   *Accelerator up 2,000 joules*

4. Position the accelerator coil assembly at the desired location, noting the direction of the projectile as it is accelerated along the flyway tube. Velocities can be reached that can break glass and do damage to objects that are directly and individually struck. Caution should be used, as the projectile can also cause injury.

You may use a flyway tube of at least several meters with a retaining stop at its end. This prevents the projectile from flying off and automatically resets its position for the next launch as it returns back to the original position by the force of gravity. Do not exceed 3,000 volts as the assembly will quickly break up due to the impact forces.

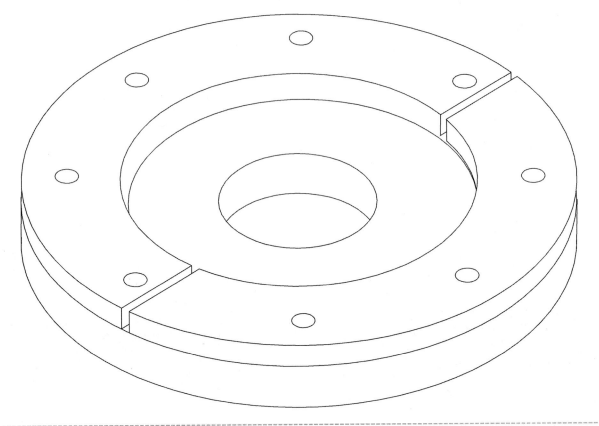

Figure 4-4 *Alternate cavity assembly for higher-powered accelerators*

5. Note the controls, as shown in Figure 3-7 of Chapter 3:

   Voltage control switch R8/S2 should be set to FCCW (OFF).

   High voltage power switch S4 should be set to DOWN (OFF).

   Charging switch S1 must be pressed to activate charging action.

6. Plug into 115 *volts-alternating current* (VAC) and turn on voltage control to verify that both *light-emitting diodes* (LEDs) will activate:

   The *red lamp LA1* indicates the system is in an "on" state.

   The *green or yellow lamp LA2* comes on but extinguishes in the charge cycle and comes on when the preset voltage level is reached, indicating the system is ready to "fire."

7. Set *charge voltage control* to 2,500 volts as marked on the panel. Plug the unit into 115 VAC and turn on the high-voltage power switch.

8. Depress the *charge switch (the green or yellow lamp now goes off) and note the increasing voltage on the panel voltmeter reading, indicating the energy-storage capacitors are charging. Charge to the preset voltage as in step 6. Note that the lamp turns on when the set level is reached, indicating that the system is ready to fire.

   Note you may bypass the voltage control settings by turning full *clockwise* (CW) and manually setting the charge voltage to the desired value as indicated on the meter.

9. Depress the fire switch and note the projectile accelerating up the flyway tube. Remove the flyway stop pin and verify the overhead clearance for higher voltages over 2,500. Repeat at a setting of 3,000 and up to 4,000 volts. Note that the system energy varies as the square of the increased voltage. For example, projectile kinetic energy will increase by 9 when voltage increases from 1,000 to 3,000.

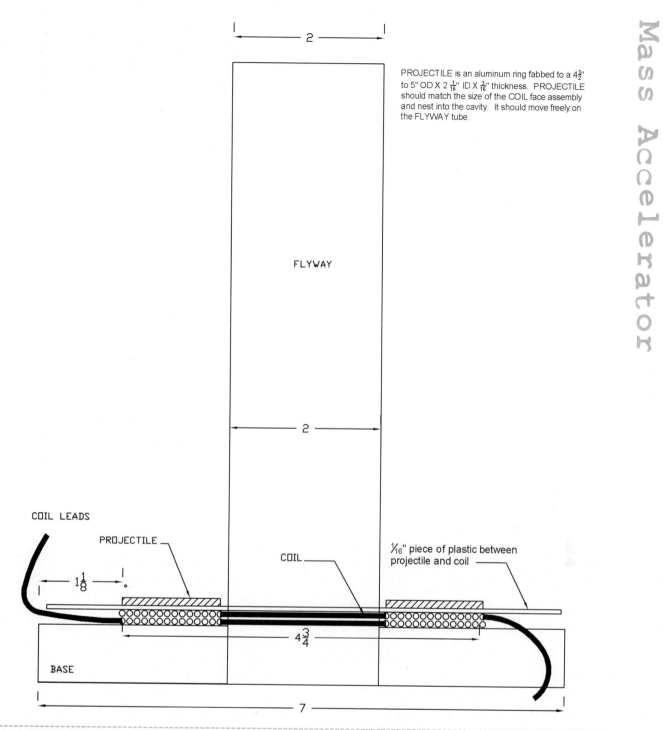

PROJECTILE is an aluminum ring fabbed to a 4½"
to 5" OD X 2 1/16" ID X 3/16" thickness. PROJECTILE
should match the size of the COIL face assembly
and nest into the cavity. It should move freely on
the FLYWAY tube.

Figure 4-5   *Simple accelerator for up to 1,000 joules*

Note that continued operation at over 3,000 volts eventually will mechanically weaken the coil and erode the spark switch. Do not go over 2,500 volts as marked on the knob, except for experimenting! The system may be test-fired up to 5,000 volts for experimental research but will require continual maintenance of the coil and cleaning of the spark switch

electrodes. Always increase in 500-volt increments to verify system integrity. Use an externally calibrated meter connected across the energy-storage capacitors to verify the panel meter calibration from time to time.

Also note that the spark gap is factory set at .1-inch for operation at 2 to 3 Kv. Lower-voltage operation may require decreasing the gap for reliable triggering. Operation over 3 Kv will require widening to prevent premature triggering. You may have to tweak the gap by loosening one of the Allen set screws for your particular range of operation. The spark switch may prematurely fire if dirty or improperly gapped.

Danger! Unplug and discharge energy-storage capacitors before making any adjustments. Verify by shorting capacitor terminals with the safety discharge probe.

Also, if system uses a push-on/off charging switch, it will automatically recharge the capacitor bank after firing. Basic units use a push-on switch for the charging switch, as charging always requires the switch to be depressed, providing safer operation at the cost of convenience.

Chapter Five

Plasma Thermal Gun

Your PPRO1 plasma gun utilizes the electrothermal energy obtained from vaporizing a metallic material, creating a pressure wave of sufficient magnitude to cause a projectile to accelerate to ballistic velocities (see Figure 5-1). The system shown on our data is a relatively low velocity device within the building capability of most hobbyists. The project may be scaled up for serious research with an obvious increase in high voltage handling and ballistic hazards. The project requires experience in the handling of high-voltage, high-energy devices.

Construction requires a working high-energy pulser as described in Chapter 3, "High-Energy Pulser." It is shown using energy-storage capacitors currently available on the surplus market. These values may be altered within limits to allow the use of available capacitors that you may have on hand.

Theory of Operation

An energy-storage capacitor is charged from a programmable, controlled current source to a selected high voltage. This capacitor is switched by a triggered spark gap, dumping all the stored energy into a small-volume explosion chamber and vaporizing a thin, aluminum wire placed inside the chamber. A pressure wave forces a projectile out the barrel to a high velocity. This project is suitable for a science class demonstration if properly supervised with qualified personnel.

Assembly Steps of the Thermal Gun

To construct the thermal gun, follow these steps:

1. Fabricate the barrel and breech plug, as shown in Figure 5-2, rounding off all edges and inner barrel ridges.

2. Fabricate the breech tube, as shown in Figure 5-3. Note that the mating of the barrel and breech plug are precision-drilled using fractional bits. The inner diameter should be .007 to .008 over the outer dimensions of the barrel

Figure 5-1 *Plasma thermal gun*

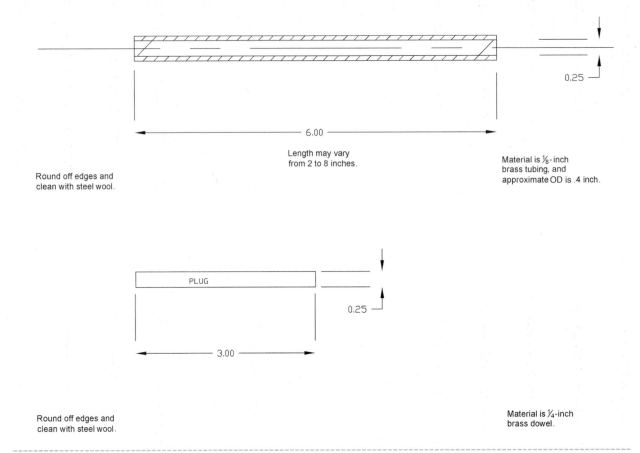

6.00

Length may vary
from 2 to 8 inches.

Round off edges and
clean with steel wool.

Material is ⅛-inch
brass tubing, and
approximate OD is .4 inch.

0.25

PLUG

3.00

0.25

Round off edges and
clean with steel wool.

Material is ¼-inch
brass dowel.

Figure 5-2 *Barrel and breech plug fabrication*

and breech plug to allow for clearance of the exploding wire, as shown in Figure 5-4.

If the clearance is too tight, it will be difficult to insert the barrel and breech plug without breaking the wire.

If the clearance is too loose, improper contact will occur on the exploding wire. Drill a bit at a time to establish the proper fitting action.

3. Solder the brass retaining washers at certain points, as shown, to allow the proper gauging of the barrel and breech plug insertion depths into the breech tube.

4. Fabricate projectiles from ¹/₂-inch pieces of ¹/₄-inch plastic, polycarbonate, or wooden dowels. Round off the end to give it a bullet-like shape, creating a streamlined effect. The final assembled gun should look like Figure 5-4 and easily fit together for the reloading of the exploding wire and projectile. You will note the metal barrel and breech plug are also the feed electrodes connected to the exploding

wire via a sandwiching action. The high-powered current pulse is now applied to these electrodes, generating an explosive, heated plasma vapor that produces a pressure wave, which forces the projectile out of the barrel.

5. Create a lab proto unit shown in Figures 5-5 and 5-6. It allows easy disassembly and provides positive electrical contact to the gun electrode ends. You may have your own ideas to this approach, but the objective must be the same: good electrical contact, easy disassembly, and reloading.

For a smaller-diameter breech plug, use the following substeps in fabricating:

a. Precisely locate the holes and center punch.

b. Drill a ¹/₈-inch pilot hole using a drill press.

c. Expand the holes to the final radius using successively larger drills.

d. Drill a ¹/₁₆-inch hole for a spring electrode sheet metal screw.

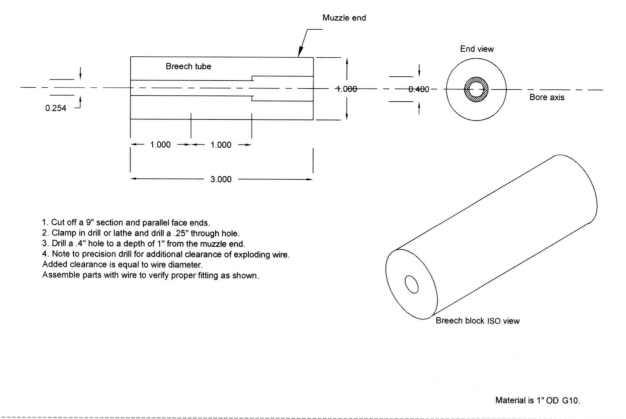

Muzzle end

Breech tube

End view

Bore axis

1.000

0.254

1.000 1.000

3.000

1. Cut off a 9" section and parallel face ends.
2. Clamp in drill or lathe and drill a .25" through hole.
3. Drill a .4" hole to a depth of 1" from the muzzle end.
4. Note to precision drill for additional clearance of exploding wire.
Added clearance is equal to wire diameter.
Assemble parts with wire to verify proper fitting as shown.

Breech block ISO view

Material is 1" OD G10.

Figure 5-3 *Breech tube fabrication*

e. Drill two $1/16$-inch holes for mounting blocks to the base section.

f. Use a band saw to cut out sections shown by dashed lines. File for final sizing and finishing.

6. Fabricate and shape the spring electrodes as shown. Use pliers and a small vise to bend them properly. These pieces must snap into the gun assembly and provide a positive contact.

7. Fabricate the base section in Figure 5-6 from a piece of $3/4$-inch finished plywood. Sand and varnish it for a good appearance. Mount the components as shown and clamp the wires in place using nylon clamps and $1/2$-inch sheet metal screws. Solder the leads of the feed wires to spring electrodes.

8. Connect up as shown in Figure 5-7. Use at least #16 wire. The leads must be short and direct.

9. Obtain a suitable target, such as a small pillow. This will prevent the projectile from rico-cheting all over the place, making it hard to find for reloading.

10. Prep the gun with exploding wire, as shown in Figure 5-4. Snap it into the holders and insert the projectile to point as shown.

It is assumed that the high-energy pulser is properly operating and connected as per the instructions and has a calibrated voltmeter and set spark switch. Familiarize yourself with the pulser operation.

11. Connect the high-energy pulser and energy-storage capacitors as shown. Note that the inductor L2 is not used in this project.

12. Allow charging up to 2,000 volts and then test the fire unit. Note the loud crack as the projectile exits the barrel. Increase the voltage in steps of 500 and note the increase in velocity. Experiment using different projectiles, targets, and so on.

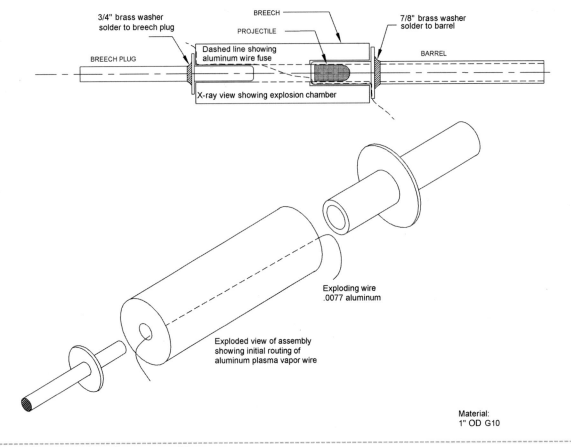

Figure 5-4   *Breech and gun assembly*

Special Notes

The system as shown was tested with two 32 *microfarad* (mfd), 4 *kilovolt* (Kv) capacitors in parallel for a total energy of 500+ joules. The projectile velocity was more than 200 *meters per second* (m/s). It is suggested to use only one capacitor for the science project approach.

Serious experimenters should consider our HEP90 higher-energy charger and additional energy-storage capacitors. The gun section must be beefed up for any significant increase in energy. Some suggestions are the following if stored energy is to exceed 1,000 joules:

- The entire assembly must be placed inside an explosion shield consisting of a half-inch of Lexan.

- Retaining washers should be replaced with multiple washers or a sleeve all securely soldered in place.

- The breech tube must be sleeved into a secondary sleeve of equal wall thickness. The entire breech assembly is now inserted into a steel tube or pipe for further reinforcement.

- The mounting blocks should be reinforced with a 1/4-inch angle of aluminum or steel.

- The mounting base should be fabricated from aluminum or steel.

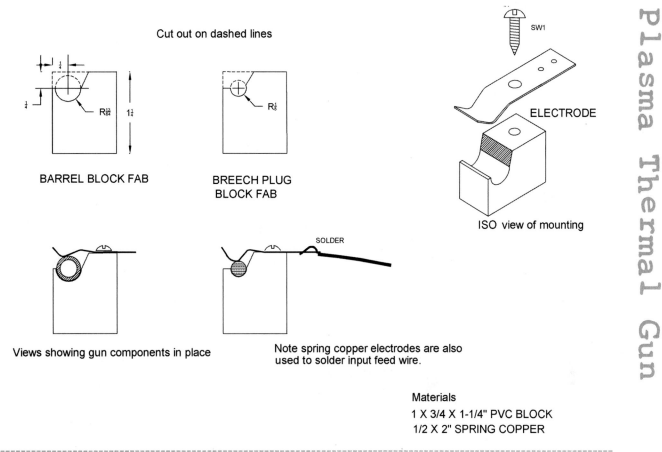

Cut out on dashed lines

BARREL BLOCK FAB

BREECH PLUG
BLOCK FAB

ELECTRODE

ISO view of mounting

SW1

SOLDER

Views showing gun components in place

Note spring copper electrodes are also
used to solder input feed wire.

Materials
1 X 3/4 X 1-1/4" PVC BLOCK
1/2 X 2" SPRING COPPER

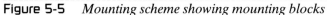

Figure 5-5 *Mounting scheme showing mounting blocks*

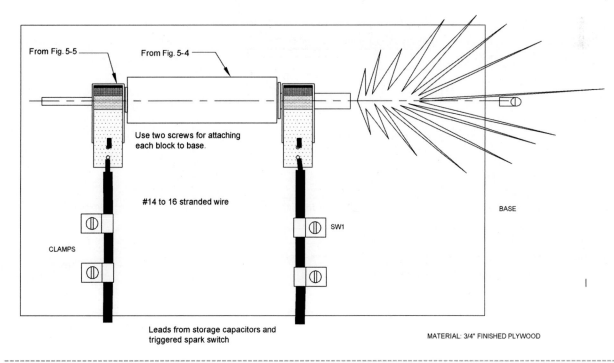

From Fig. 5-5

From Fig. 5-4

Use two screws for attaching
each block to base.

#14 to 16 stranded wire

SW1

BASE

CLAMPS

Leads from storage capacitors and
triggered spark switch

MATERIAL: 3/4" FINISHED PLYWOOD

Figure 5-6 *Plasma gun top view showing muzzle blast when firing*

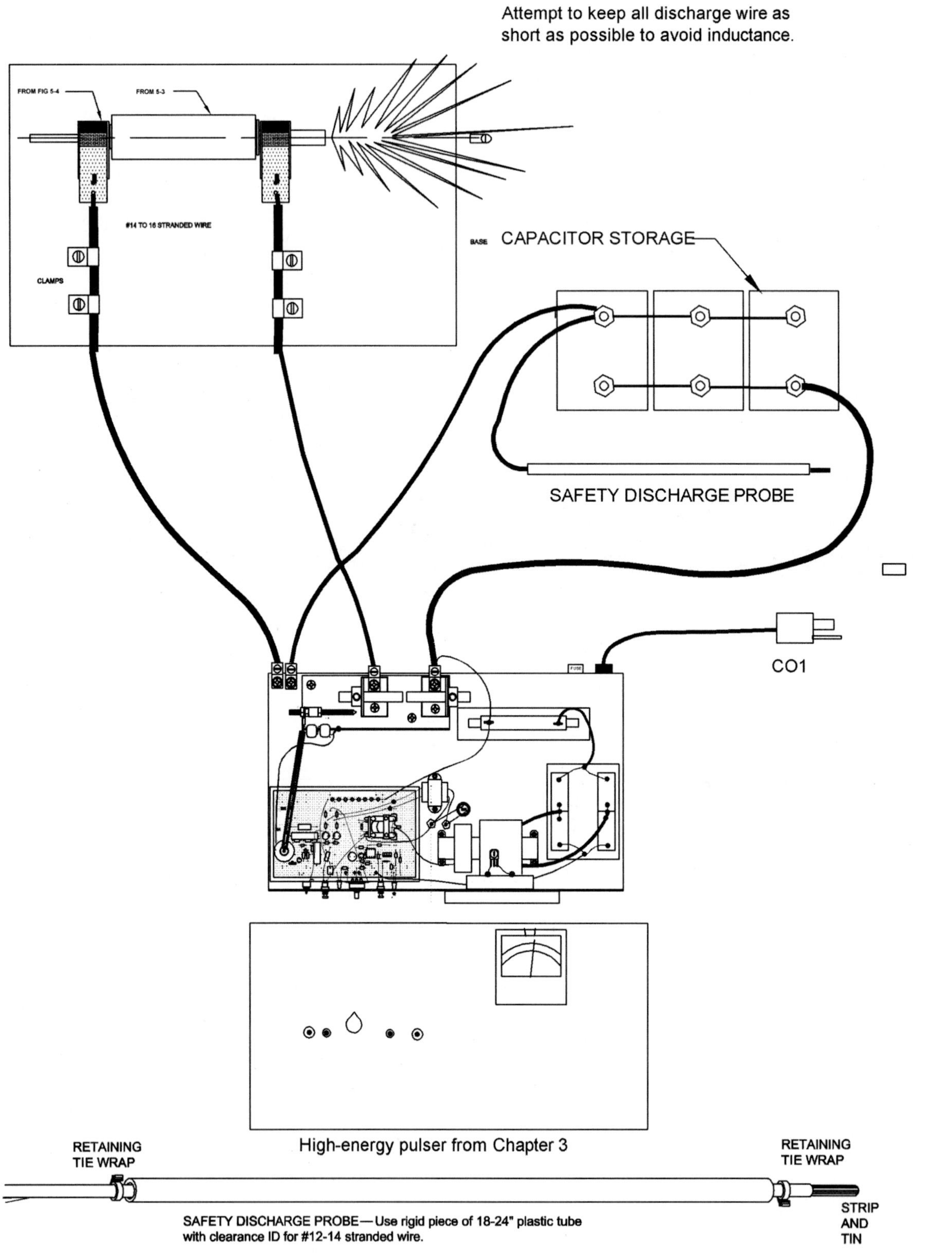

Attempt to keep all discharge wire as
short as possible to avoid inductance.

FROM FIG 5-4   FROM 5-3

#14 TO 16 STRANDED WIRE

CLAMPS

BASE   CAPACITOR STORAGE

SAFETY DISCHARGE PROBE

CO1

FUSE

High-energy pulser from Chapter 3

RETAINING
TIE WRAP

RETAINING
TIE WRAP

STRIP
AND
TIN

SAFETY DISCHARGE PROBE—Use rigid piece of 18-24" plastic tube
with clearance ID for #12-14 stranded wire.

Figure 5-7   *Setup of plasma thermal gun*

Chapter Five

When attempting to fire larger systems, it is important to increase the energy in small steps, always rechecking the system integrity for any damage resulting from prior shots. You will need the following parts to make this system:

- The fabricated parts as shown in Figures 5-4 and 5-3. These can be purchased through the factory but will require final fitting. Raw materials are available through most hardware stores.

- The energy storage caps are shown as 32 mfd/4500 volts and are available through the factory. Other values obtainable through surplus sources may be used, but we cannot guarantee performance.

- The HEP3 charger/pulser/spark switch is available through the factory as a kit or as an assembled unit. Our HEP9 higher-powered version intended for more serious experimental research is also available. It can charge banks of multiple capacitors up to 5,000 joules.

Danger! If a system uses a push-on/off charging switch, it will automatically recharge the capacitor bank after firing. Basic units use a push-on switch for the charging switch, as charging always requires the switch to be depressed, providing safer operation at the cost of convenience. The red lamp (LA1) indicates the system is in an "on" state. The green or yellow lamp (LA2) comes on but extinguishes in the charge cycle and comes on when the preset voltage level is reached, indicating the system is ready to "fire."

Chapter Six

Wire-Exploding Project

The explosion of wires or objects by the rapid discharge of high electrical currents produces some interesting phenomena. The obvious audible and visual display is not the only objective, and it does find use as a part of special effects and demonstrations. The detonation of a wire from a moderate energy source can provide a blast equal to a pyrotechnic high explosive. The detonation velocity can exceed thousands of meters per second, producing high-speed microparticles and switching high currents in the multigigawatts at nanosecond speeds for *electromagnetic pulse* (EMP) generation and electrokinetics. The project may be scaled up for serious research with an obvious increase in high voltage handling and ballistic hazards. This project requires experience in the handling of high-voltage, high-energy devices.

Construction requires a working high-energy pulser as described in Chapter 3, "High-Energy Pulser," along with the assembly of the explosion chamber. It is shown in Figure 6-1 using energy-storage capacitors currently available on the surplus market. These values may be altered within limits in order to use available capacitors that the builder may have on hand. It can be hazardous and requires protective eyewear. This project can function as an excellent museum demonstration when performed by qualified adult personnel. Safety eyewear must be worn by the operator.

Theory of Operation

An energy-storage capacitor is charged from a programmable, controlled current source to a selected high voltage. This capacitor is now switched by a triggered spark gap, discharging all the stored energy into and vaporizing a wire. A shock wave is produced by the rapid oxidization of the object. This project is suitable for a science class demonstration of the concept if properly supervised with qualified personnel.

Setup and Operation

To begin the project, follow these steps:

1. Assemble the explosion chamber, as shown in Figure 6-2.

2. Position the explosion chamber and connect it up as shown in Figure 6-3. The transparent shield section provides protection to the operator but is open in the rear and side sections. The rear section is used to place the target material where the explosive blast exits. It is suggested to *completely* shield the chamber if spectators are to be in the area.

   Note the inclusion of the trigger pulse blocking inductor in one of the leads. This part

Figure 6-1   *Explosion chamber*

provides reliable triggering when using low-impedance loads such as wires and so on.

3. Note the controls as shown in Figure 3-7 of Chapter 3.

   Voltage Control Switch (R8/S2) should be set to FCCW (OFF).

   High-Voltage Power Switch (S4) should be set to DOWN (OFF).

   Charging Switch (S1) must be pressed to activate the charging action.

4. Connect a 3- to 4-inch piece of #24 bare copper wire to the terminal block as shown. Make sure that the wire is firmly clamped by giving it a tug. Place a safety cover over the assembly if required.

5. Plug into 115 *volts-alternating current* (VAC) and turn on voltage control to verify both *light-emitting diode* (LED) lamps coming on:

The *red lamp (LA1)* indicates the system is in an "on" state.

The *green or yellow lamp (LA2)* comes on but extinguishes in the charge cycle and comes on when the preset voltage level is reached, indicating the system is ready to "fire."

Turn the voltage control off and unplug it once the previous steps are verified.

6. Set the charge voltage control to a 3,000-volt setting, as marked on the panel. Plug unit into 115 VAC and turn voltage control.

7. Depress *charge switch S1 (the green or yellow lamp now goes off) and note the increasing voltage on the panel voltmeter reading, indicating the energy storage capacitors are charging. Allow charging to the preset voltage as in step 6. Note that the LA2 lamp now turns on when the set level is reached, indicating that the system is ready to fire.

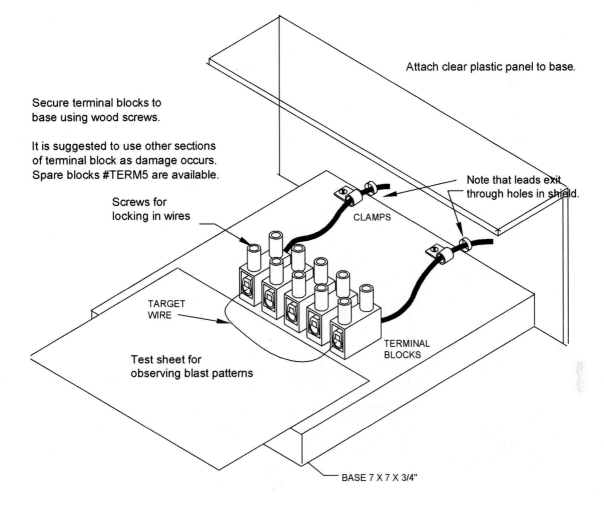

Secure terminal blocks to base using wood screws.

It is suggested to use other sections of terminal block as damage occurs. Spare blocks #TERM5 are available.

Screws for locking in wires

Attach clear plastic panel to base.

Note that leads exit through holes in shield.

CLAMPS

TARGET WIRE

Test sheet for observing blast patterns

TERMINAL BLOCKS

BASE 7 X 7 X 3/4"

Introducing all new **BLAST ART**tm
Place a piece of white paper near the target wire and note the myriad of blast patterns produced. You may experiment by placing paper at different locations and distances relative to the wire, noting the various effects of the blast. This is an excellent form of a new art that we are introducing called BLAST ART where an infinite amount of weird and bizarre blast patterns are permanently embedded onto many surface materials. Wires of different sizes and materials may be experimented with further, adding to the effects possible.

Figure 6-2 *Explosion chamber scheme*

8. Depress the fire switch and note a loud report of discharge in the spark switch. The wire should vaporize with a bright flash and a loud report similar to a firecracker. Repeat at 4,000 volts, noting a brighter flash and a louder bang.

Special Notes

System energy varies as the square of the increased voltage. For example, energy will increase by 9 when the voltage is increased from 1,000 to 3,000. You may

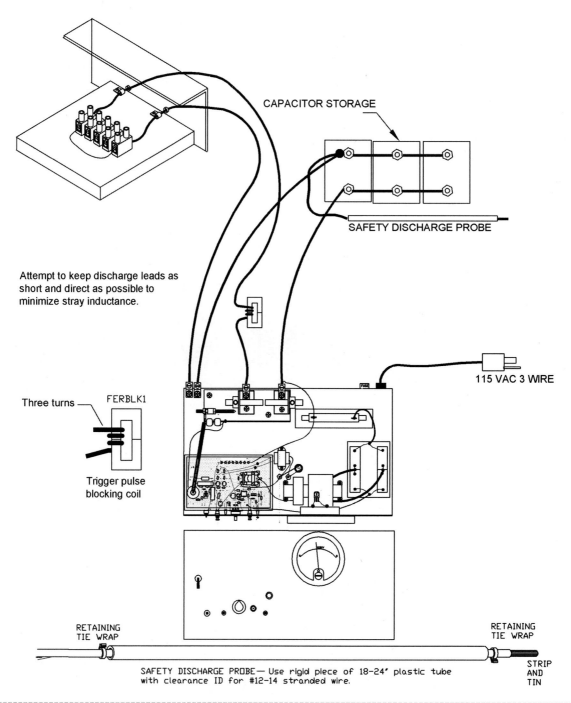

CAPACITOR STORAGE

SAFETY DISCHARGE PROBE

Attempt to keep discharge leads as
short and direct as possible to
minimize stray inductance.

115 VAC 3 WIRE

Three turns — FERBLK1

Trigger pulse
blocking coil

RETAINING
TIE WRAP

RETAINING
TIE WRAP

STRIP
AND
TIN

SAFETY DISCHARGE PROBE— Use rigid piece of 18-24" plastic tube
with clearance ID for #12-14 stranded wire.

Figure 6-3 *Setup of wire explosion project*

go to 4,500 volts for an increased effect, add more
capacity, or both. Use caution.

Also note that the continued operation at over
4,000 volts eventually will erode the spark switch. Do
not go over the 4 Kv setting except for experimenting. The system may be test-fired up to 5,000 volts for
experimental research but may require continual
maintenance of spark switch electrodes. Always

increase in 500-volt increments to verify system
integrity.

It may be a good idea to check the front panel
meter calibration from time to time. Use an externally calibrated meter connected across the storage
capacitors to verify panel voltage settings.

The spark gap is factory set at .1 inches for operation at 3 to 4 Kv. Lower-voltage operations may

require decreasing the gap for reliable triggering. Operation over 4 Kv will require widening to prevent premature triggering. You may have to tweak the gap by loosening one of the Allen set screws for your particular range of operation.

You should place a high-permeability ferrite core wound with two to three turns on one of the leads from the chamber to the G2. This part will increase the inductance of this lead to block the trigger pulse from being shorted by the low impedance of the wire yet quickly saturate at the main discharge current.

It is important that all connection leads from the capacitor bank to the pulser and the chamber be as short as possible to reduce lead inductance for optimum results.

The spark switch may prematurely fire if dirty or improperly gapped.

Danger! Always unplug and discharge energy-storage capacitors using the safety discharge probe before making any adjustments.

Place a piece of white paper adjacent to the wire and observe the interesting pattern left once it explodes.

Chapter Seven

Magnetic Can Crushing

This project is an excellent demonstration of pulsed magnetics where a normal, everyday aluminum soda can is crushed into an hourglass configuration (see Figure 7-1). Several of these systems are on interactive display at public museums where it has been reported that students would bring their own cans, placing them into the test coil and keeping them as souvenirs.

Construction requires a working high-energy pulser as described in Chapter 3, "High-Energy Pulser," along with the assembly of the shaping coil. It is shown using energy-storage capacitors currently available on the surplus market. These values can be altered within limits to allow the use of available capacitors that the builder may have on hand. The project requires experience in the handling of high-voltage, high-energy devices. This can also be an excellent museum demonstration when performed by a qualified adult.

Setup and Operation

To begin the project, proceed as follows:

1. Assemble the warping and blocking coils, as shown in Figure 7-2.

2. Position the warping coil assembly at the desired location and connect it up as shown.

Figure 7-1 *A magnetically crushed can*

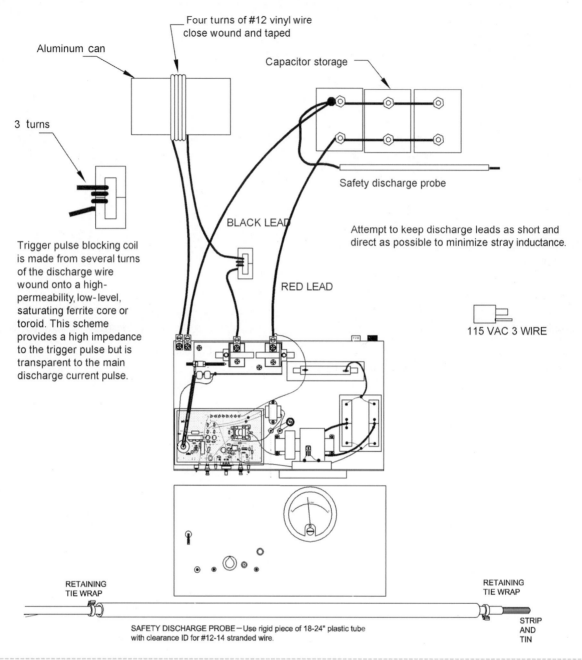

Four turns of #12 vinyl wire close wound and taped

Aluminum can

Capacitor storage

3 turns

Safety discharge probe

BLACK LEAD

Attempt to keep discharge leads as short and direct as possible to minimize stray inductance.

RED LEAD

Trigger pulse blocking coil is made from several turns of the discharge wire wound onto a high-permeability, low-level, saturating ferrite core or toroid. This scheme provides a high impedance to the trigger pulse but is transparent to the main discharge current pulse.

115 VAC 3 WIRE

RETAINING TIE WRAP

RETAINING TIE WRAP

STRIP AND TIN

SAFETY DISCHARGE PROBE—Use rigid piece of 18-24" plastic tube with clearance ID for #12-14 stranded wire.

Figure 7-2 *Setup for magnetic can crusher*

Note the inclusion of the trigger pulse blocking inductor in one of the coil leads. This part provides reliable triggering when using low-impedance loads.

You may make a suitable stand that will automatically position the pop can. Note that the coil is at a ground potential and poses a minimal shock threat when used as directed. It can be safely set up for public demonstration. Contact the factory for suggestions.

3. Set the controls as follows (reference front panel drawing, see Figure 3-7):

Voltage Control Switch (R8/S2) should be set to FCCW (OFF).

High-Voltage Power Switch (S4) should be set to OWN (OFF).

Charging Switch (S1) must be pressed to activate the charging action.

4. Plug into 115 *volts-alternating current* (VAC) and turn on the voltage control to verify both *light-emitting diode* (LED) lamps coming on:

    The *red lamp (LA1)* indicates the system is in an "on" state."

    The *green or yellow lamp (LA2)* comes on but extinguishes in the charge cycle and comes on when the preset voltage level is reached, indicating the system is ready to "fire."

5. Turn the voltage control off and unplug it once the previous steps are verified.

6. Set the charge voltage control to the 3,000-volt setting as marked on the panel. Plug the unit into 115 VAC and turn on the high-voltage safety switch S4.

7. Depress the *charge switch (S1) (the green or yellow lamp now goes off) and note the increasing voltage on the panel voltmeter reading, indicating the energy storage capacitors are charging. Allow it to charge to the preset voltage as in step 6. Note the LA2 lamp now turns on when the set level is reached, indicating that the system is ready to fire.

8. Depress the fire switch and note a loud report of discharge in the spark switch. You should see some deformation around the can under the coil. Repeat at 4,000 volts and the can deforms to about 50 percent of its original shape.

Special Notes

Note that system energy varies with the square of the increased voltage. For example, energy will increase by 9 when the voltage is increased from 1,000 to 3,000. You may go to 4,500 volts for an even greater effect, add more capacity, or both to really devastate the can.

Note that continued operation at over 4,000 volts eventually will erode the spark switch. Do not go

over the 4 *kilovolts* (Kv) setting except for experimenting. The system may be test-fired up to 5,000 volts for experimental research but may require the continual maintenance of spark switch electrodes. Always increase in 500-volt increments to verify system integrity.

It may be a good idea to check the front panel meter calibration from time to time. Use an externally calibrated meter connected across the energy storage capacitors to verify the panel voltage settings.

The spark gap is factory set at .1 inches for operation at 3 to 4 Kv. Lower-voltage operation may require decreasing the gap for reliable triggering. Operations over 4 Kv will require widening to prevent premature triggering. You may have to tweak the gap by loosening one of the Allen set screws for your particular range of operation.

You should place a high-permeability ferrite core wound with two to three turns on one of the leads from the crushing coil to the G2. This part will increase the inductance of this lead to block the trigger pulse from being shorted by the low impedance of this coil yet quickly saturate at the main discharge current.

It is important that the all connection leads from the capacitor bank the pulser and the crusher coil be as short as possible to reduce lead inductance for optimum results.

The spark switch may prematurely fire if dirty or improperly gapped.

Danger! Always unplug and discharge energy-storage capacitors using the safety discharge probe before making any adjustments.

If the system uses a push-on/off charging switch, it will automatically recharge the capacitor bank after firing. Basic units use a push-on switch for the charging switch, as charging always requires the switch to be depressed, providing safer operation at the cost of convenience.

Part Two

Handheld Burning CO₂ Gas Laser

This chapter offers an excellent, high-powered laser project that provides a full 20 watts of continuous output power and is meant to be conducted by experienced users. The system is shown constructed with all the necessary compliance for a class 4 laser device. A water-cooled, sealed laser tube is powered by a special high-efficiency, energy-conditioning circuit providing either a battery for field use or normal 115 *volts-alternating current* (VAC) for laboratory use.

This project is shown as being built in two parts (see Figure 8-1). The first is the power supply and cooling system housed in an enclosure along with all the controls, meters, and indicators. The second part is the laser head consisting of the laser tube with optics enclosed in a cylindrical tube and connected to the power supply via a 3- to 4-foot "umbilical" cable supplying coolant and electrical power. An alternative

Figure 8-1   *Carbon dioxide burning laser for laboratory use*

assembly scheme is shown for portable field and handheld applications using rechargeable batteries.

Expect to spend up to $500 for this advanced and rewarding project. The actual cost for a laser with this capability would be $5,000 or more. This project requires basic electromechanical shop skills with all parts, including protective eyewear available through Information Unlimited at www.amazing1.com. The finished device will conform to laser compliances and must be used with protective eyewear. A complete list of the necessary parts is provided at the end of the chapter in Table 8-1.

Basic Description

A gas laser tube is shown in Figure 8-1 with integral optics and a cooling jacket. This approach greatly simplifies construction, providing a neat, compact, and versatile project. The power supply used is capable of operating from a 115 VAC line or from batteries using a 200-watt sine wave inverter.

The project is shown constructed with all the necessary compliances for use in the trade or as a low-cost laboratory device. Cooling is accomplished by circulating water through the laser head cooling jacket to a forced air-cooled heat transfer radiator. A fan is used for supplying the airflow.

Theory of Operation

A carbon dioxide laser is by far one of the simplest laser devices to assemble and operate. This asset is sometimes a disadvantage as this energy can be dangerous in the hands of the inexperienced, both to the builder and his or her surroundings.

As shown in Figure 8-2, a CO_2 laser is a two-level, vibrational device emitting in the 10.6-micron infrared region. Creating lasers is accomplished by electrically exciting nitrogen gas (N_2) to an energy level close to that of the CO_2 molecule. The main dissipation of this energy is in the resonant transfer to the CO_2 molecules, causing them to change from a ground state to the excited level 2 state. You will note that the lower laser level 3 state is not prone to this

transfer of energy; hence, a population inversion now exists between levels 2 and 3.

Population inversion is an unnatural state of an atom that must be forced by an external influence. It is where higher-energy-level bands are more inhabited than lower bands occurring naturally. A population inversion is necessary for stimulating laser emission. The laser action commences as a result of the stimulated energy transition from level 2 to 3. A means of returning level 3 energy to ground level is accomplished by quenching it with a third gas being helium. Obviously, if level 3 were allowed to build up, the population inversion would soon be reduced between 2 and 3, hence terminating laser action.

It should be noted that the CO_2 gas molecule consists of many modes of vibration that contribute to the laser frequency. The main mode of vibration is around 10.6 microns and the system is often categorized in respect to this wavelength. Mother nature's atmosphere offers a natural window to this wavelength, and consequently a CO_2 laser is excellent in respect to an application requiring propagation over long distances. A CO_2 laser of sufficient power obviously is an excellent candidate for use as a direct energy-beam weapon.

The laser described in this project is referred to as a *low flow axial device*. This is the simplest approach and is nothing more than an airtight glass discharge tube with gas flowing by convection in and out of the surrounding reservoir glass enclosure. Electrodes placed into the flowing gas excite the nitrogen molecules, commencing laser action.

Circuit Operation

The power cord (CO1) supplies a 115/220 VAC from a wall receptacle for laboratory use or a sine wave inverter for portable battery operation in the field. The third green grounding lead wire must be securely attached to the metal enclosure cover (COV1) for safety.

As shown in Fiure 8-3, the key switch (SW1) energizes the primary circuits and cooling fan (FAN). A fuse (FH1/FS1) protects the main circuit from any catastrophic faults. The indicator lamp (NEON)

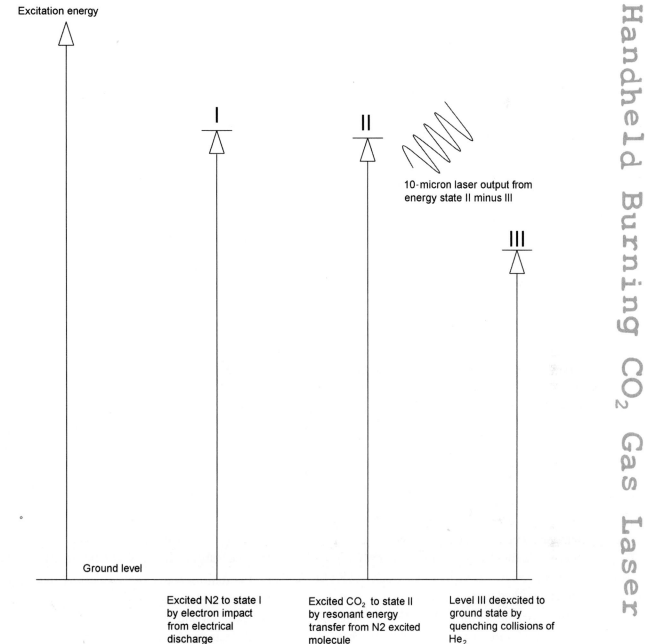

Excitation energy

I

II

10-micron laser output from energy state II minus III

III

Ground level

Excited N2 to state I by electron impact from electrical discharge

Excited CO$_2$ to state II by resonant energy transfer from N2 excited molecule

Level III deexcited to ground state by quenching collisions of He$_2$

Figure 8-2 *Carbon dioxide energy-level diagram*

requires a current-limiting resistor (R12) and lights up when SW1 is turned on. The peak charging current to C10 and C11 is limited by a rush resistor (RX). The rectifiers (D1, D2, D3, D4) are shown in a full bridge configuration when used for a 220 VAC input. For 115 VAC operation, only D1 and D3 are used along with a jumper (JMPR1) in a voltage doubler configuration. The output voltage across C10 and C11 is 300 *volts-direct current* (VDC) and is referred to as the plus and negative rails.

The secondary circuits are controlled by a switch (SW2) that energizes a low-voltage transformer (T2). Twelve VAC is a full wave rectified by diodes D30 to D33 and filtered by a capacitor (C30). A dropping resistor (R28) and zener diode (Z1) regulate output to 12 VDC. The timer (IC2) supplies a delayed voltage to the oscillator driver (IC1). It is connected as a monostable switch with a delay being determined by the time constant of capacitor C31 and a resistor R27. The delay is 10 seconds with the values as shown and

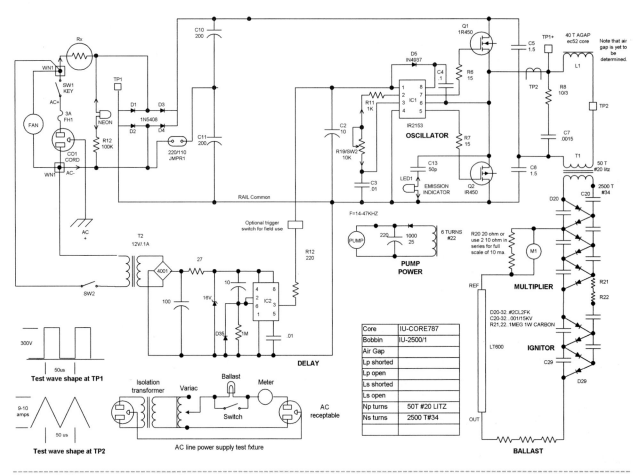

Figure 8-3 *Burning laser schematic*

is necessary for *Bureau of Radio and Health* (BRH) compliance. High voltage is now controlled by this action as IC2 supplies power to oscillator IC1. The C2 capacitor provides the high instant of current for the output pulses and must be physically close to IC1.

These drive pulses turn off and on the main *metal oxide field effect transistor* (MOSFET) switches (Q1 and Q2). The switching frequency is determined by the time constant of the power control pot (R19) and the timing capacitor (C3). A trimpot (R10) adjusts the range limits of R19. The switching circuit is in a half-bridge configuration where the MOSFET connected to the positive end (+ rail) must be driven with its source pin referenced at 150 volts above the common end (− rail). This is accomplished by biasing a bootstrap capacitor (C4) through the ultra-fast diode (D5) providing the correct DC level to fully control Q1. Resistors (R6 and R7) eliminate the high-frequency parasitic oscillation occurring as a result of rapidly switching the capacitive load of the

MOSFET gates. Emission indication is by LED1, which lights up once Q1.2 starts switching power to the laser tube. Capacitor C13 limits the current flow.

The transition time of the switched pulses across Q1 and Q2 is slowed down by a network consisting of capacitor C7 and resistor R8. The resultant time constant limits the dv/dt fast rate of voltage change that could cause a premature activation at the wrong switching MOSFET, creating a catastrophic fault mode. Capacitors C5 and 6 provide a voltage midpoint and produce the necessary storage energy to maintain the voltage level of the individual pulses.

Transformer T1 increases the 150-volt average pulses fed to the primary winding to 7,500 volts across the secondary winding. Another winding of only several turns supplies the 12 volts required for the cooling fan. This current is rectified by diode D34 and filter capacitor C33.

You will note that the output load for the power supply is a gas discharge being the laser tube. A gas discharge requires a ballasting network to control the discharge current from reaching destructive levels. This ballasting network could be a resistance or reactive impedance. The ballasting voltage is a significant proportion of the actual operating level required to provide the discharge stability.

Resistive ballast is very "lossy" and cumbersome as large resistors would be required, dissipating many watts of wasted energy. By using reactive ballasting (L1), very little energy is wasted as the now reactive current is simply switched back and forth from magnetic to electric. The reactive current still must be switched by Q1 and Q2; consequently, L1 must be carefully determined.

The now ballasted output across T1 is voltage that is tripled to over 22 *kilovolts* (Kv) by multiplier diodes (D20 to D25) and capacitors (C20 to C25). Resistors (R21 and R22) feed a parasitic multiplier consisting of diodes (D26 to D29) and capacitors (C26 to C29). This section provides the added ignition voltage to trigger the laser tube to the on state. Once the discharge current is sustained, D26 to D29 are all forward biased and the ignition section is automatically disabled. The anode of the laser tube is shown with stabilizing resistors (R23 to R25) and helps to prevent discharge jitter.

The cathode of the laser tube is returned through meter (M1) and ranging shunt resistors (R20) intended to provide 10 milliamps of full-scale deflection current.

Assembly Steps

To begin the assembly, follow these steps:

1. Fabricate the main enclosure, as shown in Figure 8-4. We used .063 black Lexan (polycarbonate) as it can be easily formed in a leaf brakemetal bending machine. Note that only overall dimensions are shown, as you should trial-fit all the components in place before drilling the holes. You could use .035 aluminum in place of the Lexan, providing proper attention is paid to the high-voltage insulating points.

2. Assemble the power supply circuit board as shown in Figure 8-5. The project uses an available *printed circuit board* (PCB) available from www.amazing1.com. More experienced builders may want to lay out their PCB or use vector board. You are, however, cautioned to observe the darkened lines on the schematic

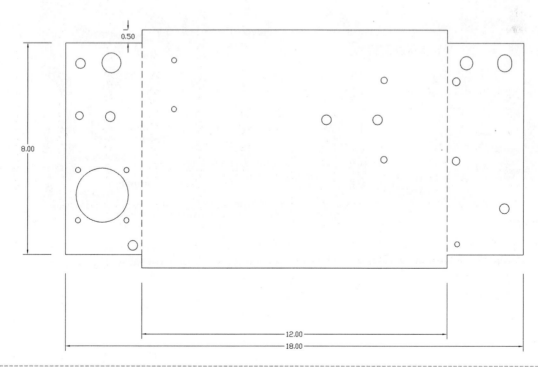

Figure 8-4 *Fabrication of power module enclosure EN1*

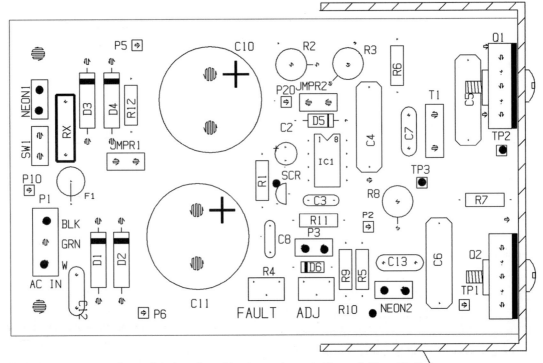

1. Black dots are connection points to external leads as shown in Figure 8-10 and 8-11. ⎫ HSB1 from Fig. 8-8

2. The PC board as shown is used in several projects and does not use all the components as indicated.

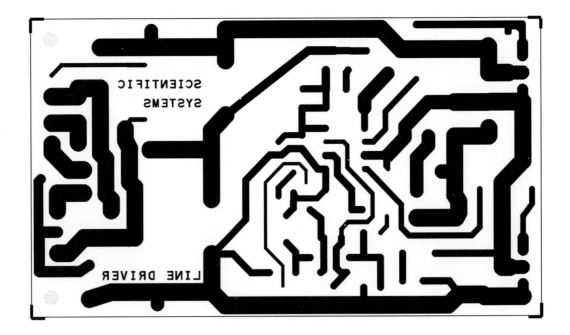

Assembly of the board assembly is shown using a printed circuit board available through www.amazing1.com Builders may use a vector board and use the component leads to duplicate the foil traces as shown.

Figure 8-5   *PCB assembly*

as these indicate high current and direct connection runs requiring the use of #18 bus wire. You will note that the PCB contains stenciling showing component locations. Only the parts shown on the schematic and parts list are to be used. Many parts will not be used on this project.

3. Assemble the delay circuit board, as shown in Figure 8-6.

4. Assemble the high-voltage multiplier circuit board, as shown in Figure 8-7. You may use vector board.

5. Assemble the heat sink bracket and the current-controlling inductor L1, as shown in Figure 8-8. It is important that the holes for attaching Q1 and Q2 are in perfect alignment when inserted into the PCB. Leads must not be stressed and the body of the transistor must be in total contact with the bracket for an efficient heat transfer.

6. Lay out the subassemblies in the main enclosure, as shown in Figure 8-9. Follow the layout locations, drilling the mounting holes as you go. High-voltage components are secured in place by large tie wraps. More conventional screws and nuts secure other assemblies.

Special note on transformer T1: This part is available assembled and ready to use. Experienced builders may want to build and wind their own. The specifications are given as the following:

The core set is a Magnetics #40658 or #40787 in P material.

The secondary winding is 2500 turns of #34 heavy coated magnet wire on a 1-inch inner-diameter bobbin. Use high-voltage TV flyback winding technique with "start" at ground potential.

For the primary wind, use 50 turns of #20 Litz wire in even layers with additional winding of 6 turns of #22 wire for pump motor power. These primary windings are wound on a diameter that will fit on the core and also slide into the 1-inch innerdiameter of the secondary.

7. Wire in first-stage components, as shown in Figure 8-10. Attempt to use existing leads and only splice using appropriate wire nuts where absolutely necessary. Leads exiting the enclosure should be 18 inches or more and will be trimmed to length when finally assembled. Use a 36-inch piece of WR20 Kv high-voltage wire for the output lead-designated HVOUT.

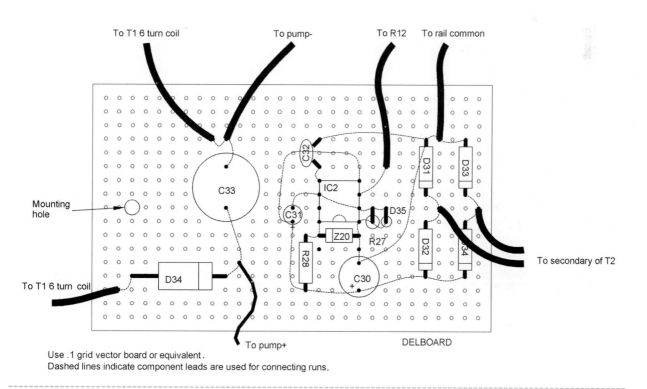

Use .1 grid vector board or equivalent.
Dashed lines indicate component leads are used for connecting runs.

Figure 8-6 *Delay control and pump power board*

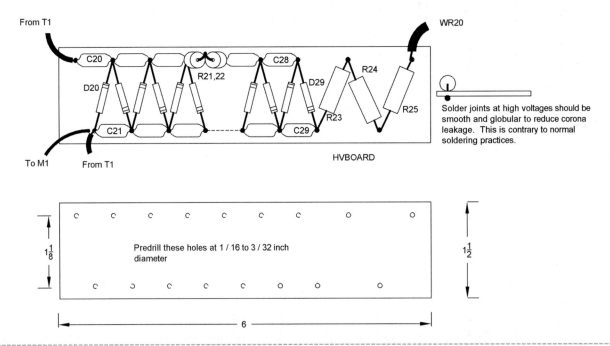

Solder joints at high voltages should be smooth and globular to reduce corona leakage. This is contrary to normal soldering practices.

HVBOARD

Predrill these holes at 1 / 16 to 3 / 32 inch diameter

$1\frac{1}{8}$

$1\frac{1}{2}$

6

Figure 8-7 *High-voltage multiplier and igniter board*

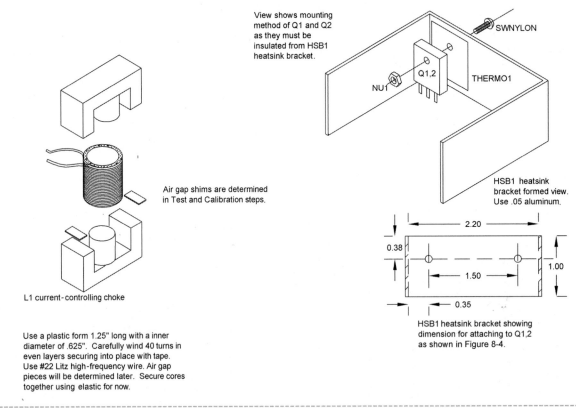

View shows mounting method of Q1 and Q2 as they must be insulated from HSB1 heatsink bracket.

SWNYLON

Q1,2

NU1

THERMO1

HSB1 heatsink bracket formed view. Use .05 aluminum.

2.20

0.38

1.00

1.50

0.35

HSB1 heatsink bracket showing dimension for attaching to Q1,2 as shown in Figure 8-4.

Air gap shims are determined in Test and Calibration steps.

L1 current-controlling choke

Use a plastic form 1.25" long with a inner diameter of .625". Carefully wind 40 turns in even layers securing into place with tape. Use #22 Litz high-frequency wire. Air gap pieces will be determined later. Secure cores together using elastic for now.

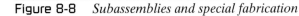

Figure 8-8 *Subassemblies and special fabrication*

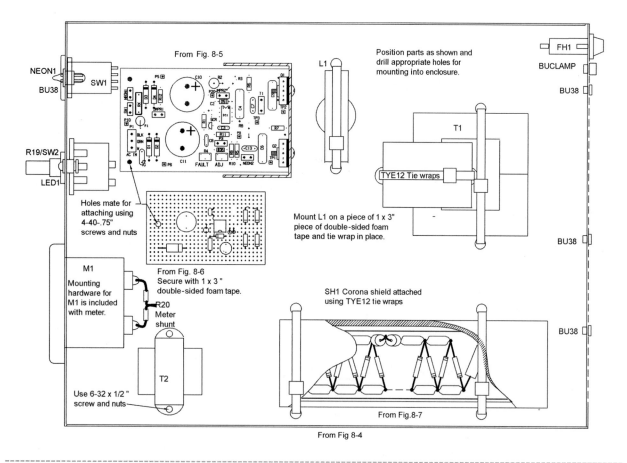

Figure 8-9 *Assembly of power module enclosure showing parts location and securing*

Use a 60-inch piece of WR5 Kv meter lead wire for the high-voltage return lead-designated HVRET. Use WR20 #20 wire for all other leads.

8. Wire in second-stage components as shown in Figure 8-11, but use WR24 #24 wire and twist where leads are paired.

9. Fabricate the cover section as shown in Figure 8-12 from .35 sheet aluminum. Use a hole saw or circle cutter for the large air flow hole or drill a series of small holes, removing the center piece and filing to a smooth contour.

10. Attach the radiator, fan, and pump to the cover, as shown in Figure 8-13.

11. Assemble the cooling system hoses and fittings, as shown in Figures 8-14 and 8-15. Stretch the .25-inch *inner diameter* (ID) latex hose onto the radiator nipples or mate it to the .25-inch fittings. You may use in-line plug and jacks for the power leads to the fan and pump so the cover can be completely sepa-

rated. Nonetheless, leads must be sufficiently long so the cover can be moved out of the way for access to the circuitry. Note the safety-grounding lug for the metal cover.

12. Examine the LTUB1 laser tube and verify it with Figure 8-16. Pay heed to this data, as the tube is fairly delicate and must be operated properly.

Laser Head Construction

The following section shows the assembly of the laser head:

1. Create the laser head enclosure from a 22-inch length of 2.5-inch ID × .125 clear polycarbonate wall tubing, as shown in Figure 8-17. The clear plastic allows visual access to the innards yet is opaque to the 10.6-micron laser energy.

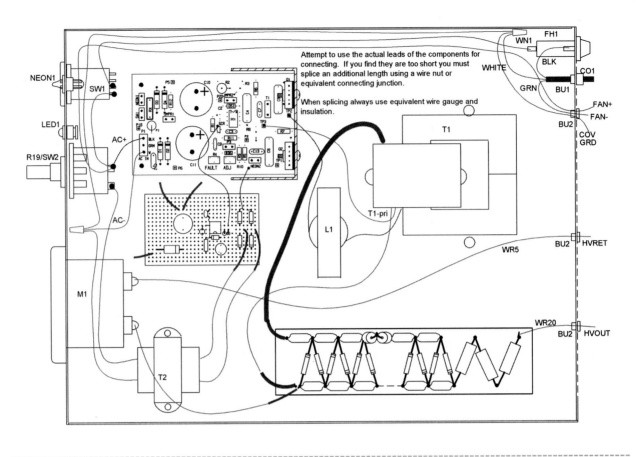

Attempt to use the actual leads of the components for connecting. If you find they are too short you must splice an additional length using a wire nut or equivalent connecting junction.

When splicing always use equivalent wire gauge and insulation.

Figure 8-10 *First-stage wiring aid*

2. Create DISKFRONT and DISKREAR from a solid 2.5-inch polycarbonate rod as shown. The front output disk has a 1-inch aperture for the beam output or as a sleeve for a focusing lens. The rear disk is fabricated with four .25-inch holes for feeding through the cooling hoses and the power leads. You may strain or relieve these feed points using tie wraps or clamp bushings. This is not shown and left to the discretion of the builder.

3. Fabricate the laser head base (HEADBASE) from a 22-inch length of a .035-inch-wide piece of aluminum sheet, as shown in Figure 8-17. Note the acute angle bend for contact to the sides when sitting flush and screwed to the enclosure.

4. Fabricate the aperture flap shutter (FLAP) from a 1.25 × 1.75 × .035 piece of aluminum. Round the corners and drill the pivot hole, as shown in Figure 8-17.

5. Measure out the hoses and power leads to the laser tube and thread through the respective holes in the REARDISK.

6. Assemble everything, as shown in Figure 8-18, but do not screw on the cover at this time. Note the instructions for labeling.

Testing Your Laser

It assumed that your system is accurately and properly wired per the instructions. To test the laser , follow these steps:

1. Obtain a container of clean drinking water and insert the fill and drain hoses, as shown in Figure 8-18. With your mouth, suck in water through the drain tube until the laser tube water jacket is reasonably filled. Insert the

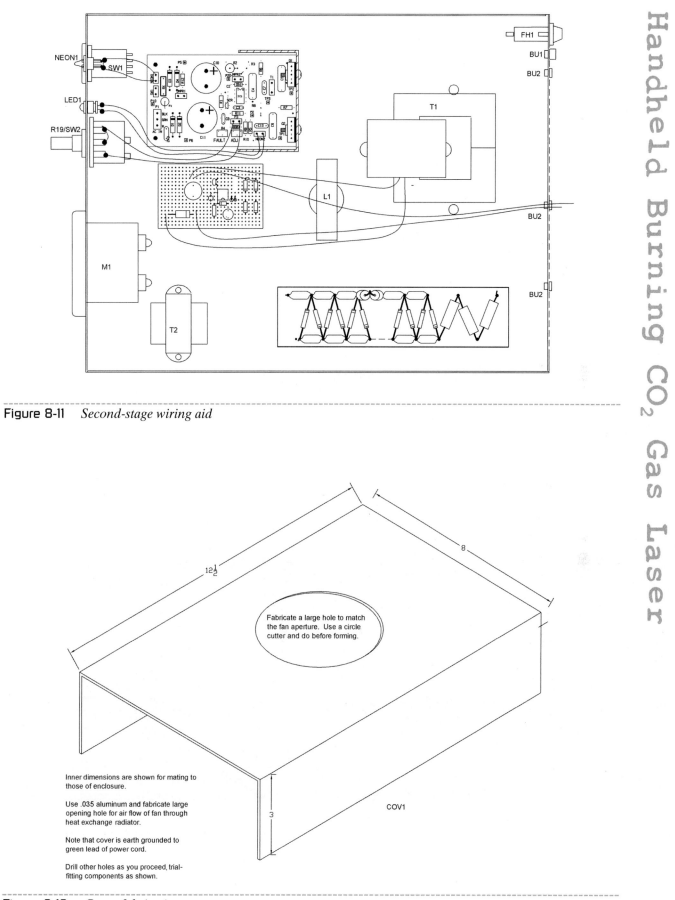

Figure 8-11 *Second-stage wiring aid*

Fabricate a large hole to match the fan aperture. Use a circle cutter and do before forming.

Inner dimensions are shown for mating to those of enclosure.

Use .035 aluminum and fabricate large opening hole for air flow of fan through heat exchange radiator.

Note that cover is earth grounded to green lead of power cord.

Drill other holes as you proceed, trial-fitting components as shown.

Figure 8-12 *Cover fabrication*

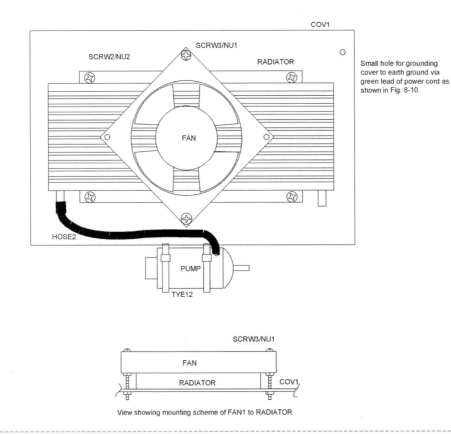

View showing mounting scheme of FAN1 to RADIATOR

Figure 8-13   *Cover and cooling components*

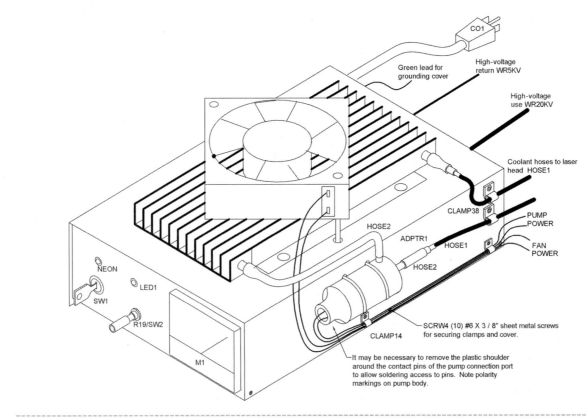

Figure 8-14   *Final view of the assembled controller and cooling system*

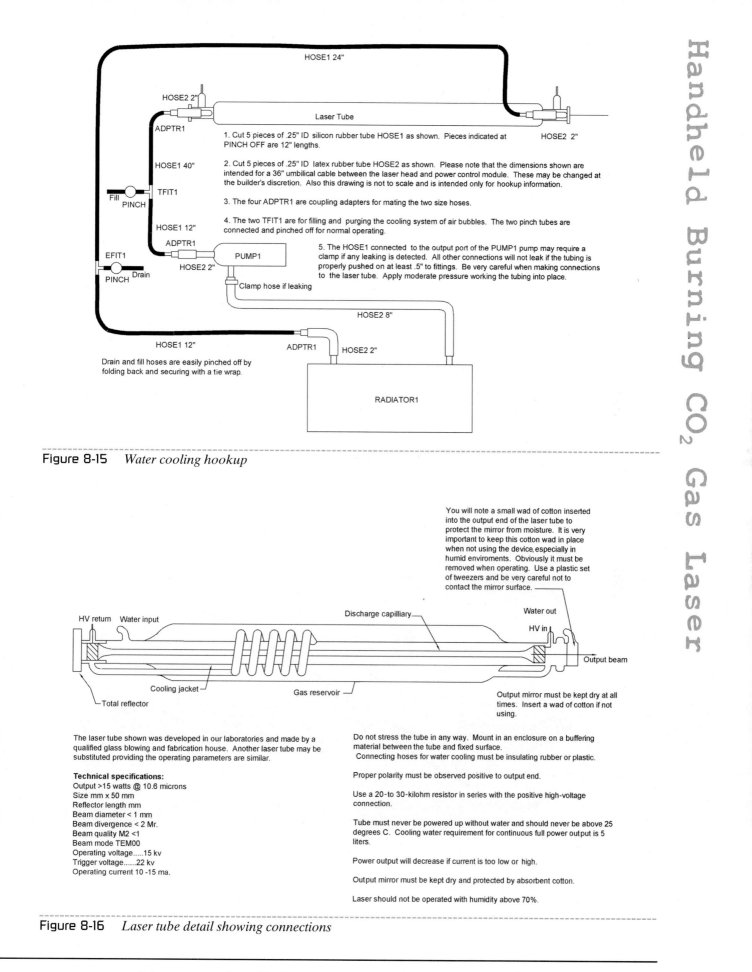

Figure 8-15   *Water cooling hookup*

HOSE1 24"

HOSE2 2"

ADPTR1

Laser Tube

HOSE2 2"

1. Cut 5 pieces of .25" ID silicon rubber tube HOSE1 as shown. Pieces indicated at PINCH OFF are 12" lengths.

HOSE1 40"

2. Cut 5 pieces of .25" ID latex rubber tube HOSE2 as shown. Please note that the dimensions shown are intended for a 36" umbilical cable between the laser head and power control module. These may be changed at the builder's discretion. Also this drawing is not to scale and is intended only for hookup information.

Fill   TFIT1
PINCH

3. The four ADPTR1 are coupling adapters for mating the two size hoses.

HOSE1 12"

4. The two TFIT1 are for filling and purging the cooling system of air bubbles. The two pinch tubes are connected and pinched off for normal operating.

ADPTR1

EFIT1   PUMP1
PINCH Drain
HOSE2 2"

5. The HOSE1 connected to the output port of the PUMP1 pump may require a clamp if any leaking is detected. All other connections will not leak if the tubing is properly pushed on at least .5" to fittings. Be very careful when making connections to the laser tube. Apply moderate pressure working the tubing into place.

Clamp hose if leaking

HOSE2 8"

HOSE1 12"   ADPTR1   HOSE2 2"

Drain and fill hoses are easily pinched off by folding back and securing with a tie wrap.

RADIATOR1

You will note a small wad of cotton inserted into the output end of the laser tube to protect the mirror from moisture. It is very important to keep this cotton wad in place when not using the device, especially in humid enviroments. Obviously it must be removed when operating. Use a plastic set of tweezers and be very careful not to contact the mirror surface.

HV return   Water input   Discharge capillary   Water out

HV in

Output beam

Total reflector   Cooling jacket   Gas reservoir

Output mirror must be kept dry at all times. Insert a wad of cotton if not using.

The laser tube shown was developed in our laboratories and made by a qualified glass blowing and fabrication house. Another laser tube may be substituted providing the operating parameters are similar.

Technical specifications:
Output >15 watts @ 10.6 microns
Size mm x 50 mm
Reflector length mm
Beam diameter < 1 mm
Beam divergence < 2 Mr.
Beam quality M2 <1
Beam mode TEM00
Operating voltage.....15 kv
Trigger voltage......22 kv
Operating current 10 -15 ma.

Do not stress the tube in any way. Mount in an enclosure on a buffering material between the tube and fixed surface.
Connecting hoses for water cooling must be insulating rubber or plastic.

Proper polarity must be observed positive to output end.

Use a 20-to 30-kilohm resistor in series with the positive high-voltage connection.

Tube must never be powered up without water and should never be above 25 degrees C. Cooling water requirement for continuous full power output is 5 liters.

Power output will decrease if current is too low or high.

Output mirror must be kept dry and protected by absorbent cotton.

Laser should not be operated with humidity above 70%.

Figure 8-16   *Laser tube detail showing connections*

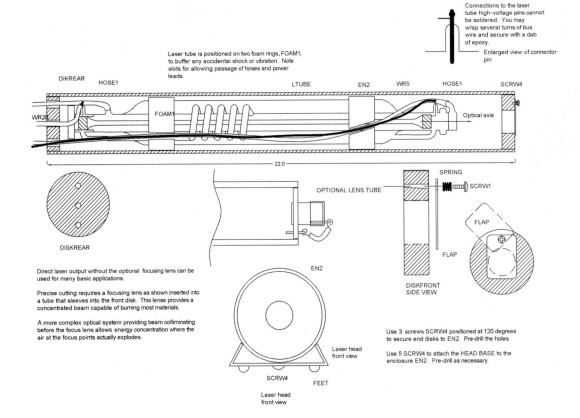

Connections to the laser tube high-voltage pins cannot be soldered. You may wrap several turns of bus wire and secure with a dab of epoxy.

Enlarged view of connector pin

Laser tube is positioned on two foam rings, FOAM1, to buffer any accidental shock or vibration. Note slots for allowing passage of hoses and power leads.

DIKREAR   HOSE1   LTUBE   EN2   WR5   HOSE1   SCRW4
WR20   FOAM1   Optical axis

22.0

DISKREAR

Direct laser output without the optional focusing lens can be used for many basic applications.

Precise cutting requires a focusing lens as shown inserted into a tube that sleeves into the front disk. This lense provides a concentrated beam capable of burning most materials.

A more complex optical system providing beam colliminating before the focus lens allows energy concentration where the air at the focus points actually explodes.

OPTIONAL LENS TUBE

SPRING
SCRW1
FLAP
FLAP
DISKFRONT
SIDE VIEW

EN2

Laser head front view

SCRW4   FEET
Laser head front view

Use 3 screws SCRW4 positioned at 120 degrees to secure end disks to EN2. Pre-drill the holes.

Use 5 SCRW4 to attach the HEAD BASE to the enclosure EN2. Pre-drill as necessary.

Figure 8-17   *Laser head side and end views*

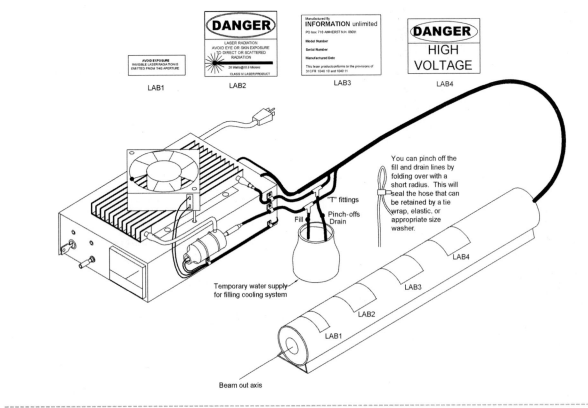

AVOID EXPOSURE
INVISIBLE LASER RADIATION IS EMITTED FROM THIS APERTURE.

LAB1

DANGER
LASER RADIATION
AVOID EYE OR SKIN EXPOSURE TO DIRECT OR SCATTERED RADIATION
26 Watts@10.6 Microns
CLASS IV LASER PRODUCT

LAB2

Manufactured By
INFORMATION unlimited
PO box 716 AMHERST N.H. 09031
Model Number
Serial Number
Manufactured Date
This laser product conforms to the provisions of 31CFR 1040 10 and 1040 11

LAB3

DANGER
HIGH VOLTAGE

LAB4

You can pinch off the fill and drain lines by folding over with a short radius. This will seal the hose that can be retained by a tie wrap, elastic, or appropriate size washer.

"T" fittings
Fill   Pinch-offs   Drain

Temporary water supply for filling cooling system

LAB4
LAB3
LAB2
LAB1

Beam out axis

Figure 8-18   *Laser system for 115 VAC laboratory use showing labels*

drain hose in the water container and verify that both hoses are secured fully submerged.

2. Connect a variable DC voltage source capable of several amps across C33, shown in Figure 8-6, observing polarity. Slowly turn up the voltage until the pump starts to circulate water. Allow it to continue to run while tapping and tipping the laser head and radiator section, purging out all the air bubbles. Let it run for 30 minutes and verify that the system is not leaking and there are no air bubbles. Note that you may have to suck the air on the line several times to get the pump to prime. Disconnect the DC supply.

3. Obtain a variable voltage transformer (a variac and an isolation transformer, and ballast the input through a 60-watt bulb). Figure 8-3 shows a schematic of this very useful testing system for line-powered circuits. Verify switches SW1 and SW2/R19 are in their off position. Connect a dual-input scope to test point TP1 to measure the voltage and a current probe at test point TP2. Short the output terminals of secondary T1 and temporarily assemble L1 without any air gaps.

4. Verify that the laser tube is electrically connected with the right polarity and that the pump and fan are connected. Turn the key switch on and slowly turn up the variac, noting the neon indicator lamp comes on with the cooling fan.

5. Rotate R19 to an on position and note that the LED lamp has a 5- to 10-second delay.

6. Note a wave shape, as shown in Figure 8-3, at TP1. Rotate R19 full *clockwise* (CW) and note the increasing period. Adjust trimpot R10 for a full period of 50 microseconds (20 kHz). Note a very small current at TP2.

7. Shut the power down and put a layer of approximately 10 mils (.01 inch)-thick tape or an equivalent for the first air gap setting, as shown in Figure 8-8. Repeat this, increasing the air gap until you get the current waveform value as shown in TP2. You should have a peak current rise at TP2 of 7 to 8 amps with the 60-watt ballast in the line. This will increase to around 10 amps with the ballast removed (shorted). Our lab units require a gap thickness of .03 inches.

8. You are now ready to activate the laser tube obtaining output power. Secure the tube to a nonconductive surface and verify the power leads are the right polarity and are securely attached to the pins. Remove the shorting lead across TI output.

Caution: Do not attempt to solder to these pins, as the tube will be destroyed. Obtain a 3- to 4-inch piece of wood and place it securely 10 to 12 inches in front of the laser tube output end. Remove the protective cotton swab and gently clean the mirror with a Q-tip and some acetone.

9. Connect a current meter that can accurately measure 10 milliamps and temporarily insert it in a series with the laser tube ground lead.

10. Turn off all switches and controls. Plug the system into the unballasted line with an AC ammeter. With the laser tube pointing at the wooden target and your safety glasses on, turn on the key switch, noting the neon lamp lighting and the fan coming on.

11. Click on R19 and note the LED lighting with about a 5- to 10-second delay. If you did your homework and Murphy's laws are lenient, you should detect a burning spot on the wooden target as you rotate the control clockwise. Adjust to 5 milliamps as indicated on the external current meter. The laser output should be burning the wood with an almost white-hot impact point. Note the pump running and water circulating through the tube. Measure 5 to 6 volts DC across the pump terminals. Calibrate the panel meter to midscale for 5 milliamps, giving a full-scale value of 10. The meter is calibrated by shunting with a resistor. We use a 20-ohm basic value and trim with other larger values to bring the meter reading into the proper range.

12. Verify all functions, air bubbles, the excessive heating of parts, potential high-voltage breakdown points, and loose connections and correct as necessary.

13. Finally assemble and label everything. Study the following notes.

Special Notes

Included is some useful information about the laser system.

Optional Focusing Lens

Your laser system may be used with a focusing lens for close work such as etching, wood and plastic engraving, materials testing, precision cutting and drilling, and other obvious fine fabrication. The focused energy will cause a surface to burn like the sun! The focusing lens can be mounted to a tube that sleeves into the front disk aperture. Use your own ingenuity to secure it in place.

The raw beam from the laser itself without any optional optics will project a burning spot of about .3 to .4 inches at a distance of 10 feet. Obvious caution is required when projecting the beam without a known target stop, as the energy can be a serious fire hazard.

Battery Operation

Portable field use simply involves using a rechargeable battery pack. A garden tractor or motorcycle battery can provide up to 1 hour of continuous operation when the beam current is set to 5 milliamps. Our shop field model uses four 12-volt, 4-amp gel cells with a 200-watt sine wave converter, as shown in Figure 8-19. Again, the device is a two-part system with the batteries, converter, and power supply packaged as a side pack with a shoulder strap. The laser head is built into a gun configuration with a pushbutton trigger switch, as shown in optionally wired Figure 8-3.

Cleaning Output Optics

It is important to protect the output lens with a cotton swab to absorb moisture that would damage the output optics. The output mirror should be cleaned with a Q-tip immersed in acetone periodically.

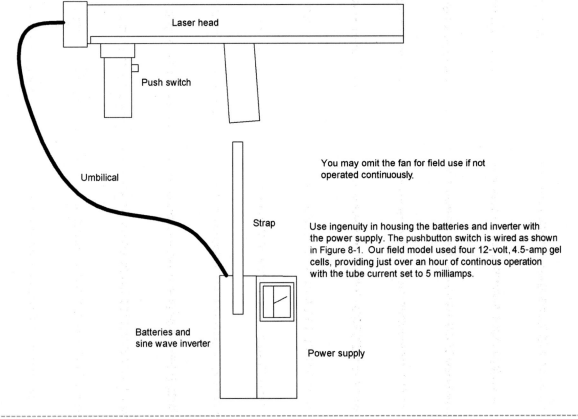

Laser head

Push switch

Umbilical

Strap

You may omit the fan for field use if not operated continuously.

Use ingenuity in housing the batteries and inverter with the power supply. The pushbutton switch is wired as shown in Figure 8-1. Our field model used four 12-volt, 4.5-amp gel cells, providing just over an hour of continous operation with the tube current set to 5 milliamps.

Batteries and sine wave inverter

Power supply

Figure 8-19   *Portable battery-operated system*

Always remember to remove the cotton before powering up or serious damage to the mirror will result.

Purging the Cooling System

It may be necessary to purge out the air bubbles from the cooling system from time to time. Do not allow large air bubbles to form, as the mirrors may not be adequately cooled due to these air pockets.

Operating Time

The system may be left on indefinitely, providing there is no excessive heating of the electronics and the cooling system adequately keeps the laser tube at ambient or below.

Increasing Power Output

The system is shown operating at a tube current of 5 milliamps, resulting in around 10 to 15 watts of output. You may further decrease the air gap in L1 to obtain more tube current of up to 10 milliamps. This will result in up to 20 watts but may overheat the electronics if left on for an extended period.

General-Purpose Test Fixture for Line-Powered Circuits

A universal test fixture schematic is shown in the inset drawing in Figure 8-3. This fixture is strongly recommended for the serious hobbyist or builder constructing line-powered circuitry. Not only does it provide some electrical safety, but it also can help prevent catastrophic component burnouts due to prototype and project wiring errors. Our labs use several of these fixtures ranging in line currents of 3 amps at 115 VAC to 100 amps, 3 phase, at 220 VAC.

The suggested unit for most projects up to 500 watts is to obtain an isolation transformer and a variac good for up to 5 amps. Wire as shown using #18 wire and observe line cord polarity relative to the hot side. You can also monitor line current by connecting in a series with the hot lead to the output.

The earth-grounding wire is directly connected from an input cord to an output receptacle and to the case ground of the fixture. The output lead is also in series with a lightbulb socket with a shorting switch. This scheme allows you to set the maximum short-circuit fault current by ballast wattage selection. Once it is determined that the circuit is reasonably functional, the ballast is shorted out by the switch.

Table 8-1 Handheld burning CO_2 laser parts

Ref. #	Qty.	Description
		Main board assembly from Figure 8-5
R1		220-ohm, $1/4$-watt resistor (red-red-br)
R6, 7	2	15-ohm, $1/4$-watt resistor (br-grn-blk)
R8		10-ohm, 3-watt resistor *metal oxide* (MOX)
R10		10K vertical trimpot
R11		1K, $1/4$-watt resistor (br-blk-red)
R12		100K, $1/4$-watt resistor (br-blk-yel)
Rx		In rush current limiter #CL190
C2		10 *microfarad* (mfd)/25-volt vertical electrolytic capacitor
C3		.01 mfd/100-volt plastic capacitor
C4		.1 mfd/600-volt *metalized polypropylene* (MPP) capacitor
C5, 6	2	1.5 mfd/250-volt MPP capacitor
C7		.0015 mfd/600-volt MPP capacitor
C10, 11	2	220 to 330 mfd/200-volt vertical electrolytic capacitor
C13		50 *picofarad* (Pfd), 1 Kv ceramic capacitor
D1, 2, 3, 4	4	IN5408 1 Kv, 3-amp rectifier
D5		1N4937 1 Kv fast-switching diode
IC1		IR2153 *dual in-line package* (DIP) driver
Q1, 2	2	IRF450 MOSFET
PC1		PCB and wire as shown in Figure 8-5, #PCLINE
THERMO1	2	Insulating thermo pads for Q1 and Q2
SWNYLON	2	6-32 $\times$ $1/2$-inch nylon screws and metal nuts

Table 8-1 Continued

Ref. #	Qty.	Description
Remaining electrical parts		
FH1/3 A		Panel-mounted fuse holder and 3-amp slow-blow fuse
R19/SW2		10K pot and switch
R20		Meter shunt resistor start, 20 ohms and $1/4$ watt; see text
R21, 22	2	1-meg, 1-watt resistors
R23, 24, 25	3	10K 5-watt MOX resistors
R27		1-meg, $1/4$ -watt resistor
R28		27-ohm, $1/2$ -watt resistor
C20–29	10	.001 mfd, 15 Kv ceramic capacitors, #.001/15 Kv
C30		100 mfd, 25-volt vertical electrolytic capacitor
C31		10 mfd, 25-volt vertical electrolytic capacitor
C32		.01 mfd, 50-volt disc capacitor
C33		1000 mfd, 25-volt vertical electrolytic capacitor
D20–29	10	10 Kv, 100-milliampere, 100-nanosecond, high-voltage diode, #VG10HC
D30–33	4	1N4001 50-volt, 1-amp diode
D34		1N5401 50-volt, 3-amp diode
D35		IN914 small signal diode
Z1		16-volt, 1-watt zener diode
IC2		555 DIP timer
CO1		3-wire #18 power cord
NEON1		Neon indicator lamp with leads
SW1		Key switch—removable in off position only
T1		High-voltage ferrite switching transformer, #LABURT10; see text
T2		12-volt, 100 milliampere power transformer, #12DC/.1
L1		Current-controlling inductor shown in Figure 8-8, #LABURL10
M1		Panel meter with 50-microamp movement
LED1		Green ultra-bright LED
WN1	5	Small wire nuts for #20 wires
WN3	2	Medium wire nut for #18 wires

Ref. #	Qty.	Description
WR24	3 feet	#24 red and black pieces of hookup wire
WR20	6 feet	#20 red and black pieces of hookup wire
WR5KV	6 feet	5 Kv flexible test lead wire
WR20KV	5 feet	20 Kv silicon high-voltage wire
Cooling section parts		
FAN1		4-inch-square 110 *cubic feet per minute* (CFM) 115 VAC muffin fan
RADIATOR		Transmission cooler, JC Whitney, #COOLER
PUMP		12-volt automotive window washer pump, #PUMP1
LTUBE		20-watt CO_2 laser tube, #COTUB2; see text
HOSE1	10 feet	$1/8$ -inch ID silicon rubber tubing or equivalent
HOSE2	2 feet	$1/4$ -inch ID latex rubber tubing or equivalent
TFIT1	2	$1/8$ -inch T fitting
ADPTR1	4	$1/8$ - to $1/4$ -inch hose adapter
Fabricated parts		
EN1		Enclosure as shown in Figure 8-4
DEL BOARD		$.1 \times .1$ vector board cut 3×1.6 inches as shown in Figure 8-6
HVBOARD		$1/16$ -inch lexan 6- $\times$ 1.5-inch fabricate as shown in Figure 8-7
HSB1		Heatsink bracket as shown in Figure 8-8
COV1		Aluminum cover as shown in Figure 8-12
SH1		Corona shield fabricate from 7- $\times$ 1.5-inch *polyvinyl chloride* (PVC) tubing
HEADBASE		Aluminum base for laser head support as shown in Figure 8-17
EN2		Laser head tube 22- $\times$ 2.5-inch *innerdiameter* (ID) $\times$ .125-inch clear Lexan shown in Figure 8-17
DISKFRONT		Output end .5 $\times$ 2.5 *outerdiameter* (OD) Lexan disk as shown in Figure 8-17
DISKREAR		Feed end .5 $\times$ 2.5 OD Lexan disk as shown in Figure 8-17
FLAP		Aperture flap as shown in Figure 8-17
Hardware and small mechanical parts		
SCRW1	6	6-32 $\times$ .5-inch machine screw
SCRW2	4	8-32 $\times$ .5-inch machine screw
SCRW3	2	6-32 $\times$ 1.5-inch machine screw

Chapter Eight

Ref. #	Qty.	Description
SCRW4	20	6 × .25-inch sheet metal screw
SCRW5	3	4-40 × .75-inch machine screw
SWNYLON	2	6-32 × .5-inch nylon screw
NU1	10	6-32 machine hex nut
NU2	4	8-32 machine hex nut
NU3	6	4-40 machine hex nut
SPRING		.25- × .5-inch medium compression spring
TYE12	7	12-inch nylon tie wraps
LUG3		#8 solder lug for cover grounding
LUG4	2	Slip on lug for laser tube connection pins as shown in Figure 8-17
WN1	5	HI3 wire nuts
BUCLAMP		Clamp bushing for power cord

Ref. #	Qty.	Description
BU38	8	.25-inch feedthrough bushings
FEET	8	Rubber feet for power supply and laser head
CLAMP25	2	.25-inch nylon cable clamp
CLAMP38	2	.375-inch nylon cable clamp

Optional support items
150-watt sine wave inverter for 12 *volts-direct current* (VDC) operation, #SINE150
12-volt, 4.5-amp rechargeable battery 2.75- × 4- × 3.6-inch, #BAT12
12-volt, 1-amp recharger kit, #BC12K
5-inch focusing lens with a 1- diameter× 5-inch fl. zinc selenide, #MENC5

Handheld Burning Diode
Laser Ray Gun

This project is a very rewarding one for the laser experimenter with the result being a tool that can burn and cut many different materials (see Figure 9-1). The continuous output of 1 watt at wavelengths from .7 to 1 micron requires class IV labeling and other compliances that are clearly shown and explained. Project data is highly detailed, as errors in assembly can be very costly, owing to replacing the laser diode; they are not cheap! Protective eyewear must be worn and is available at www.amazing1.com.

Expect to spend $50 to $100 for all the components, minus the laser diode. Sometimes these are available on the surplus market ranging from $100 to $200. Required test equipment involves only DC-reading meters. All parts are available on the previously mentioned web site and are listed in Table 9-1.

The laser will operate most laser diodes from 10 *milliwatts* (MW) to 1 watt or more. The circuit is not temperature compensated and is intended to operate near ambient temperature. This is usually not a problem for normal room temperature applications.

Applications

The system, when operated with a .5- to 1-watt diode, is capable of cutting, engraving, trimming, and performing many small-scale fabricating applications.

When operated at around 1060 *nanometers* (nm), output may be frequency doubled to an intense, visible green beam at 532 nm. Output in the green can be a conservative third of the input where a 1-watt diode will easily produce 300 milliwatts of visible, intense green light.

Diodes operating in the 800 to 900 nm range produce intense, *infrared* (IR) illumination for night vision devices. Low-powered devices can invisibly illuminate windows that act as a microphone, converting internal sounds and voices to mechanical vibrations. The reflected light now spatially varies and is processed back into sounds by a special optical

Figure 9-1   *Diode burning laser*

receiver. A laser window bounce listening system is described in Chapter 13, " Laser Window Bounce Listening Device." Your laser will operate most laser diodes from 10 milliwatts to 1 watt or more. The circuit is not temperature compensated and is intended to operate near ambient temperature. This is usually not a problem for normal room temperature applications.

Basic Theory

Your laser diode requires a constant current source of clean, stable, ripple-free power. The output results when a forward-biased semiconductor junction emits optical energy due to stimulated emission. A *light-emitting diode* (LED), on the other hand, produces optical output via spontaneous emission and lacks coherence properties.

A forward-biased diode junction must be current controlled as this junction resistance is not linear and has a tendency to draw large increments of current for small, incremental changes in voltage. The diode is also sensitive to temperature and must be operated with a feedback loop to control the drive current if used in widely varying temperatures. This is not a built-in feature of this system as operation is intended to be around room temperature. The circuit provides the necessary control and conditioning of the diode current for most room temperature applications. The diode junction is also very thin and cannot take any voltage transients without a catastrophic failure. Therefore, all transients and surges must be minimized to protect this junction.

Circuit Theory

Input to the system may be a battery pack capable of supplying $1\frac{1}{2}$ times the rated laser diode current or an equivalent wall adapter for direct 115 *volts-alternating current* (VAC) use. We suggest a nonregulated type as the switching units can cause transient damage under certain conditions. The input power is initially controlled by a toggle switch (S1) that immediately turns on the blower (BLW1) and the

indicating LED (LA1). The actual laser output is controlled by the action of a key switch (S2) that initiates a timer (I1) providing the necessary delay required for a class IV laser. Output from the timer now turns on a DC switch consisting of Q1 and Q2 through a slow turn-on by the charging action of start capacitor (C7), as shown in Figure 9-2.

This DC switch now feeds current to a voltage regulator (VR1) with its output set to 7 to 8 volts by the trim pot (R5). LED LA2 now indicates that the laser is emitting energy.

The regulated output of VR1 is conditioned for transients by transorb PZ1 and a tantalum capacitor C9 and is then fed to a current-controlled module (CC1). The laser diode is connected to the module through a resistor (R7) junctioned at the solder terminal strip (TE4). The laser diode current is monitored by reading the voltage at pins 3 and 4 of TE4 or via an external jack (J2) connected to these points. The diode current is carefully controlled by the multi-turned trimpot for current adjusting mounted on the CC1 module and it must be closely monitored in order to not exceed the manufacturer ratings. You will note that all switches and contacts are bypassed by a .1-microfarad capacitor to minimize noise from contact bounce.

Assembly Steps

The figures show our layout and should be followed reasonably closely. It is suggested that you obtain the necessary parts and then trial-position and fabricate the chassis as you proceed. The steps for assembly are as follows:

1. Insert the components into the perforated board as shown in Figure 9-3, using the component leads for connection points, as indicated by the dashed lines. Attempt to follow the layout using the hole locations as shown. Note the component polarity, where shown, avoids short circuits, cold solder, solder bridges, and wire shorts. Figures 9-6, 9-7, and 9-8 show various stages of wiring.

2. Create the chassis per Figure 9-4 from .063 aluminum as shown. Note that important dimensions are shown, such as the overall

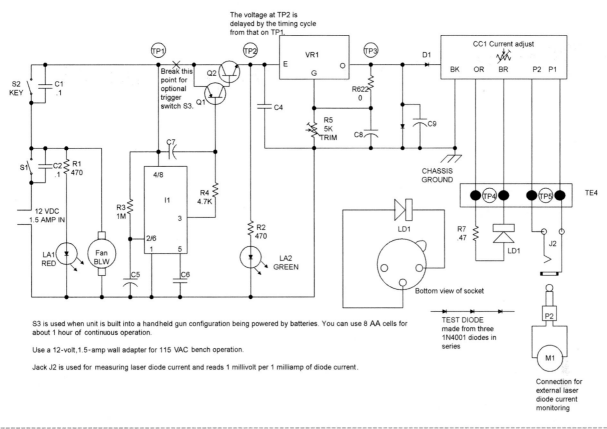

Figure 9-2 *Burning diode laser schematic*

length, width, and so on. Make the actual holes for components as shown. Dimension them as needed for particular component size and in the proper locations according to the drawings.

3. Assemble the bracket as shown in Figure 9-4 from a 4- × 2-inch piece of .063 aluminum and mount it to the chassis via two #8 × ³⁄₈ sheet metal screws. Note the holes for the MK1 mounting socket of the voltage regulator (VR1). These holes must be free from any burrs or sharp edges that could puncture the insulating mica washer.

Drill a single hole for the single screw mounting of the CC1 current controller.

4. Mechanically assemble the chassis as shown in Figure 9-5:

   a. Attach the terminal lug strip (TE6) via pieces of bus wire secured under the nuts of the two grounding screws, SW1 and NU1. These pieces of support wires connect to the two outer lugs of this strip grounding these points.

   b. Snap in BU1, two plastic bushings.

   c. Attach the front panel components.

5. Connect the remaining interconnecting wires and connections as shown in Figures 9-6, 9-7, and 9-8.

6. Verify the wiring accuracy and check for shorts, bad solder joints, and general assembly errors.

7. Assemble TUBE1 and 2 as shown in Figures 9-9 and 9-10. Note these pieces must mate with one another in a telescope-like structure.

   It may be necessary to ream out the outer tube to obtain a proper sliding fit.

8. Glue the lens to the end of TUBE2. Use a *room temperature vulcanizing* (RTV) adhesive sparingly, avoiding touching the lens.

9. Carefully solder the umbilical lead WIREZIP to the respective connections on the SOCK1 socket. Observe the polarity when connecting the other end to TE6.

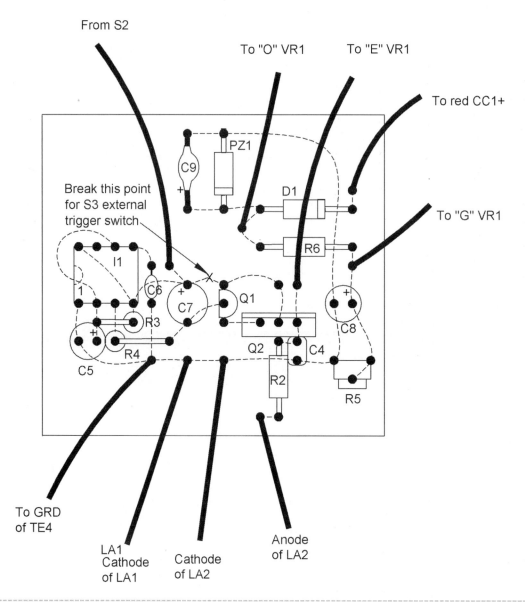

From S2

To "O" VR1   To "E" VR1

To red CC1+

PZ1

C9

To "G" VR1

D1

Break this point
for S3 external
trigger switch

I1

R6

C6

Q1

C7

C8

R3

Q2

C4

R4

R2

C5

R5

To GRD
of TE4

LA1
Cathode
of LA1

Cathode
of LA2

Anode
of LA2

Figure 9-3   *Assembly board blowup*

10. Assemble the LDIODE laser diode to the HSINK heat sink via a retaining washer and two screws. Note the sandwiching action of the assembly with a diode nesting in the recess of the heat sink. Do not insert the laser diode into the socket at this time.

Circuit Pretest

To conduct a pretest on the circuit, follow these steps:

1. Assemble a test diode using three 1N4001 connected in a series to simulate the laser diode as shown in Figure 9-2.

2. Connect the leads of the test diode to the pins of the laser diode socket using small alligator clips, observing the polarity.

3. Preset all switches to "off" positions and connect a voltmeter to TP3.

4. Turn on S1 and note the blower BLW1 and LED LA1 coming on.

5. Turn on S2 and note a 5- to 10-second delay before LA2 comes on and the voltmeter indicates a voltage. Note that the indicator LED on the CC1 module also lights up. Adjust trimpot R5 for 7 volts on the meter. Adjust input voltage from 10 to 14 and note the meter not changing from 7 volts. This verifies proper voltage regulation.

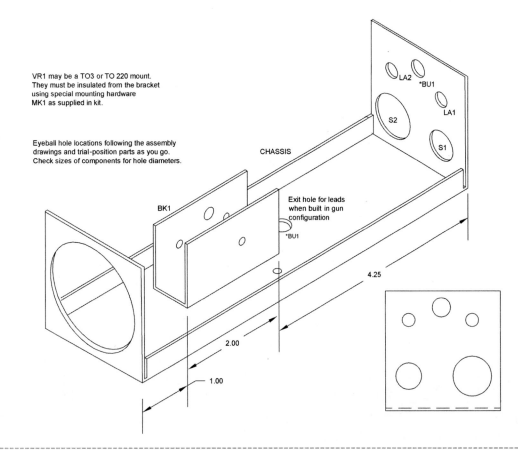

VR1 may be a TO3 or TO 220 mount.
They must be insulated from the bracket
using special mounting hardware
MK1 as supplied in kit.

Eyeball hole locations following the assembly
drawings and trial-position parts as you go.
Check sizes of components for hole diameters.

CHASSIS

BK1

Exit hole for leads
when built in gun
configuration
*BU1

LA2
*BU1
LA1
S2
S1

4.25

2.00

1.00

Figure 9-4 *Chassis fabrication showing bracket BK1*

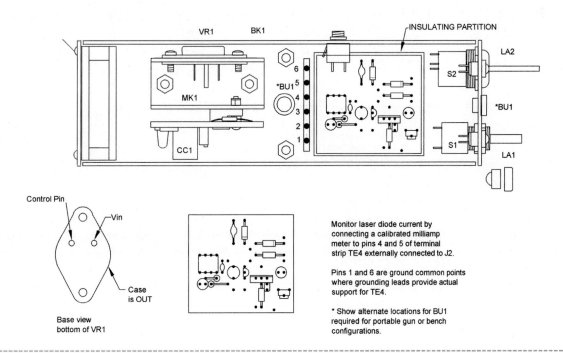

HHH

VR1   BK1   INSULATING PARTITION

LA2
S2
*BU1
S1
LA1

6
*BU1 5
4
3
2
1

MK1

CC1

Control Pin
Vin
Case
is OUT

Base view
bottom of VR1

Monitor laser diode current by
connecting a calibrated milliamp
meter to pins 4 and 5 of terminal
strip TE4 externally connected to J2.

Pins 1 and 6 are ground common points
where grounding leads provide actual
support for TE4.

* Show alternate locations for BU1
required for portable gun or bench
configurations.

Figure 9-5 *Mechanical layout*

Twist wire pairs together wherever possible
for neatness and mechanical integrity.

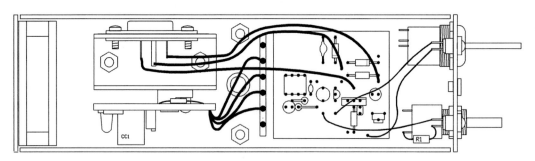

Figure 9-6 *Wiring aid first level*

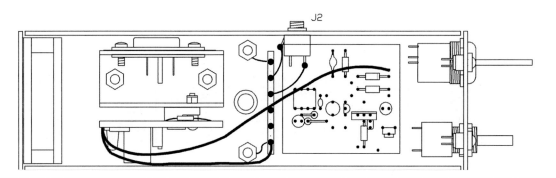

Figure 9-7 *Wiring aid for level two*

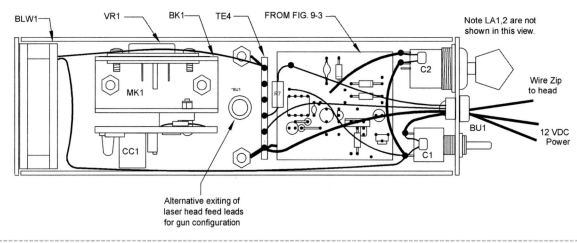

Figure 9-8 *Wiring aid for level three*

6. Connect a digital millivoltmeter to TP5 and adjust the multiturn trimpot on CC1 for exactly 1 volt. This corresponds to a current of 1 amp flowing through the test diode. You will note that a reading of 1 millivolt on the digital meter equals 1 milliamp of diode current.

7. Check the output regulation by shorting out one of the test diodes and note that the milli-voltmeter does not change. This verifies the current control.

8. You may now connect in a real laser diode. The system output will be dependent on the

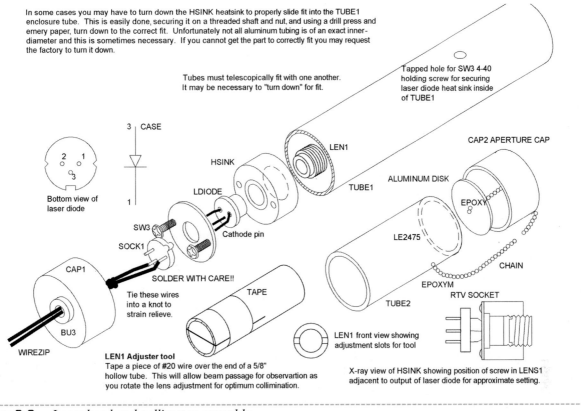

In some cases you may have to turn down the HSINK heatsink to properly slide fit into the TUBE1 enclosure tube. This is easily done, securing it on a threaded shaft and nut, and using a drill press and emery paper, turn down to the correct fit. Unfortunately not all aluminum tubing is of an exact inner-diameter and this is sometimes necessary. If you cannot get the part to correctly fit you may request the factory to turn it down.

Tapped hole for SW3 4-40 holding screw for securing laser diode heat sink inside of TUBE1

Tubes must telescopically fit with one another. It may be necessary to "turn down" for fit.

3 | CASE

2 | 1
3

Bottom view of laser diode

1

CASE

HSINK

LEN1

CAP2 APERTURE CAP

LDIODE

ALUMINUM DISK

TUBE1

EPOXY

SW3

LE2475

CHAIN

SOCK1

Cathode pin

EPOXYM

RTV SOCKET

CAP1

SOLDER WITH CARE!!

TAPE

TUBE2

Tie these wires into a knot to strain relieve.

BU3

WIREZIP

LEN1 front view showing adjustment slots for tool

LEN1 Adjuster tool
Tape a piece of #20 wire over the end of a 5/8" hollow tube. This will allow beam passage for observartion as you rotate the lens adjustment for optimum collimination.

X-ray view of HSINK showing position of screw in LENS1 adjacent to output of laser diode for approximate setting.

Figure 9-9 *Laser head and collimator assembly*

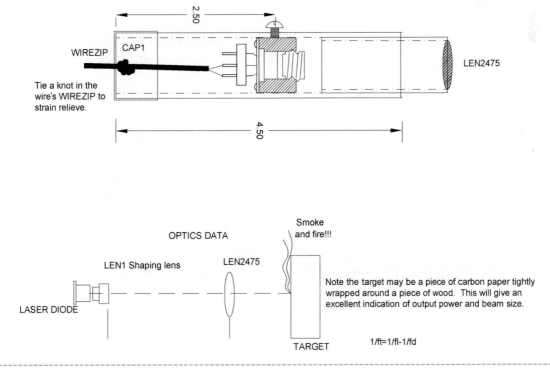

X-ray view of HSINK showing position of screw in LEN1 adjacent to output of laser diode for approximate setting.

2.50

WIREZIP | CAP1

LEN2475

Tie a knot in the wire's WIREZIP to strain relieve.

4.50

Smoke and fire!!!

OPTICS DATA

LEN1 Shaping lens

LEN2475

LASER DIODE

Note the target may be a piece of carbon paper tightly wrapped around a piece of wood. This will give an excellent indication of output power and beam size.

TARGET

1/ft=1/fl-1/fd

Figure 9-10 *X-ray view of laser head and collimator*

power, wavelength, and beam profile of the actual laser diode. It is assumed that you are familiar with these properties, their applications, and the proper safety precautions.

Before you open up the box containing your laser diode, it is imperative to read the following pages.

Study and carefully follow all of the manufacturer's instructions usually supplied with the diode.

Safety

High-power diode lasers emit radiation in the visible and infrared region of the spectrum. When in use, safety precautions must be taken to avoid the possibility of eye damage. For Class IIIb lasers, do not look into the beam or allow direct eye exposure. For Class IV lasers, do not expose the eye or skin to direct or scattered radiation. If viewing is required, the beam should be observed by reflection from a matte surface, utilizing an image converter or a suitable fluorescent screen. Class IIIb lasers operate at powers less than 500 milliwatts. Class IV lasers operate at powers greater than 500 milliwatts. Safety glasses for this project are available at www.amazing1.com.

General Handling

The window of the 9-millimeter hermetically sealed laser diode packages is quite thin (~250 micrometers) in order to minimize the optical path length. Do not push on the window when inserting the laser diode into a socket. Push the device into place by pressing on the base of the package.

Static Electrical Damages (ESD)

Laser diodes are very reliable under normal operating conditions, but like most semiconductor devices,

they can be easily damaged or destroyed by inadvertent electrical or static discharges. A static-free environment is mandatory. Grounded tweezers, a grounded wriststrap on the user, a grounded work surface, antistatic floors, and a case ground for the laser diode all reduce the risk of damaging static discharge through the diode. Retain the laser diode in a static fine environment when not in use (such as the shipping container). Short the pins on packaged diodes at all times when not in use. Insert pins into conductive foam or wrap wire from pin to pin. (Note: A nonshorted laser can be damaged by ESD even without touching it.)

Excessive Forward Voltage

A forward current greater than the specified limit or any reverse voltage may damage the diode junction. Putting a clamping diode across the output of the power supply solves this problem. Commonly, these conditions occur from a static discharge or from activation or range-changing voltage transients in laboratory power supplies. Many power supplies, even current-regulated one, exhibit fast voltage spikes a when switched on or off. The following precautions are recommended to minimize the risk of destructive electrical transients occurring:

- Reduce static charge accumulation by wearing a grounded wriststrap when handling laser diodes.

- Use a grounded work area and store the laser diodes in their original shipping packages when not in use.

- Eliminate transient power supply spikes by using a power supply specifically designed for laser diodes or another "slow-start" power supply. This criterion is built into this project.

Excessive Forward Current

An excessive forward current can cause operation at optical power levels, which may damage the output facet in less than 1 *microsecond* (usec). Laser action may continue after this damage at lower efficiency and lower power, or only spontaneous emission may remain. The allowable current depends on the pulse width in pulsed or quasi-clockwise operation, and peak optical power must be reduced as the pulse width increases.

Reverse Currents

Reverse currents may also damage the diode. Forward or reverse transients may be caused by energy reflections in pulsed systems, capacitance in fixtures or cables, or output capacitors in constant-current supplies operated with no load connected. Drive levels on drivers for moderate-power, continous-wave laser diodes may be tested by using a dummy load, such as two 1N4001 diodes and a 1-ohm resistor in series.

High Temperature

For normal, room-temperature operation, the laser diode should be firmly mounted on a heat sink, whose temperature should be maintained at 20° to 25° C. The maximum operating and storage limits are shown on the product data sheet.

Here is some helpful information on how to operate your laser:

1. Note that the optical diagram on the inset of Figure 9-10 shows the shaping and collimating lens (LEN1) along with the focusing lens (LEN2475). Lens L1 is screwed into the mating threads of the heat sink (HSINK1) to a point that provides optimum collimation. Once set, it may be left alone except for occasional cleaning. It also need not be perfectly set for acceptable operation.

2. Obtain some black carbon paper and wrap it around a 1- × 2-inch piece of wood. Place the laser diode with only the shaping lens installed securely in a mound of clay. Position the carbon paper jig about 1 to 3 inches away. Apply power and slowly adjust the current trimpot on the CC1 module to 1.3 amps. You

will note smoke and fire as the jig is moved across the output beam.

3. The focusing lens is adjusted for proper beam delivery as a burning spot the size of a pin-point or wider. Note that the lens tube may be inserted with LEN2 at the output end, providing a shorter focal length than when the lens tube is inserted with LEN1 on the internal end adjacent to the laser diode. Focal lengths may range from approximately 2 to around 50 centimeters. You may use a longer focal length for a far-field effect.

System output will be dependent on the power, wavelength, and beam profile of the actual laser diode. It is assumed the user is familiar with these properties, their applications, and safety precautions.

4. Once proper operation is verified, you may assemble and label everything, as shown in Figure 9-11.

Table 9-1 Handheld burning diode laser parts

Ref. #	Qty.	Description	DB #
R1, 2	2	470 ohms, $1/4$ watt (yel-pur-br)	
R3		1 meg, $1/4$ watt (br-blk-grn)	
R4		4.7K, $1/4$ watt (yel-pur-red)	
R5		5K trimpot	
R6		220 ohm, (red-red-br) $1/4$ watt	
R7		.47 ohm, 3 watt	
DIODES	3	IN4001 used for test laser diode	
C1, 2, 3, 4	4	.1 mfd/50-volt plastic cap.; note C3 on S3	
C5, 8	2	10 mfd/25-volt vert. electrolytic cap.	
C6		.01 mfd/50-volt disc cap.	
C7		100 mfd/25-volt vert. electrolytic cap.	
C9		3.3 mfd/20-volt tantalum cap.	
S1		SPST toggle switch	
S2		Key switch nonremovable key in "on" position KEYSW	
S3		Optional switch for trigger	
BLW1		2- × 2-inch 12-volt blower fan	

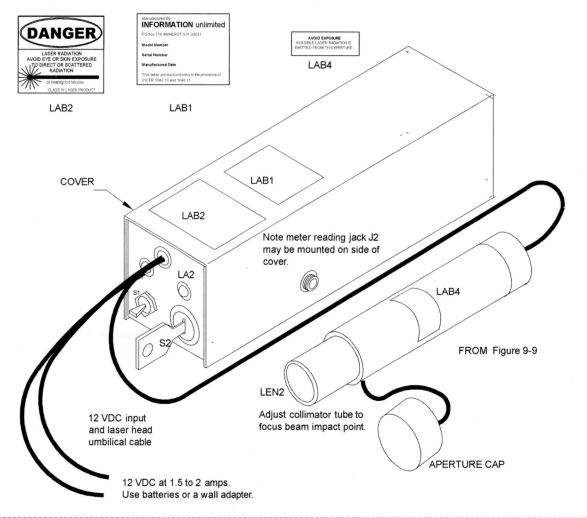

DANGER
LASER RADIATION
AVOID EYE OR SKIN EXPOSURE
TO DIRECT OR SCATTERED
RADIATION
20 Watts@10.6 Microns
CLASS IV LASER PRODUCT

LAB2

Manufactured By
INFORMATION unlimited
PO box 716 AMHERST N.H. 03031
Model Number
Serial Number
Manufactured Date
This laser productconforms to the provisions of
31CFR 1040.10 and 1040.11

LAB1

AVOID EXPOSURE
INVISIBLE LASER RADIATION IS
EMITTED FROM THIS APERTURE

LAB4

COVER

LAB1

LAB2

LA2

S1

S2

Note meter reading jack J2
may be mounted on side of
cover.

LAB4

FROM Figure 9-9

LEN2

12 VDC input
and laser head
umbilical cable

Adjust collimator tube to
focus beam impact point.

APERTURE CAP

12 VDC at 1.5 to 2 amps.
Use batteries or a wall adapter.

Figure 9-11   *Final assembly with laser head and labels*

Table 9-1 Continued

Ref. #	Qty.	Description	DB #
LA1		Red LED indicator for primary power on	
LA2		Green LED indicator for laser emitting	
D1		1N5401 power diode	
PZ1		PKE15 transorb	
CC1		Adjustable current driver module	CC1
VR1		317 voltage regulator, shown as TO3	
Q1		PN2907 PNP general purpose transistor	
Q2		MJE3055 NPN GP transistor TO220 case	
I1		555 dual in-line package IP timer	

Ref. #	Qty.	Description	DB #
J2/P2		3.5 mm stereo jack and plug for external meter	
J1		2.5 mm DC jack or use direct wiring	
TE6		6-lug solderable terminal strip	
PB1		$1^{3}/_{4} \times 2.1 \times .1$ perforated circuit board	
COVER		Shown in Figure 9-11	
CHASSIS		Shown in Figure 9-4	
BK1		Shown in Figure 9-4 inset	
MK1		Mounting kit for TO3 transistor	
BU1		$^{3}/_{8}$-inch clamp bushing	
WIRE22	2 feet	22-gauge, vinyl-stranded hookup wire	

Ref. #	Qty.	Description	DB #
WIRE20	6 feet	20-gauge, vinyl-stranded hookup wire	
SW1/NU1	4	6-32 × ¹/₂-inch screws and nuts	
SW2	2	4-40 × ¹/₂-inch screws for mounting BLW1	
FEET	4	¹/₂-inch rubber stick-on feet for chassis bottom	
PLASTIC		2- × 2-inch piece of insulating plastic	

Laser Head and Collimator

Ref. #	Qty.	Description	DB #
WIREZIP	4 feet	#18 zip lead, parallel stranded wire	
TUBE1		4¹/₂ × 1 × .062 wall aluminum tube; see Figure 9-9	
TUBE2		2¹/₄ × ⁷/₈ × .062 wall aluminum tube; see Figure 9-9	
HSINK1		Heatsink, washer, and screws combination per Figure 9-9	HSINK

Ref. #	Qty.	Description	DB #
SOCK1		Three-pin TO5 transistor socket	
CAP1, 2	2	1-inch plastic caps; see Figure 9-9	
BU3		Small, ¹/₄-inch Heyman clamp bushing	
LEN1		Special threaded lens fixture	LENS13
LEN2475		24- × 75-millimeter extending lens	LE2475
SW3	2	4-40 × ³/₈-inch small screws for securing HSINK	
LDIODE		See text of options and instructions on laser diode.	LD0098
LAB1		Laser certification label	
LAB2		Laser classification label	
LAB4		Aperture label	

Chapter Ten

Long-Range Optical Laser Project

This relatively low cost kit, shown in Figure 10-1, shows how to build a laser gun that can project a red dot over several miles. Someone looking toward the laser would see it as the brightest object on the horizon from distances of tens of miles. The project is intended for beginners who may have more mechanical than electrical skills. The enclosure and fabrication are readily available from most hardware stores, but it will require some work using files, saws, drills, and other small hand tools. Expect to spend about $30, and that includes the laser module.

This project is intended for the laser experimenter who desires to add a simple optical system that allows for far-field focusing, beam collimation, or beam expanding. The kit contains all the crucial parts, including a working 3- to 5-milliwatt, 650 *nanometer* (nm) red laser module.

Construction parts such as tubing, handles, brackets, and hardware are all readily available at your local hardware store. Our suggested lab approach is to acquire some telescopic tubing of metal or plastic, or a combination of both, and fabricate the laser and lense holders from a suitable sized plastic rod or the equivalent. The center holes of the holders must be as accurate as possible or the optical system integrity will be greatly degraded.

The grip or handle is fabricated from a 1-inch-thick piece of finished pine that has a section carefully removed for the battery holder and pushbutton switch. Passage holes for the power leads are drilled as shown in Figure 10-2. A clear piece of lexan (polycarbonate) sheet is formed with a flange at a 90-degree bend that is the surface for mounting the switch. This cover is screwed to the wooden handle and has a neat appearance if done with care and precision. A contoured shape may be fabricated into the handle to mate to the curvature of the tube. See Figure 10-3 for assembly details.

The operation of the laser can be used without any of the optics, providing a similar performance to a higher-quality laser pointer. The addition of the lens system allows far-field focusing, close-range expan-

Figure 10-1 *Optical laser electronic assembly*

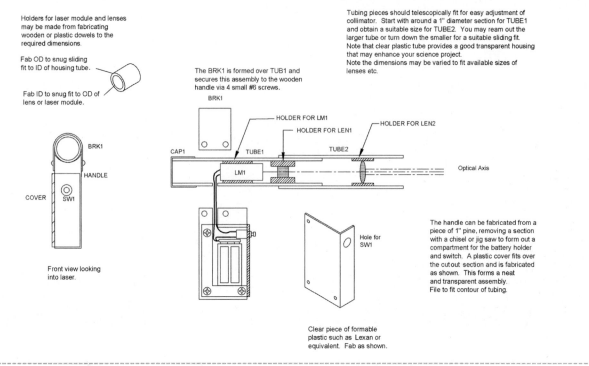

Holders for laser module and lenses may be made from fabricating wooden or plastic dowels to the required dimensions.

Fab OD to snug sliding fit to ID of housing tube.

Fab ID to snug fit to OD of lens or laser module.

Tubing pieces should telescopically fit for easy adjustment of collimator. Start with around a 1" diameter section for TUBE1 and obtain a suitable size for TUBE2. You may ream out the larger tube or turn down the smaller for a suitable sliding fit. Note that clear plastic tube provides a good transparent housing that may enhance your science project.
Note the dimensions may be varied to fit available sizes of lenses etc.

The BRK1 is formed over TUB1 and secures this assembly to the wooden handle via 4 small #6 screws.

BRK1

HOLDER FOR LM1
HOLDER FOR LEN1
HOLDER FOR LEN2

CAP1
TUBE1
TUBE2
Optical Axis
LM1

BRK1
HANDLE
COVER
SW1

Front view looking into laser.

Hole for SW1

The handle can be fabricated from a piece of 1" pine, removing a section with a chisel or jig saw to form out a compartment for the battery holder and switch. A plastic cover fits over the cutout section and is fabricated as shown. This forms a neat and transparent assembly. File to fit contour of tubing.

Clear piece of formable plastic such as Lexan or equivalent. Fab as shown.

Figure 10-2 *Long-range optical laser gun*

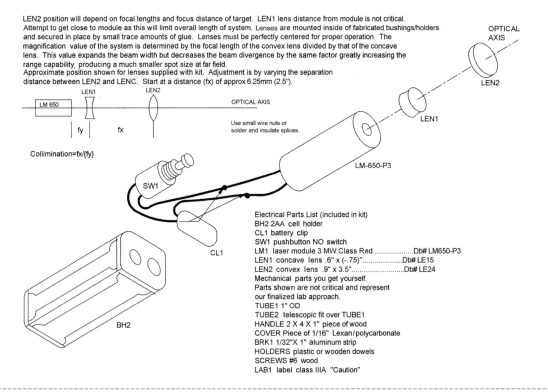

LEN2 position will depend on focal lengths and focus distance of target. LEN1 lens distance from module is not critical. Attempt to get close to module as this will limit overall length of system. Lenses are mounted inside of fabricated bushings/holders and secured in place by small trace amounts of glue. Lenses must be perfectly centered for proper operation. The magnification value of the system is determined by the focal length of the convex lens divided by that of the concave lens. This value expands the beam width but decreases the beam divergence by the same factor greatly increasing the range capability, producing a much smaller spot size at far field.
Approximate position shown for lenses supplied with kit. Adjustment is by varying the separation distance between LEN2 and LENC. Start at a distance (fx) of approx 6.25mm (2.5").

OPTICAL AXIS

LEN1
LEN2
LM 650
OPTICAL AXIS
fy | fx

Use small wire nuts or solder and insulate splices.

Collimination=fx/{fy}

LEN2
LEN1
LM-650-P3
SW1
CL1
BH2

Electrical Parts List (included in kit)
BH2 2AA cell holder
CL1 battery clip
SW1 pushbutton NO switch
LM1 laser module 3 MW Class Red...................Db# LM650-P3
LEN1 concave lens .6" x (-.75)"...................Db# LE15
LEN2 convex lens .9" x 3.5".........................Db# LE24
Mechanical parts you get yourself.
Parts shown are not critical and represent our finalized lab approach.
TUBE1 1" OD
TUBE2 telescopic fit over TUBE1
HANDLE 2 X 4 X 1" piece of wood
COVER Piece of 1/16" Lexan/polycarbonate
BRK1 1/32"X 1" aluminum strip
HOLDERS plastic or wooden dowels
SCREWS #6 wood
LAB1 label class IIIA "Caution"

Figure 10-3 *Isometric assembly drawing*

sion, and far-field collimation where the beam spot effect will be reduced by the collimating power of the system. As an example, the beam diameter without optics at 300 meters will be approximately a 50-centimeter diameter spot. With a 10× collimator, the diameter will be reduced to 5 centimeters.

Chapter Eleven

Handheld, Ultrabright Green Laser

This excellent visual effects laser produces a visible green beam of light. The device is shown constructed as a fully compliant 20-milliwatt class 3b laser (see Figure 11-1). This tool has many useful applications, and the range can be up to tens of miles.

Expect to spend up to several hundred dollars on this reasonably simple construction project. All parts are available through Information Unlimited at www.amazing1.com and are listed in Table 11-1.

Applications

The green diode laser project requires protective eyewear if viewing the direct beam or reflection. It is intended for star pointing by amateur astronomers, spooking light-sensitive animals, long-range pointing, target intimidation, special effects, or just general optical research and experimentation.

Figure 11-1   *Handheld, ultrabright green laser*

The laser operates from a 6-volt battery pack and includes an adjustable collimator where the beam spot can be expanded and converged for extra-long-range applications. The system is in full compliance for a class 3b laser project and is built using easy-to-fabricate plastic parts.

Circuit Description

Figure 11-2 shows a key switch (S1) applying main power to the circuit and starting timer I1, initiating the activation cycle. The delay time of I1 is determined by the time constant of R2 and C2. The LED1 *light-emitting diode* (LED) indicator now energizes, indicating the system is ready to fire via S2. Resistor R1 controls the current to the LED. Current to the laser module (LM1) is sent through transistor switch Q1 by controlling its base current from pin 3 of I1 through resistor R3. You will note that the beam control pushbutton switch S2 is used to control the laser output and can only apply power after the emission delay circuit times out and pin 3 of I1 goes high.

Resistors R4 and R5 are selected to allow the required current through the laser module. The Z1 zener diode maintains a 3-volt level at the laser module input connections.

Assembly Steps

To begin the project, follow these steps:

1. Identify all the parts and pieces and verify them with the bill of materials.

2. Insert the components, starting from one end of the perforated circuit board, and follow the locations shown in Figure 11-3, using the individual holes as guides. Use the leads of the actual components as the connection runs. These are indicated by the dashed lines. It is a good idea to trial-fit the larger parts before actually starting to solder.

   Always avoid bare wire bridges, messy solder joints, and potential solder shorts. Check for cold or loose solder joints. Pay attention to the polarity of the capacitors with polarity signs and all the semiconductors.

3. Cut, strip, and tin 4-inch wire leads for connecting to S1, S2, and the LED. Solder to these components along CL1 and LM1, as shown in Figure 11-3. Note the polarity of the laser module. Red is positive, and black is negative.

4. Fabricate rear cap CA1 from a 1⅝-inch plastic cap. Note the large hole for the S1 key switch and the small hole for the LED.

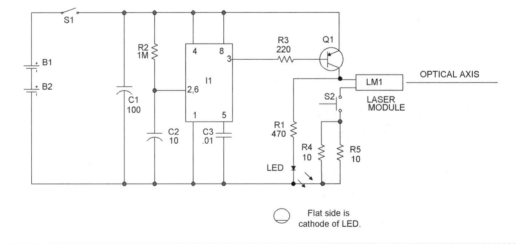

Figure 11-2   *Laser model schematic*

Chapter Eleven

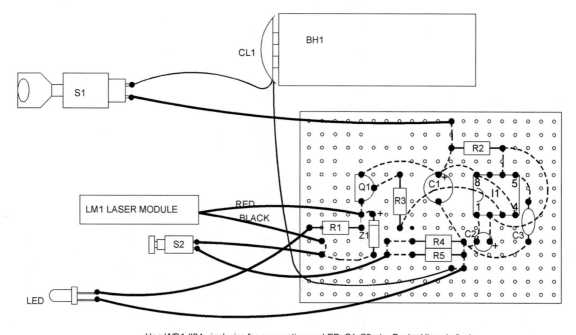

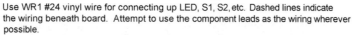

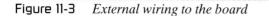

Use WR1 #24 vinyl wire for connecting up LED, S1, S2, etc. Dashed lines indicate the wiring beneath board. Attempt to use the component leads as the wiring wherever possible.

Figure 11-3   *External wiring to the board*

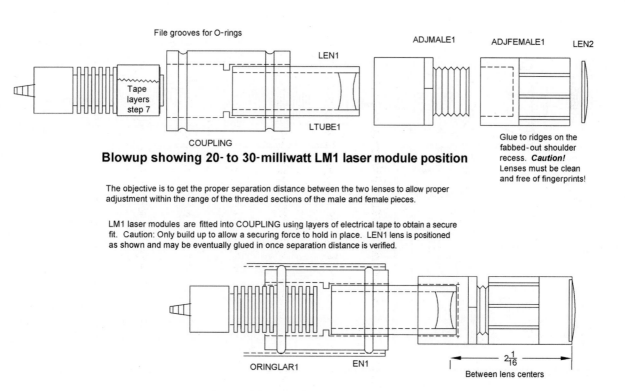

Blowup showing 20- to 30-milliwatt LM1 laser module position

The objective is to get the proper separation distance between the two lenses to allow proper adjustment within the range of the threaded sections of the male and female pieces.

LM1 laser modules are fitted into COUPLING using layers of electrical tape to obtain a secure fit. Caution: Only build up to allow a securing force to hold in place. LEN1 lens is positioned as shown and may be eventually glued in once separation distance is verified.

Assembled collimator x-ray showing laser and lenses position

Figure 11-4   *Collimator assembly*

Electrical Pretest

To test out the electrical system, follow these steps:

1. Position the laser module so that it is pointing in a safe direction.

2. Turn the key switch off (full *counterclockwise* [CCW]) and connect it to the 6-volt battery pack (BH1).

3. Turn the key switch on one notch and note that the emission indicator comes on within a 10-second period. This is the turn-on delay necessary for compliance with the Bureau of Radiological Health (BRH).

4. Activate pushbutton switch S2 and note a bright green output. Allow several seconds for it to brighten. You may check the input current to the laser module by connecting a current meter in series with one of the leads to the laser module and measure 250 milliamps for the total current.

Collimation and Final Assembly

Follow these steps for collimator and final assembly

1. File some grooves onto the coupling for large O-rings. Trial-fit them into the EN1 enclosure until a secure fit is obtained. You should file a bit at a time to avoid overdoing (refer to Figure 11-4).

2. Shim up the LM1 laser module with several layers of electrical tape to provide a positive friction fit with the coupling piece. Fully insert it until the module abuts to the internal center shoulder in the coupling.

3. Clean the larger lens LEN2 and then glue it onto ADJFEMALE1, as shown, using silicon rubber glue or an equivalent.

4. Insert LEN1 into the laser output end of the LTUB1 assembly as shown. Do not glue at this time.

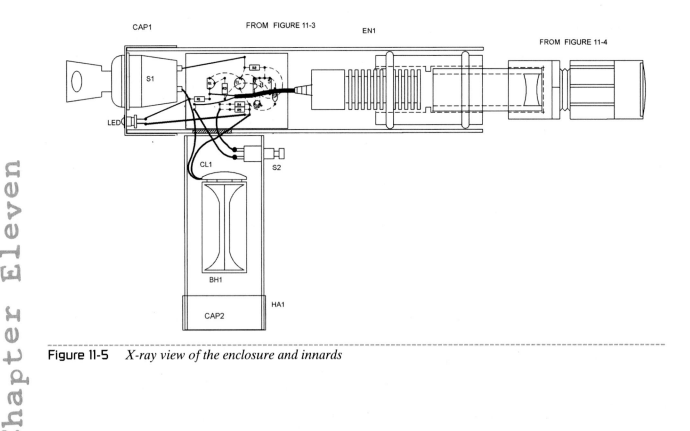

Figure 11-5 *X-ray view of the enclosure and innards*

5. Fully insert the LTUB1 into the other end of the coupling to the internal center shoulder. You will now have an assembly as shown in Figure 11-5.

6. Finally, assemble and label everything, as shown in Figure 11-6. Note the clearance hole in the enclosure adjacent to the handle for routing through the connecting leads. You may want to extend the laser module leads. Use caution as the reversing polarity will destroy the laser.

7. Insert 4 AA batteries into the BH1 holder in the handle and slide on CAP2. Turn on the key switch and verify proper operation as done in an electrical pretest section.

8. Point the laser at a target around 100 meters away and adjust the collimator to the smallest spot. Do this in low light or preferably at night. It may take several seconds for the laser to come up to the required brightness. Note that the adjustment should be about a $^1/_2$ to $^3/_4$ turn from being fully tight for a mechanically stable operation. If too far out, the assembly will be loose. You may have to wire brush threads to clean and lube with dry Teflon for smooth action. You may compensate for the distance between lenses by changing the length of TUBE1. You may now glue LEN1 in place as noted previously.

The collimator will expand the beam width but decrease the divergence by the same ratio. This greatly reduces the far field beam diameter.

Note that the laser modules are at a wavelength of 532 nanometers and will be much brighter to the eye than an equivalent powered red device at 630 to 650 nanometers. Higher-powered green lasers are available at www.amazing1.com. Also note that most of the mechanical parts used in this project can be obtained through most hardware stores.

The pointing range on a clear night can be in excess of many miles. The target detection can only be limited in the line of sight by the curvature of the earth. On a clear night, when looking toward the laser, it will appear as a very bright source from up to tens of miles.

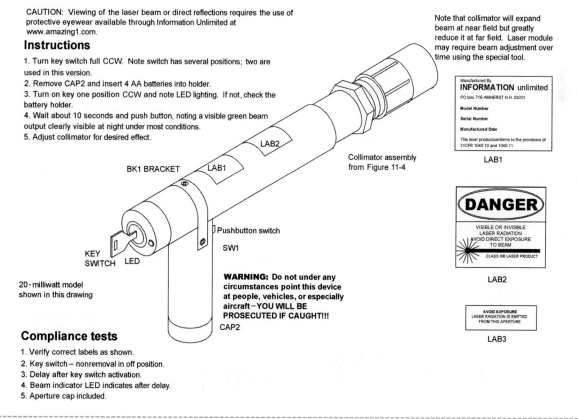

Figure 11-6 *Final assembly view and labels*

Table 11-1   Handheld, ultrabright green laser parts

Ref. #	Qty.	Description	DB#
R1	1	470-ohm, $\frac{1}{4}$-watt (yel-pur-br) resistor	
R2	1	1-megohm, $\frac{1}{4}$-watt (br-blk-grn) resistor	
R3	1	220-ohm, $\frac{1}{4}$-watt (red-red-br) resistor	
R4, 5	2	10-ohm, $\frac{1}{4}$-watt (br-blk-blk) resistor	
C1	1	100-microfarad, 25-volt vertical electrolytic capacitor	
C2	1	10-microfarad, 25-volt vertical electrolytic capacitor	
C3	1	.01-microfarad, 50-volt polyester capacitor	
LED	1	High-brightness LED	
I1	1	555 dual in-line integrated circuit chip timer IC	
Z1	1	3-volt, $\frac{1}{2}$-watt zener diode 1N5225	
Q1	1	PN2907 PNP general purpose transistor	
*LM1	1	Class IIIBiib 10- to 30-milliwatt 532 nanometer laser module	LM532-20
S1	1	Panel-mount key switch	
S2	1	Pushbutton NO switch	
PB1	1	1.5- × 2- × .1-inch grid vector board	
BH1	1	4 AA cell holder	
WR1	24 inches	#24 hookup vinyl wire	
EN1	1	7- to 12-inch × 1.5-inch ID × .058 wall plastic tube	
HA1	1	5 × 1.5 ID × .058 wall plastic tube	
CAP1, 2	2	1 $\frac{5}{8}$-inch plastic caps, fabricated as shown	
CL1	1	Battery clip lead	
BK1	1	6 × $\frac{1}{2}$ × .035 aluminum bracket	
LEN1	1	.6 × −1 negative lens	LE15-25
LEN2	1	24- × 75-millimeter double convex lens	LE2475
*LTUBE	1	.625- ID × 2-inch sked 40 *polyvinyl chloride* (PVC) tube	
*COUPLIG	1	PVC sked 80 $\frac{1}{2}$-inch coupling; use for lm 10—Spears #02467	
ORINGS	2	$\frac{1}{2}$ × $\frac{5}{8}$ × .0625 rubber O-ring #014	
ORINGL	2	1 $\frac{3}{16}$ × 1 $\frac{7}{16}$ × .125 rubber O-ring #217	
ADJMALE	1	$\frac{1}{2}$-inch PVC sked 40 slip fit to male thread, Genova 30405	
ADJFEM	1	$\frac{1}{2}$-inch PVC sked 40 slip fit to female thread, Genova 30305	
LAB1	1	Certification label	
LAB2	1	Compliance class IIIB label	
LAB3	1	Aperture label	
SW1	3	#6 × $\frac{3}{8}$-inch sheet metal screw	

Chapter Eleven

Chapter Twelve

Laser Property-Protection Fence

A low-cost property-protection device provides a laser beam perimeter that, when broken by an intruder, will sound an alert or an intrusion deterrent. It consists of a laser beam generator, optical detector, controller, and front-surface mirrors to reflect the beam around the perimeter (see Figure 12-1). The device can be interfaced with our painful sound field deterrent system described in Chapter 28, "Pain Field Property-Protection Guard."

Simple-to-build modules require basic installation skills in setup and alignment. This project may require readily available hardware depending on the particular installation.

The system is intended to protect a given area against unauthorized intrusion by producing a wall of ultrasonic pain. The system can be used for small inside areas or larger outside areas. The system utilizes a laser that bounces between several strategically placed front-surface mirrors before impacting an optical light detector and controller. Once the beam is broken, the controller sends a signal to the sonic pain field generator, discouraging the intruders.

The total range of the light path depends on the alignment, quality of the laser beam, and the number of mirrors used. Expect a range of 1,000 feet with three mirrors and up to 2,000 feet with a single mirror.

Power for both the laser and controller comes from low-voltage adapters using batteries for backup or short-term installations. The light source is a visible

Figure 12-1   *Property protection system*

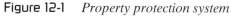

red laser but may be invisible infrared. The laser section must be securely mounted on a sturdy pole and bracket. A tree or existing structure may be used, providing it allows a clear shot to the first mirror and is out of the way of normal traffic.

The optical receiver and controller section requires 12 *volts-direct current* (VDC) as supplied by the included wall adapter transformer. It must be mounted like the laser section and avoid any exposure to bright sunlight. This section also provides the control to other external deterrents, as well as having a buzzer and *light-emitting diode* (LED) indicating lamp.

The device is shown housed into a plastic cylindrical tube (EN1) fitted with a focusing tube and a lens for long-range applications. You may use an aperture cap with its opening customized to size depending on the range and collimation of the reflected laser light. A diffusion plate can also be made from two milky white plastic caps assembled together. This method may be necessary when being used with the highly coherent beam of light from a nearby laser.

Special Note

The optical light detector controller receiver section can also be used to cause an event when light is detected and exceeds a certain value. The threshold detection level is set by the adjustment of a trimpot. This mode of operation is designed to be used in conjunction with any laser, light gun, or similar device to control any external function. It can provide a means of improving handgun steadiness by holding the laser beam on the aperture as long as possible. A means for the instinctive point firing of such weapons as shotguns and machine guns is also realized. Long-range optical remote control is possible up to a half-mile or over.

Circuit Description

Light entering the focus tube (FT1) through the lens (LE1) is detected by phototransistor Q1. From Figure 12-2 Q1 is placed so that it picks up the laser light

and causes its collector to conduct through the R1 base sensitivity trimpot resistor that biases transistor Q2 into conduction. A positive voltage is now developed across the R3 collector resistor. This voltage remains across coupling capacitor C1 until the light beam is broken. When this occurs, a negative-going voltage appears at the junction of biasing resistors R4 and R5 that are connected to the control pin 2 of monostable timer I1. The timing cycle now turns "on," causing pin 3 to go positive. It turns on transistor Q3 and energizes the sounding buzzer (BZ1), the LED, and control relay (RE1). The relay is capable of switching noninductive loads of up to 5 amps at 120 *volts-alternating current* (VAC) or of providing the detection level to trigger our PPG10 pain-field generator or other deterrent.

Construction Steps for the Receiver and Controller

Follow these steps in constructing the optical receiver and controller:

1. Lay out and identify all parts.

2. Assemble the board as shown in Figure 12-3. Note the PNP and NPN transistors. Use the leads of components as connecting wires whenever possible. Avoid bare wire bridges and short-circuit points.

3. Attach leads and external components, as shown in Figure 12-4.

4. Create the focus tube FT1 from a 6-inch piece of $1^7/_8$-inch *polyvinyl chloride* (PVC) tube, shown in Figure 12-5. Create EN1 from a 12-inch length of $2^3/_8$-inch tubing. Glue the lens to the shoulder of FT1. Note that the tubes should provide a secure telescoping fit.

5. Make the final assembly as shown with leads exiting through CAP1 using the clamp bushing BU1.

6. Test your unit by presetting both trimpots to midrange. Note the 1-megohm pot R6 controls the "on" timer once the light beam is broken. The 100K pot R1 controls the light necessary to keep the unit from triggering.

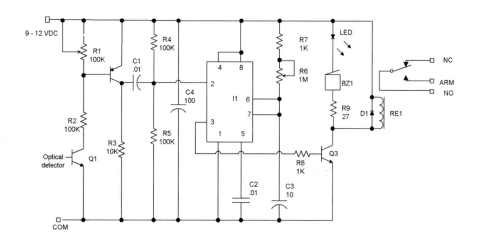

This circuit is intended as an alarm using a laser to continuously activate phototransistor Q1. When the beam is broken by an intrusion, the I1 timer is activated, turning on the buzzer, the LED, and the relay for the external of our #PPG100 Phasor Shockwave Blaster described in Chapter 26.

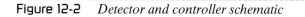

Figure 12-2 *Detector and controller schematic*

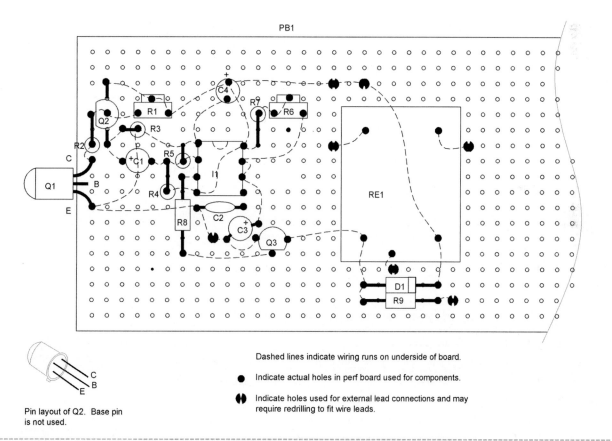

Dashed lines indicate wiring runs on underside of board.

● Indicate actual holes in perf board used for components.

◖◗ Indicate holes used for external lead connections and may require redrilling to fit wire leads.

Pin layout of Q2. Base pin is not used.

Figure 12-3 *Magnified view showing wire routing on the underside of the board*

7. Place the unit in a brightly lit area and connect it to a 12-volt power supply or wall adapter. Point the unit to a light source and then place your hand over the aperture. The LED and buzzer should respond for a period of time determined by the R6 setting. Adjust R1 for reliable triggering and R6 for the desired response time. The relay also activates,

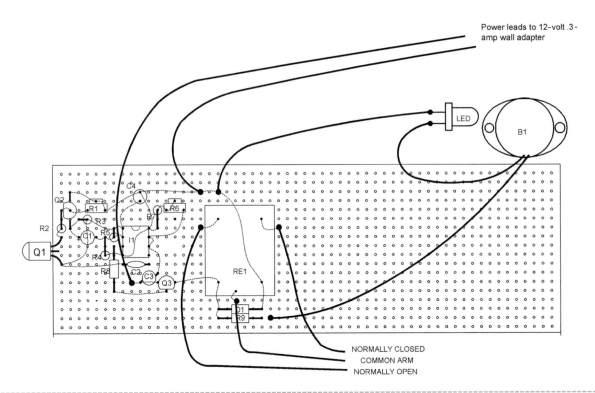

Figure 12-4   *Board assembly showing external wiring*

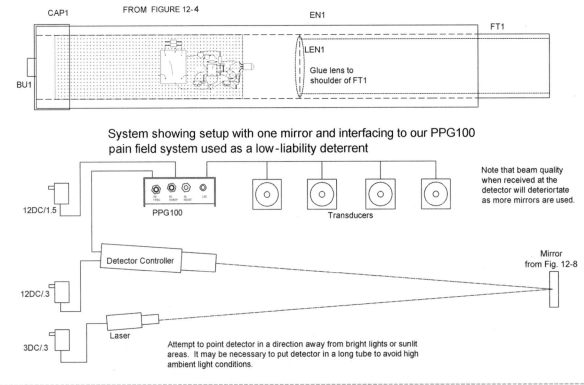

Figure 12-5   *X-ray view optical detector and controller in enclosure*

closing the contacts that now will turn on the pain field generator or other device. This now verifies the proper operation of the receiver controller.

Construction Steps for the Laser Beam Generator

Follow these steps in constructing of the laser beam generator:

1. Mechanically assemble the laser section, as shown in Figure 12-6. You may wire the 3-volt

adapter directly, and strain relieves the power leads, which can be done using a clamp bushing or tying a knot. Verify that the far-field beam spot is fairly condensed when using the collimator.

2. Proceed to set up the system as outlined in Figures 12-5, 12-7, and 12-8. Stabilize the laser, detector, and all the mirrors. Use your own ingenuity in these steps, using available materials and any possible mounting points for components.

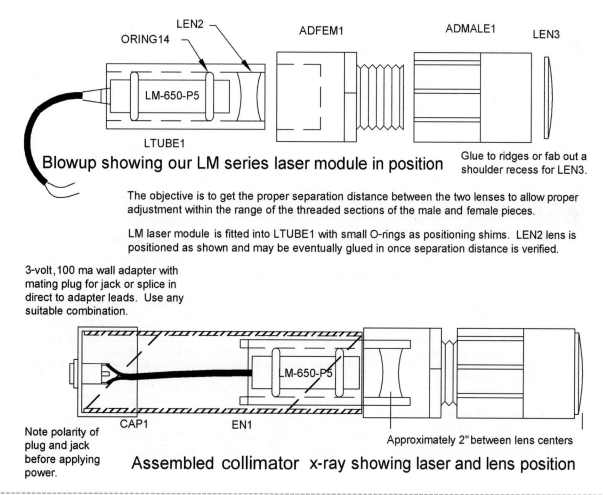

Blowup showing our LM series laser module in position

Glue to ridges or fab out a shoulder recess for LEN3.

The objective is to get the proper separation distance between the two lenses to allow proper adjustment within the range of the threaded sections of the male and female pieces.

LM laser module is fitted into LTUBE1 with small O-rings as positioning shims. LEN2 lens is positioned as shown and may be eventually glued in once separation distance is verified.

3-volt, 100 ma wall adapter with mating plug for jack or splice in direct to adapter leads. Use any suitable combination.

Note polarity of plug and jack before applying power.

Approximately 2" between lens centers

Assembled collimator x-ray showing laser and lens position

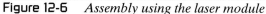

Figure 12-6   *Assembly using the laser module*

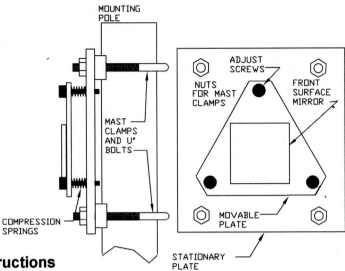

MOUNTING POLE

MAST CLAMPS AND 'U' BOLTS

COMPRESSION SPRINGS

NUTS FOR MAST CLAMPS

ADJUST SCREWS

FRONT SURFACE MIRROR

MOVABLE PLATE

STATIONARY PLATE

Basic instructions

Your laser protector fence is designed to provide an intrusion-detection perimeter around your home or target area. It uses a Class iiia diode laser to minimize but not eliminate any optical hazard. A low-liability sonic shock field generator such as the system shown in Chapter 26 is suggested as a deterrent. Caution: check local laws for proper posting of this equipment on your property. Remember if you injure a criminal, even if he is robbing you, it can result in a stiff penalty in certain states for violating his rights.

System setup

1. Obtain the following equipment listed in the latest Information Unlimited catalog:

A. Class iiia diode laser or a HeNe Laser

B. Optical light detector and controller with built-in alarm or

C. PPG10 low liability sonic shock pain field generator from chapter 26

2. Examine the area you want to protect. Attempt to position beam travel over level terrain to eliminate possible "easy-to-sneak-under"points.

3. Establish corner points of property where mirrors are to be positioned. Note that a clear view of adjacent mirrors, laser, and receiver must be maintained in all weather conditions for reliable operation.

4. Determine the corner you want to position the actual laser and laser detector at. Note there should be access to these devices for powering and control.

5. Construct mirror mounts using your own ingenuity; allow for adjustment and stability. Roughly align mirror to approximate position.

6. Turn on laser and position it to hit center of first mirror. Secure laser in place.

Adjust first mirror to hit center of second and repeat for remaining mirrors all the way back to the optical receiver. Secure them in place as you go.

Special note - Large areas may require an optical collimator at the laser output to reduce beam width at longer ranges. The collimator can be a rifle scope, telescope, binoculars etc., positioned in axial alignment with the laser. Beam divergence will be reduced by the magnification factor of the system used; however, the actual beam will be expanded by the same factor. The net result is a smaller cross-section at longer ranges.

Figure 12-7   *Scheme showing adjustable mirror mounts*

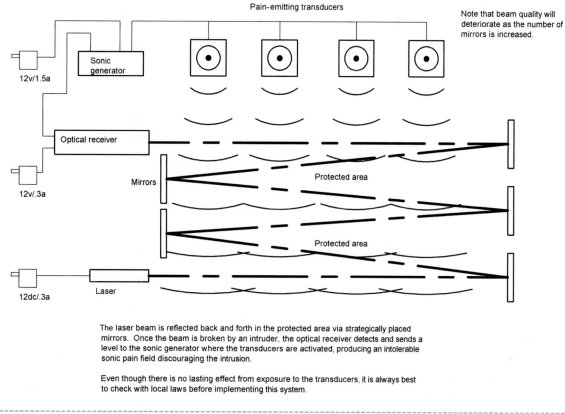

Note that beam quality will deteriorate as the number of mirrors is increased.

The laser beam is reflected back and forth in the protected area via strategically placed mirrors. Once the beam is broken by an intruder, the optical receiver detects and sends a level to the sonic generator where the transducers are activated, producing an intolerable sonic pain field discouraging the intrusion.

Even though there is no lasting effect from exposure to the transducers, it is always best to check with local laws before implementing this system.

Figure 12-8 *Example of system layout using multiple mirrors*

Table 12-1 Laser property-protection fence

Ref. #	Qty.	Description	DB#
R1		100K vertical trimpot resistor	
R2, 4, 5	3	100K, 1/4-watt (br-bl-yel) resistor	
R3		10K, 1/4-watt (br-bl-or) resistor	
R6		1 m, 1/4-watt (br-bl-gr) resistor	
R7, 8	2	1K, 1/4-watt (br-bl-red) resistor	
R9		27, 1/4-watt (red-pur-blk) resistor	
C1, 3	2	10-microfarad, 25-volt vertical electrolytic capacitor	
C2		.01-microfarad, 25-volt disk capacitor (103)	
C4		100-microfarad, 25-volt vertical electrolytic-capacitor	
Q1		L14 phototransistor	L14P2
Q2		PN2907 PNP general-purpose transistor	
Q3		PN2222 NPN general-purpose transistor	
D1		IN4007 1 Kv diode	
I1		555 DIP dual in-line integrated circuit timer LED1	Ultrabright LED
BZ1		4- to 8-volt, 20-milliampere, 90-decibel, 1-inch OD buzzer	

Table 12-1   Continued

Ref. #	Qty.	Description	DB#
RE1		12-volt SPDT relay 115/5a contacts	
PB1		$3 \times 1 .75$ of $.1 \times .1$ grid perforated board	
WR24	60 inches	#24 vinyl hookup wire	
EN1		12 inches of $2^{3}/_{8}$-inch schedule 80 PVC tubing	
FT1		6 inches of $1^{7}/_{8}$-inch schedule 40 PVC tubing	
CAP1		$2^{3}/_{8}$-inch plastic cap	
BU1	2	Small clamp bushing	
LEN1		45×90 focusing lens	LE4590
12DC/.3		12-volt, .3-amps adapter for detector	12DC/.3
PPG100		Optional pain field deterrent	PPG100
Laser section			
3DC/.3		3-volt, .3-amp wall adapter for laser	3DC/.3
LM1		3-milliwatt laser module	LM650P3
LTUBE1		$2^{1}/_{4} - \times ^{7}/_{8}$-inch OD PVC tubing	
EN2		$3^{1}/_{2} - \times 1$-inch OD plastic tubing	
CAP2		1-inch plastic cap	
LEN2		$.6 \times -1$-inch negative lens	LE15
LEN3		$.9 \times 3$-inch F/L double convex lens	LE24
ADMALE		$^{1}/_{2}$-inch PVC slip to male Genova #30405	
ADFEM		$^{1}/_{2}$-inch PVC slip to female Genova #30305	
ORING	2	$^{1}/_{2} - \times ^{5}/_{8}$-inch O-rings #014	

Chapter Twelve

Chapter Thirteen

Laser Window Bounce Listening Device

This chapter offers an excellent demonstration in which a laser illuminates a window or another reflecting surface and is then reflected back to a specialized optical receiver. Information is retrieved from the window laser as a result of vibrations on the reflecting surface. This device (see Figure 13-1) makes many potential tasks possible from a safe distance, such as surveillance or monitoring radioactive pipes in reactors, high-voltage insulators, and other objects in hostile environments.

Complete detail on the construction of the required electronic modules is provided in this chapter, allowing you to set up and demonstrate the concept without a precision machined optical housing lab. This approach is excellent for science projects,

and a sample housing is shown for those wanting to build a medium-performance field-usable device lab.

Expect to spend around $100 for the basic parts. Most are readily available, whereas specialized parts are available through Information Unlimited at www.amazing1.com. The parts for the laser transmitter and receiver are listed in Tables 13-1 and 13-2.

Special Note

The builder must realize that using a system for acoustical surveillance could violate Federal Law #90-352 (which states that it is illegal to covertly

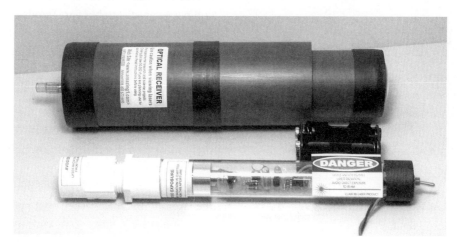

Figure 13-1   *Laser bounce listening system*

listen to a conversation without permission from one of the parties) and may also be in further violation of certain state and local laws. Remember that you need permission from one of the parties being listened to. Our advice on this point is when in doubt, consult your state attorney general as to your bottom-line objective. Using the device for a science fair project, research and development, or individual experimenting usually is not a legal issue.

Basic Theory

This project shows how to interface any optical laser and a sensitive light with a sound detector/receiver to create a system capable of listening to vibrations from a laser-illuminated reflecting surface. This device makes it feasible to listen to mechanical abnormalities of moving machinery, dangerous bursting pipes, volcanic activity, high-voltage insulator failures, and nuclear reactors, all from a safe distance. Monitoring and acoustically listening to a certain premises without ever illegally entering, as would be necessary for purposes of installing a bug or a similar device, now are possible. The principle is sound, but to achieve optimum results it may sometimes be difficult due to the many variables often encountered. These variables strongly depend on the optical and mechanical properties of the reflecting surface, the positioning of the *laser transmitter* relative to the *laser receiver*, and the quality of the target sounds in intensity, wavelength, and background interference.

Project Description

Your system requires three parts: *the laser transmitter* for surface illumination, the *laser receiver*, and the *sound-to-optical surface*. The laser transmitter projects a coherent, collimated beam of light that is reflected off the sound-to-optical surface. This reflected or scattered light is detected by the laser receiver and is converted into sound through a headset. The system uses two methods of receiving the reflected or scattered light.

The *direct reflection* (DR) method is where the laser transmitter and receiver are at two defined locations dictated by Snell's Law, which states that the angle of reflection equals the angle of incidence in all planes. This approach can be cumbersome and is limited due to positioning problems, as both the transmitter and receiver must be optically aligned in both the horizontal and vertical planes. It does provide a maximum range of operation. The DR method provides a high level of signal that may require optically attenuating with polarized filters. A shortcut for demo purposes is to use a reflective surface such as a mirror or window pane, and position it for proper alignment. This surface can be secured using your own ingenuity.

The preferred method is the *scattered reflections* mode, using the signal provided by the light from where the laser hits the surface. This is the spot you usually see at any relative position and is many times weaker than the DR, but it reduces the problem of positioning. The laser transmitter and receiver can be mounted on the same frame but must be aligned so both may view the target point. Obviously, large optical lenses on the receiver and a higher-power laser with a collimator can greatly enhance this method by collecting and providing more light for the signal.

All systems use our laser light receiver, utilizing a "light-biased" phototransistor for operation at low light signal levels. A voice enhancement bandpass filter limits high- and low-frequency sounds, further utilizing performance.

The surface of reflection plays a major role in performance, as well as the actual position of the laser spot. Double- and triple-glazed windows further complicate the situation. We can only suggest patience and perseverance to become familiar with the tricks and sometimes "black magic" of optimizing the performance of this system.

You will notice a peculiar property of the reflected laser light, such as from a window containing interference bands. This is primarily due to the phase interference occurring from a relatively flat surface. A slight motion or distortion of this surface will cause these interference bands to vary in position. This quality we use in modulating the phototransistor of the laser light receiver. The proper position of the phototransistor relative to its view of these bands can be a very sensitive adjustment. A slight change can cause a tremendous difference in reception.

You will note that it is *not* necessary to view the beam at an exact right angle to the reflection surface. Any reasonably acute angle will produce the interference bands since they are a function of the differential flatness of the reflecting surface. The detector wants to view the interface of these interference bands for optimizing performance. The importance of good, sturdy tripods for the system cannot be overstressed.

An ideal condition for testing is to position both the laser transmitter and the laser receiver so that the reflected beam has a small angle or the reflected path is near that of the incident path. This might be difficult in actual use because you have two planes (azimuth and elevation) to deal with. It is suggested when experimenting to attempt to position the "test" reflecting or vibrating surface to fine-tune these nearly coincidental axes.

Project Construction

The project is divided into three sections consisting of the laser transmitter, the laser receiver, and the completed assembly.

Laser Transmitter

First, a little laser theory. "Laser" is an acronym for *light amplification by stimulated emission of radiation.* Visible lasers are used in many applications, including gun sites, pointers, printers, construction and surveying aids, compact disc players, barcode readers, light shows, and many others. The helium-neon gas laser is one of the most familiar types, with its bright red directional beam. It's been a workhorse for years despite its fragile glass laser tube and its requirements for costly high-voltage power supplies. Infrared diode lasers are used in telemetry, security detection, night vision, robotics, covert target designation, and laser bounce surveillance.

The recently developed laser diodes emit coherent laser light in the visible and invisible spectrum, and do not require a high-voltage power supply. Because they're small, low cost, and fairly rugged, laser diodes are well suited for many new applications.

Before reviewing some basic laser theory, we must first talk about regular light for a minute. When you turn on a lightbulb, light energy is emitted in what is referred to as a spontaneous form. It is an integration of many individual atomic-energy-level changes, each producing its own little "packet" or photon of light energy, with each photon having a particular phase and frequency. In the case of a lightbulb, electrical energy "pumps" the filament electrons to higher than normal atomic energy levels shown in the 13-2a section of Figure 13-2.

Photons are emitted when the electrons return to their initial states and give up that energy in the form of light. The frequency of the light is dependent on the difference between the previously excited and the normal energy-level states: the larger the difference in energy levels, the lower the wavelength of light. The light produced by the process of spontaneous emission is incoherent or random, as shown in the 13-2b section of Figure 13-2.

Unlike spontaneous emission, laser light is highly directional. The radiant energy is released in steps, or in synchronism, resulting in coherent, reinforced light where all the waves are in phase. In other words, all the rays are parallel and at the same wavelength. This requires that the number of excited atoms in the higher energy state exceeds that of the initial or rest state. This condition, referred to as *population inversion,* normally doesn't occur in nature and must be "forced."

Given a population inversion, each energized atom is then "stimulated" to return to its lower energy state by the emission energy or the incident light of an adjacent atom, as shown in 13-2c. The result is coherent lightwaves, as shown in 13-2d. An optical cavity with mirrored ends is usually necessary to provide the right amount of stimulated energy for laser light. As shown in 13-2e, the light is reflected back and forth within its confines until it is a powerful beam that is allowed to exit the cavity as useful laser light energy.

A laser diode is similar to an ordinary *light-emitting diode* (LED) in that both are composed of a semiconductor PN junction, as shown in Figure 13-2f.

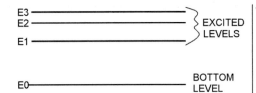

Fig 13-2a   Light is the result of radiation produced
within an individual atom by an electron being
"pumped" to a higher than normal energy level
by an external energy source.

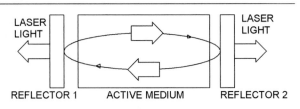

LASER
LIGHT

LASER
LIGHT

REFLECTOR 1   ACTIVE MEDIUM   REFLECTOR 2

Fig 13-2e   An optical cavity having mirrored ends provides the
right amount of stimulated energy for laser light. Light is
reflected back and forth within its confines until it is a powerful
beam that is allowed to exit the cavity as useful laser radiation.

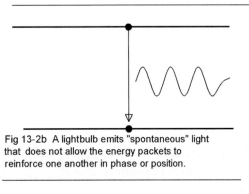

Fig 13-2b   A lightbulb emits "spontaneous" light
that does not allow the energy packets to
reinforce one another in phase or position.

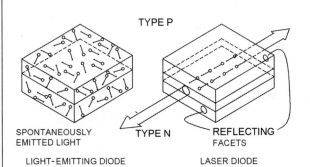

TYPE P

SPONTANEOUSLY
EMITTED LIGHT

TYPE N

REFLECTING
FACETS

LIGHT-EMITTING DIODE   LASER DIODE

Fig 13-2f   A laser diode is similiar to an ordinary LED except
that the LED produces spontaneous light, while the laser emits
light by stimulated emission where the wavelengths and
temporal relation are coherent. A laser diode also contains two
reflecting mirrors that form a cavity and permit the emitted light
to be highly directional.

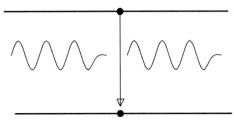

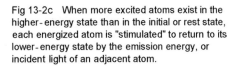

Fig 13-2c   When more excited atoms exist in the
higher-energy state than in the initial or rest state,
each energized atom is "stimulated" to return to its
lower-energy state by the emission energy, or
incident light of an adjacent atom.

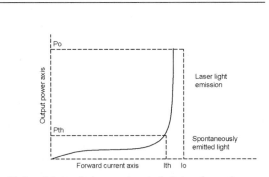

Fig 13-2g   A laser diode operates similarly to a forward-
biased diode. The vertical axis corresponds to optical output
while the horizontal axis is the forward diode current. Io is
the operating current, and anything below Ith will produce the
effects of an LED.

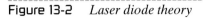

LASER LIGHT (COHERENT RADIATION)

Fig 13-2d   A laser beam is the result of an
"in lockstep" train of coherent lightwaves.

Figure 13-2   *Laser diode theory*

An electrical potential causes a flow of holes and
electrons that, upon recombination, emit light. The
LED produces spontaneous light, whereas the laser
emits light by stimulated emission. The laser diode
also contains two reflecting mirrors that form what's
called a Fabry Perot cavity and permit the emitted

light to be highly directional, an important laser
property. The 13-2g section of Figure 13-2 shows the
characteristic diode curve showing the slope sensitivity of the device.

In spite of a laser diode's apparent physical
ruggedness, it is very sensitive to temperature

changes, electrical transients, and operating-current parameters. It is totally unforgiving of errors, so our circuitry and construction techniques must take that into consideration. The curve shows a steep slope where laser operation takes place and the input-current "window" on the horizontal axis is very narrow; consequently, the driver circuit must operate within those limits or you'll end up with one of the world's most expensive medium-powered LEDs.

It is possible in certain cases to cool your laser diode down to a low temperature and obtain more output. Caution is advised for two reasons: You may damage the diode's built-in optics and you may exceed the laser classification rating.

The laser in this circuit uses a diode that can produce up to 10 milliwatts of optical power at a wavelength of 880 nanometers. The output level must be kept to fewer than 5 milliwatts to comply with safety issues when using this application.

Circuit

The schematic of the handheld laser is shown in Figure 13-3. The laser diode (LD1) is actually an assembly that contains a laser-emitting section (LD) and a photodiode section (PD). The photodiode enables the circuit to monitor the laser diode's output and produce the feedback necessary to control the circuit and protect the diode from excess drive current.

The laser diode is connected in series with current-limiting resistors R3 and R4 and the collector of transistor Q2. The current through Q2 is controlled by transistor Q1. The zener diode (Z1) maintains the

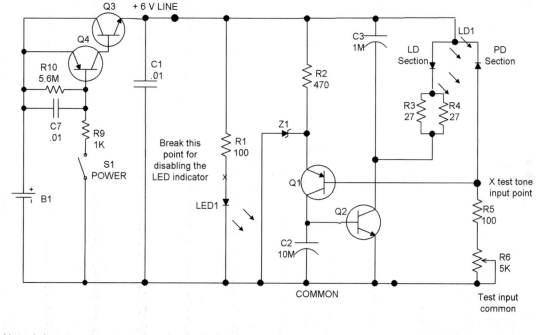

Note 1 Laser output power may be controlled to some extent by physical pressure to contact probes connected in place of S1 switch.

Note 2 Do not adjust R6 to limits as the resistance jump could damage the laser diode.

Note 3 Power to the module is shown using a 6-volt battery pack of 4 AA cells. You may use any clean source of 6 volts, providing it can deliver a current of 150 ma without a voltage drop. Use caution when setting the power on weak batteries as when new ones are replaced the laser may be overpowered.

Note 4 The test tone oscillator and voice modulator are connected to the points shown across R5,6. A separate battery of 9 volts is recommended for power to these external circuits. Connecting wires to these circuits are twisted and routed out through rear cap CAP1.

Figure 13-3 *Laser schematic*

voltage across Q1, and resistor R2 limits the zener current. The collector current of Q1, also the base current of Q2, is controlled by its own base, which is connected across resistors R5 and R6. Current through the photo diode PD develops a voltage across these resistors that is proportional to the optical output energy and constitutes the feedback required for output stabilization. Increased output causes Q1 to conduct less base current to Q2, resulting in less laser diode current. Potentiometer R6 presets the value of the quiescent current. Capacitors C2 and C3 limit transients at the base of Q2, whereas capacitor C1 limits them from the 6-volt line. The system turns on when transistor Q3 starts to conduct and eventually reaches saturation.

The leads to switch S1 may be touch-control electrodes consisting of small pieces of metallic tape that, when bridged by finger contact, cause a small amount of base current to flow into transistor Q4. The collector current of Q4 flows into the base of Q3, causing it to saturate and supply current to the laser diode circuitry. The base current to Q4 is limited by resistors R9 where R10 and capacitor C7 reduce the circuit's sensitivity to stray AC or static fields that could cause premature activation. This scheme would only be used in certain applications and may be eliminated by a simple switch.

The laser is powered by four external AA batteries or any convenient source of 6 volts. S1 controls the power to the circuit and may be a key switch, push button, toggle, or slider switch rated at .5 amps. LED1 is intended for emission indication, and resistor R1 controls the diode current.

Construction of the Electronics Assembly Board

To construct the board, follow these steps:

1. Identify all the parts and pieces and verify them with the bill of materials.

2. Insert the components, starting from one end of the circuit board, and follow the locations shown in Figure 13-4. If you are using a perforated vector board, it is a good idea to trial-fit the larger parts before actually starting to solder. Always avoid bare wire bridges, messy

solder joints, and potential solder shorts. A *printed circuit board* (PCB) is available for this project.

3. Check for cold or loose solder joints. Pay attention to the polarity of the capacitors with polarity signs and all the semiconductors.

4. Carefully solder the bus wire extension pieces to the laser socket as shown in Figure 13-5. The socket must abut to the top side of the assembly board when soldered in place. It is suggested that the socket be secured in place using *room temperature vulcanizing* (RTV) silicon rubber when the proper assembly is verified.

5. Solder two 6-inch switch leads and a CL1 battery clip to the assembly board. You may also connect leads to the optional test-tone circuit if building the laser for window bounce experiments. Note the dual connections to pins 1 and 3. These are the anode of the diode laser and the cathode of the monitor diode.

Assembly Is Ready for Pretest

Do not install the laser diode in the circuit at this time. The circuit must be checked and calibrated beforehand. To test the construction, follow these steps:

1. Make sure that the batteries are at full capacity before you proceed with the following. Check on the lowest meter range; any current flowing into the circuit above a fraction of a micro-amp with the system off will cause premature discharging of the batteries. Check for defective components, flux paths, excessive moisture, and so on if any current reading is detected in this step.

2. Set up a meter in series with one of the battery leads to monitor the system current in the range of 10 to 250 milliamperes.

3. Temporarily unsolder the LED from R1. Short the switch leads together and measure a current of approximately 15 milliamperes. Bridge R1 to the LED and note the current going to 40 milliamperes and the LED lighting. Still

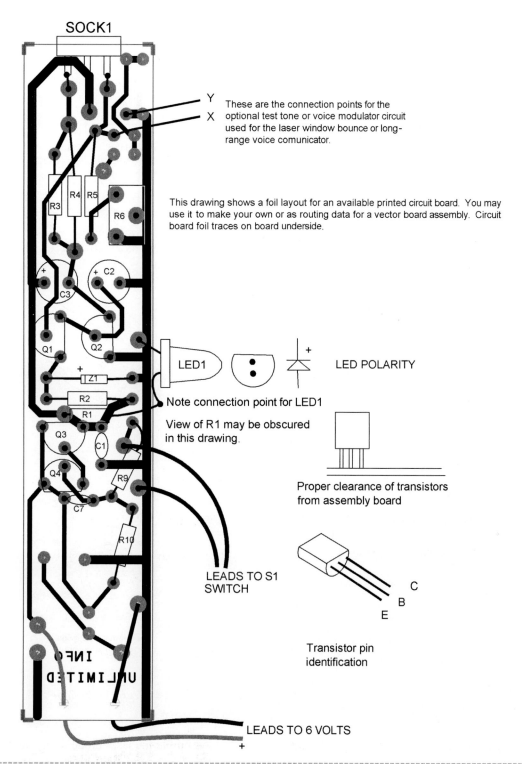

SOCK1

Y
X
These are the connection points for the optional test tone or voice modulator circuit used for the laser window bounce or long-range voice comunicator.

R3 R4 R5
R6

This drawing shows a foil layout for an available printed circuit board. You may use it to make your own or as routing data for a vector board assembly. Circuit board foil traces on board underside.

+ C2
C3
+ C2
Q1 Q2
LED1
LED POLARITY
+
Z1
R2
R1
Note connection point for LED1
Q3
C1
View of R1 may be obscured in this drawing.
Q4
R9
Proper clearance of transistors from assembly board
C7
R10
LEADS TO S1 SWITCH
C
B
E
UNLIMITED INFO
Transistor pin identification

LEADS TO 6 VOLTS
+

Figure 13-4 *Assembly circuit board*

leave it disconnected. This verifies the LED1 lighting necessary for emission indication.

4. Obtain a 1N4001 diode and connect it in place of the LD laser diode section of LD1. You can connect the anode to the 6-volt line and the cathode

end to the junction of the R3 four-socket end. Rotate R6 to full *counterclockwise* (CCW).

5. Turn the system on and note the input current going to 250 milliamperes. Add a 10K resistor in place of the PD section of LD1 and note

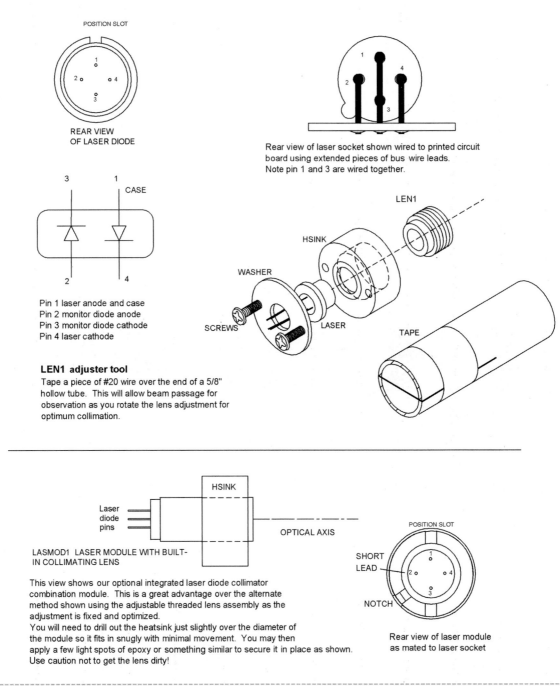

POSITION SLOT

REAR VIEW
OF LASER DIODE

Rear view of laser socket shown wired to printed circuit
board using extended pieces of bus wire leads.
Note pin 1 and 3 are wired together.

LEN1

HSINK

WASHER

SCREWS

LASER

TAPE

3 1

CASE

2 4

Pin 1 laser anode and case
Pin 2 monitor diode anode
Pin 3 monitor diode cathode
Pin 4 laser cathode

LEN1 adjuster tool
Tape a piece of #20 wire over the end of a 5/8"
hollow tube. This will allow beam passage for
observation as you rotate the lens adjustment for
optimum collimation.

HSINK

Laser
diode
pins

OPTICAL AXIS

LASMOD1 LASER MODULE WITH BUILT-
IN COLLIMATING LENS

This view shows our optional integrated laser diode collimator
combination module. This is a great advantage over the alternate
method shown using the adjustable threaded lens assembly as the
adjustment is fixed and optimized.
You will need to drill out the heatsink just slightly over the diameter of
the module so it fits in snugly with minimal movement. You may then
apply a few light spots of epoxy or something similar to secure it in place as shown.
Use caution not to get the lens dirty!

POSITION SLOT

SHORT
LEAD

NOTCH

Rear view of laser module
as mated to laser socket

Figure 13-5 *Laser socket wiring for a separate diode and a diode with an integral collimator*

the input current dropping to around 15 milliamperes.

6. Rotate R6 through its range and note the
input current varying between 15 and 250 milliamperes. This is *not* a linear control.

Note that an integrated laser diode with optics
(LASMOD1) is available as shown in Figure
13-5, making optical alignment simpler.

7. Assemble the laser diode to the heat sink, as
shown in Figure 13-5, and very carefully install
the laser diode into the socket. Observe the
position slot shown. Verify that R6 is fully
CCW. Apply power and carefully adjust the
input current to 100 milliamperes. The diode
should be producing around 2 milliwatts of
output and is detectable using infrared-sensi-
tive paper or a night vision device. Many video
cameras will respond to this laser wavelength.

Chapter Thirteen

You can operate the diode up to 125 milliamperes and obtain almost 10 milliwatts of output. This will exceed the class 3a rating and the system must be labeled accordingly. Safety glasses must be used at this power level.

Once you set the desired power level, you can reconnect LED1, noting that it draws 20 milliamperes of additional current. The optional test-tone or voice modulator can be connected and will be clearly heard when used with the optical detector by just picking up the scatter. Note these optional circuits can operate up to 12 VDC.

The output of this laser is also an excellent invisible illumination source for starlight night vision scopes, allowing long-range viewing in total darkness.

At this point you should have a fully functional board-level laser system capable of producing up to 10 milliwatts of optical output power at 820 nanometers. The laser must now be housed or contained in a suitable enclosure dependent on its application.

Special Note

Unfortunately, the output of this system is in the infrared range and is invisible to the naked eye as the radiation is at 820 nanometers and appears as a dull, red glow. The actual viewing of the laser energy will require a night vision device. Many video cameras will pick up this radiation.

Lumitek supplies infrared sensor paper that will glow, indicating the beam profile. It is available on their web site at lumintek.com. Look for Q11 sensor paper or sticks.

Final Enclosure and Optics

Insert the small lens (LEN1) with the small, plastic-threaded retainer into the heat sink, as shown in Figure 13-5. If your heat sink is threaded, simply screw it in until it bottoms out. If it is not threaded, you must shim up the assembly with some thin tape until it fits into the heat sink reasonably securely. Adjustment is much simpler using the threaded heat sink as you simply screw it into the right position to obtain the smallest far-field beam profile as indicated by the

detection equipment used. The unthreaded version will require trial-and-error positioning of the lens to achieve the correct position. This can be very cumbersome and requires patience and perseverance to find the optimum position. The objective is to obtain the minimum beam divergence at far field (several meters).

Fabrication

Fabricate the enclosure tube (EN1) from a 1- × $^{1}/_{16}$ - × 7-inch polycarbonate tube, as shown in Figure 13-6. We use clear tube to enhance the science project approach, showing the innards.

Create the lens tube (LTUB1) from a .835- OD × .6- × 2-inch schedule 40 *polyvinyl chloride* (PVC) tube. Note these are the nominal dimensions for $^{1}/_{2}$-inch PVC tubing. Insert a negative concave lens (LEN2) into the LTUB1.

Make sure the lens is clean without fingerprints. Use a clean eraser on a pencil to position it. This should fit smoothly into this tube. PVC tubing can vary in dimension and you may have to shim or actually bore out the tube for proper fitting of this part.

Final Assembly

Finally, assemble the attaching lens (LEN3) to the fitting (ADJFEMALE) with glue as shown. Press the lens tube completely into the mating fitting (ADJMALE). The collimator assembly now fits into the enclosure tube. You may have to shim the LTUB1 section to obtain a secure fit.

The laser module fits inside the enclosure tube and is secured by a small nylon screw (SW1) mating to the clearance hole on the board. S1 is mounted to the rear cap (CAP1). A clearance hole for power adjust trimpot R6 must be drilled in the enclosure tube.

Now the ultimate trick is to get the two-lens separation distance approximately to 2 inches or where the best collimation occurs at the midrange of the threaded adjustment of the ADJMALE and FEMALE fittings. Once this is verified you can secure it using a forgiving adhesive such as RTV silicon rubber.

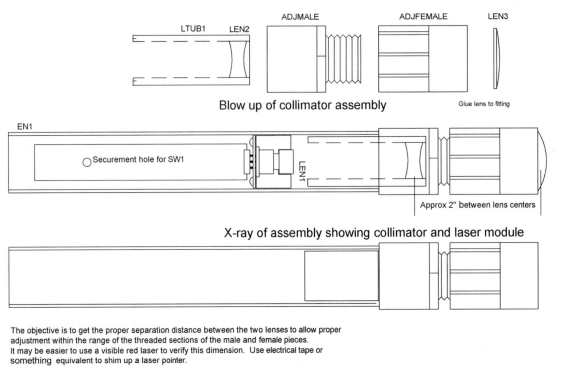

Blow up of collimator assembly

Glue lens to fitting

X-ray of assembly showing collimator and laser module

Approx 2" between lens centers

The objective is to get the proper separation distance between the two lenses to allow proper adjustment within the range of the threaded sections of the male and female pieces.
It may be easier to use a visible red laser to verify this dimension. Use electrical tape or something equivalent to shim up a laser pointer.

Beam spot without collimator will be approx. 12" at 500 feet. The collimator will reduce this width by a factor of 5 to 7.

Figure 13-6 *Housing and collimator*

Experiment by checking the basic lens position in the heat sink and by resetting the collimation adjustment, noting the objective of obtaining minimal beam divergence at far field. You may also require an expanded beam for night viewing. The final assembly should resemble that shown in Figure 13-7.

Test-Tone Module Construction

To assemble the test-tone module, follow these steps:

1. Identify all parts and pieces and verify them with the bill of materials.

2. Insert the components, starting from one end of the perforated circuit board, and follow the locations shown later in Figure 13-14, using the individual holes as guides. The board is cut $1 \times 1.5 \times .1$.

   Use the leads of the actual components as the connection runs. These are indicated by the

dashed lines. It is a good idea to trial-fit the larger parts before actually starting to solder.

Always avoid bare wire bridges, sloppy solder joints, and potential solder shorts. Check for cold or loose solder joints.

Pay attention to the polarity of the capacitors with polarity signs and all the semiconductors.

3. Cut, strip, and tin the wire leads for connecting to the diode laser module points X and Y. These should be #24 leads to fit into the hole pads of the laser module.

4. Connect a scope across output X and Y and connect a 9-volt battery to the CL1 snap clip. Note the circuit draws 10 milliamperes when activated.

5. Turn on S1 and note a 1,000 to 1,500 Hz test tone of 2 volts peak.

6. Connect the output leads to the laser board, as shown at points X and Y. Preset the laser diode current to 100 milliamperes as directed. Aim the laser at a light-colored surface. You

CAUTION: Viewing of laser beam or direct reflections require the use of protective eyewear available through KENTEK in N.H. tel# 1 603 435 7201

Note that collimator will expand the beam at near field but greatly reduce it at far field. LEN1 basic lens in heastsink may require beam adjustment over time using a special tool.

Instructions:

1. Note switch is off and insert batteries.
2. Mount or secure laser assembly and point to target.
3. Power up by turning on S1. Note emission LED coming on.
4. Adjust optics for desired effects.

Note that the collimator's adjustable piece is shown as the male part. This is a user preference. We prefer the female section with the flutes.

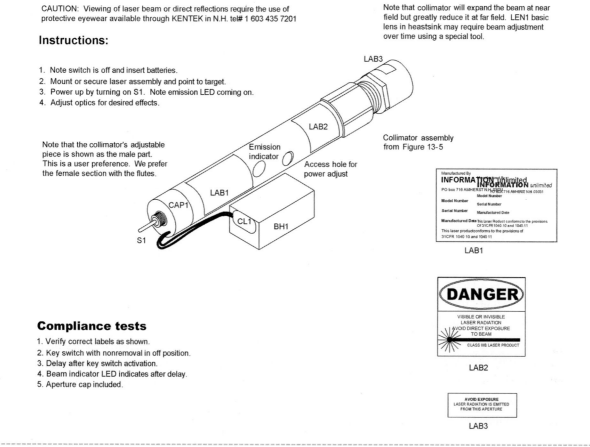

Collimator assembly from Figure 13-5

Compliance tests

1. Verify correct labels as shown.
2. Key switch with nonremoval in off position.
3. Delay after key switch activation.
4. Beam indicator LED indicates after delay.
5. Aperture cap included.

Figure 13-7   *Final assembly view and labels*

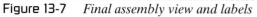

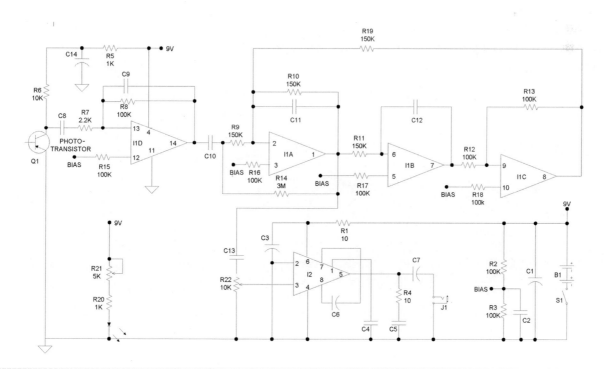

Figure 13-8   *Optional receiver schematic*

should clearly and loudly hear the test-tone signal when pointing the laser receiver in the direction of the scattered reflections.

Laser Receiver

This section shows how to construct the electro-optical receiver capable of detecting and reproducing the modulated information placed on optical beams of laser light energy. It also enables you to listen to any varying, periodic source of light, such as calculator displays, TV sets, normal lighting, the light produced from a fire, lightning, infrared sources, and of course intentionally modulated beams for voice or other analog communications. It functions as an excellent detector for this laser listening project by detecting the vibrations on a window or other similar surface when illuminated by laser light. These signals are clearly reproduced via a speaker or headset.

The device is housed in a round PVC enclosure that is easily mounted to a camera tripod when used

for sensitive positioning applications. The gain control power switch and headphone jacks are mounted on the rear of the enclosure. A pistol grip configuration is suggested for pointing and listening to random sources.

Circuit Theory

Figure 13-8 shows the output from a sensitive photo-transistor (Q1) amplified by a low noise amplifier (I1D) to a gain of 50. The output is pass-band filtered by voice filter I1ABC at 3-decibel rolloff points between 500 and 2,000 Hz. This provides maximum performance at most voice frequencies. The filtered output is further amplified to a usable level by IC2 for headsets or a small speaker. The J1 jack is connected to allow individual volume controls when using high-performance headsets (HS30). System gain is controlled via a pot and switch combination (R22/S1). S1 controls battery (B1) power to the circuit. A network consisting of resistors (R2 and R3) provides the necessary midpoint biasing for I1ABCD.

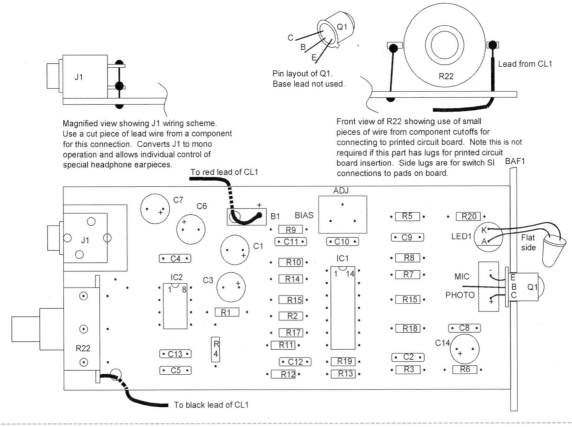

Figure 13-9 *Optical receiver assembly board*

A special circuit consisting of LED1 provides a light bias to phototransistor Q1. This is especially useful in a low light signal condition when viewing the scattered reflection mode of operation. Resistor biasing Q1 creates bothersome noise, limiting the full performance of the system. Trimpot R21 controls the emission output of LED1 and must be carefully adjusted in final testing for optimum effect.

Construction Steps

To build the laser receiver, follow these steps:

1. Lay out and identify all the parts and pieces with the parts list—note the color identification on the resistors. Capacitors are easily identified by markings with alternative value codes in parentheses. Assembly is shown on a PCB with a foil layout, as shown in this data. Construction may be on a vector board if layout is followed.

2. Insert the resistors as shown in Figure 13-9. Start with R1 and progress until all the resistors are mounted. Solder and clip off the excess leads. It is important to avoid solder bridges, shorts, and cold solder joints. Figure 13-10 shows the foil traces on the underside of the PCB for those wanting to etch their own.

3. Repeat with capacitors, observing the polarity of C1, C3, C6, C7, and C14.

4. Insert I1 and I2; note the position marks.

5. Insert Q1 phototransistor. Note the polarity as shown in Figure 13-7.

6. Insert B21, R22, and J1. Solder in place.

7. Connect the CL1 battery clip. Note the color-coded leads and dedicated strain relief holes in the PCB.

8. Fabricate the baffle plate (BAF1), as shown in Figure 13-11. This piece is used to view the position of the reflected laser or focused light as gathered by the extender lens. It is necessary to position this light onto the lens of the phototransistor. The baffle should be small enough to slide into the enclosure tube without obstruction. Drill two small holes for the leads of LED1 and center the hole for the phototransistor. Insert the LED1 leads, observing the polarity through the holes, and solder in place. Note the leads must be as long as possible. Position LED1 off to the side relative to the Q1 phototransistor so the emission signal can obliquely be seen yet does not block out any signal along the optical axis. Position it as shown in Figure 13-12 and secure it to the PCB with an RTV adhesive or an equivalent.

9. Create SPCR1 from a $1^{1}/_{2}$ - × 2-inch piece of plastic, cardboard, or any suitable resource. This piece is sandwiched between the battery and the bottom of the PC board for insulating

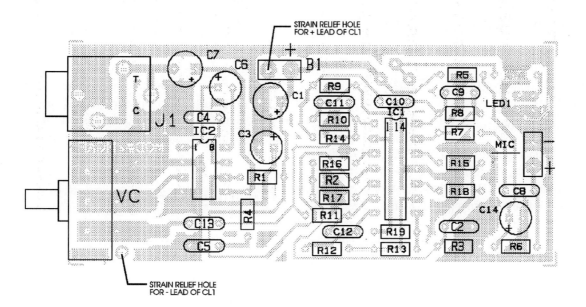

Figure 13-10 *Assembly board layout foil traces*

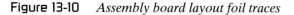

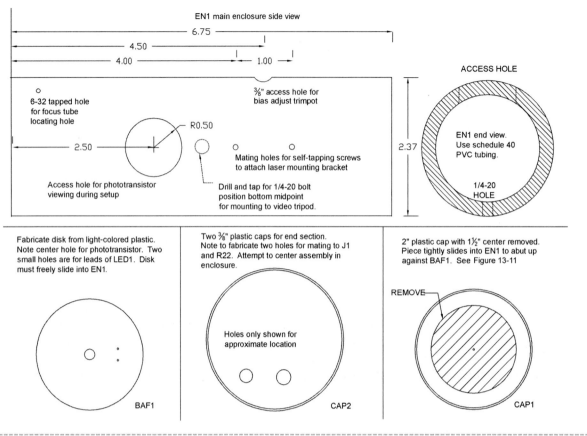

EN1 main enclosure side view

6.75
4.50
4.00 — 1.00

ACCESS HOLE

6-32 tapped hole
for focus tube
locating hole

⅜" access hole for
bias adjust trimpot

R0.50

2.50

Mating holes for self-tapping screws
to attach laser mounting bracket

EN1 end view.
Use schedule 40
PVC tubing.

2.37

Access hole for phototransistor
viewing during setup

Drill and tap for 1/4-20 bolt
position bottom midpoint
for mounting to video tripod.

1/4-20
HOLE

Fabricate disk from light-colored plastic.
Note center hole for phototransistor. Two
small holes are for leads of LED1. Disk
must freely slide into EN1.

Two ⅜" plastic caps for end section.
Note to fabricate two holes for mating to J1
and R22. Attempt to center assembly in
enclosure.

2" plastic cap with 1½" center removed.
Piece tightly slides into EN1 to abut up
against BAF1. See Figure 13-11

Holes only shown for
approximate location

REMOVE

BAF1

CAP2

CAP1

Figure 13-11   *Parts fabrication sketches*

purposes, as shown in Figure 13-12. The battery can simply be held in place with suitable elastic or an O-ring as shown. Create the collimator as shown in Figure 13-13.

10. Check all the wiring and the integrity of the solder points. Check for solder bridges and shorts on the assembly circuit board.

Unit Is Ready for Pretest

To test the unit out, follow these steps:

1. Connect a headset to J1 and clip in a 9-volt battery or connect to a suitable DC bench supply. Preset R21 FCCW. Click on R22/S1 and advance until a hum is heard in the headphones. Back off to a comfortable listening level. Note the hum is from the 60 Hz lighting.

2. Obtain a calculator with a clocked display and point unit to pick up the display emissions. Note a loud tone indicating signal pickup.

3. Verify the filter operation (which is not necessary unless circuit problems are encountered) by inserting a variable sine wave signal through an attenuator network across Q1. Preset the signal generator frequency to 1 kHz and turn up the level until a 1-volt peak-to-peak signal is measured via a scope connected across J1. Slowly vary the signal frequency and note the output response. The signal should start to roll off at 300 Hz with 3 kHz peaking in the center of these extremes. The battery current should read approximately 10 to 20 milliamperes with the headphones connected, detecting a normal signal.

Final Assembly

The final steps are as follows:

1. Fabricate EN1, CAP1, and CAP2, as shown in Figure 13-11.

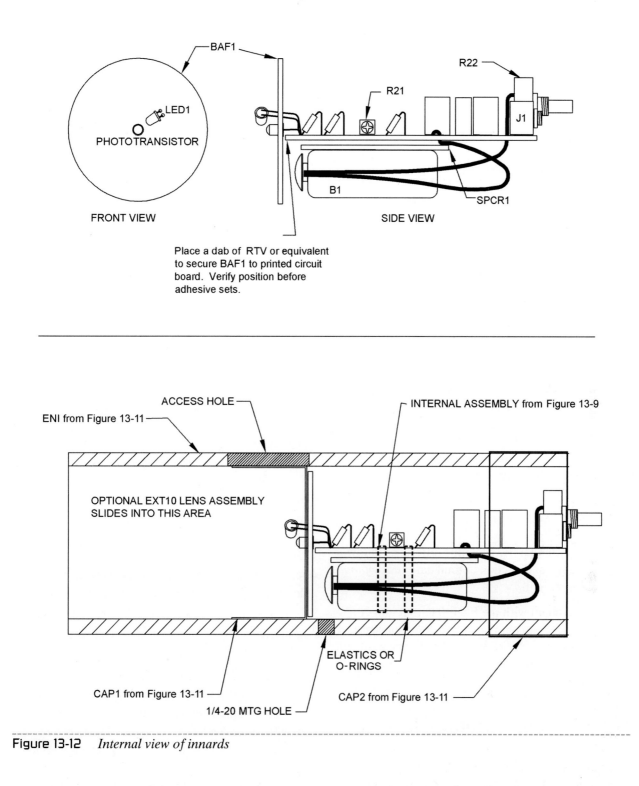

FRONT VIEW

SIDE VIEW

Place a dab of RTV or equivalent
to secure BAF1 to printed circuit
board. Verify position before
adhesive sets.

Figure 13-12 *Internal view of innards*

2. Assemble as shown in Figure 13-12. Note the
   positioning of CAP1. This keeps BAF1 in
   alignment.

Laser Receiver Is Ready for Final Testing

To conduct the final test, follow these steps;

1. Attach the unit to a video tripod via the $\frac{1}{4}$-20
   threaded hole.

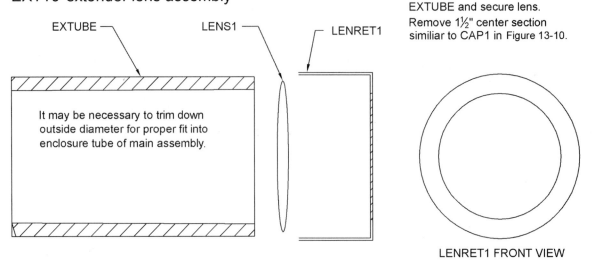

EXT10 extender lens assembly

EXTUBE ———

LENS1 ———

LENRET1

Lens retainer slides onto
EXTUBE and secure lens.
Remove 1½" center section
similiar to CAP1 in Figure 13-10.

It may be necessary to trim down
outside diameter for proper fit into
enclosure tube of main assembly.

LENRET1 FRONT VIEW

Figure 13-13   *Extender lens assembly*

2. Obtain a source of signal such as the clocked
    display of a hand calculator or a similar
    device. Note the liquid crystal will not work.
    Point the unit in the general direction of the
    display at a distance of 10 to 15 feet. Do not
    use the EXT10 lens at this time.

3. Allow for total darkness and attempt to
    obtain a signal by panning the tripod. Adjust
    the bias light control R21 for the best signal to
    background noise. Note the difference in low-
    level signal detection in total darkness *without*
    the bias light.

4. Experiment with the EXT10 lens at longer
    distances and appreciate the detection sensi-
    tivity of this device. Note that alignment may
    be difficult due to a narrow field of view.

System Application, Safety, and Legality

These fully working modules may be packaged by the
builder into a compatible housing that will allow
micro-mechanical adjustments. You can use a visible
red laser module for short-range demonstration or
sighting and alignment purposes. The illuminating
laser transmitter must be precisely in alignment with
the optical axis of the laser receiver and be mechanically fine-tuned via micrometer head screws to

receive and process the scattered reflections. The
laser transmitter described in this project uses the
test-tone circuit that greatly simplifies rough optical
alignment to a far-field surface. Output of the laser
transmitter may be preset to 2 to 4 milliwatts for class
3a or at 8 to 10 milliwatts for class 3b. A collimator is
included that greatly extends the potential range of
the system. Optional use of the telephoto lens retrofitted to the optical receiver will greatly enhance distance and performance. The prebuilt modules may be
set on tripods to verify performance integrity or for
demonstration purposes before being enclosed.

The lasers used can be class 3a to class 3b visible
or invisible. Protective eyewear is positively required
in case you look into the direct reflection. Using the
scattered reflection mode is less dangerous.

Get permission of those parties you are listening
to! Experimental demonstration of this system
should not pose a legal problem, nor should it be
used for applications not involving oral interception.

Setup Using Direct Reflection

The direct reflection method requires the following
steps:

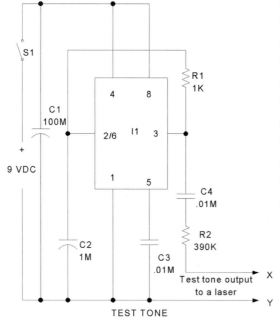

Bill of Materials

R1 1K 1/4 W (BR-BLK-RED)

R2   (1) 390K 1/4 W (OR-WH-YEL)

C1   (1) 100 M / 25V vert electro capacitor

C2   (1) 1 M / 25V vert electro capacitor

C3,4   (2) .01 M / 50V disc (103)

I1   (1) 555 DIP timer IC

S2   (1) Slider SPDT switch

CL1   (1) Battery snap clip

WR1   (30") #24 vinyl hookup wire

PB1   (1) 1 X 1.5" .1 grid perf board

Assembly board layout

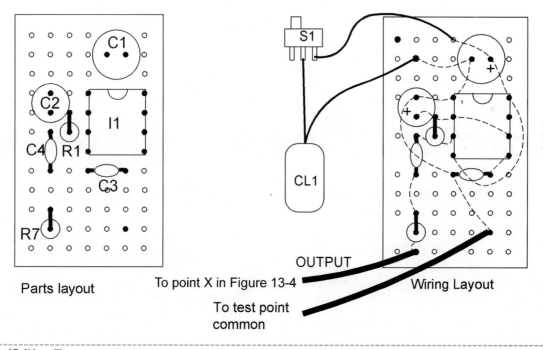

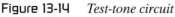

Parts layout   To point X in Figure 13-4   Wiring Layout

To test point common

OUTPUT

Figure 13-14 *Test-tone circuit*

1. Obtain two video camera tripods and secure the laser transmitter to one and the laser receiver to the other. Use duct tape, bungee cord, electrical tape, and your own ingenuity.

2. Remove the rear cover of the laser receiver and install a 9-volt battery into the clip.

3. Determine the target window. Select an easy one that is nearly "normal" and on the same

level where you are located. Place a loud radio on the opposite side of the window.

4. If your laser is a pointer or gun sight, it will be necessary to apply pressure to the trigger switch. This can be accomplished using a paper clip or clothespin to clamp the switch.

5. Position the laser transmitter tripod so the angle is as close as possible to the normal, reflected surface. This will allow minimal separation between the transmitter and receiver.

   Note that this is not necessary for proper performance but is easier until you are familiar with overall system alignment.

6. Locate the position of the reflected "laser spot" resulting from the direct reflection of the laser beam as it bounces back from the window. This will depend a lot on the relative position as in step 5, since the angle of reflection will equal the angle of incidence (Snell's Law).

7. Carefully adjust the positioning of the laser receiver so that it intercepts the spot from the direct reflection. The final position where the reflected signal is incident on the phototransistor as viewed through the view hole.

8. If you are using the extender focal lens, adjust it so that the reflected signal is about the size of a penny as viewed on the phototransistor and the white baffled disc. This lens is not necessary for ranges below 50 feet.

9. Turn the amplifier on via the control and adjust it to a comfortable audio level. Optimum results may require "tweaking" to an actual signal. A rough adjustment requires detecting a weak optical signal source in total darkness and adjusting for the best performance/noise figure. Note that the unit will not work correctly if not properly set.

10. Carefully adjust the position of the laser receiver for maximum clarity and volume. Note that only a slight adjustment can make a world of difference in performance. Experiment with the lens assembly when using ranges over 50 feet. Note the laser beam spot profile on the surface of the white baffle disc. You will see interference bands or fringes consisting of light and dark sections.

Note that clarity, volume, and general performance depend on many factors. The size of the window; the setting of the pane; and even the vibration picked up from window air conditioners, motors, pumps, oil burners, and so on can seriously degrade a usable signal.

Serious experimenters may want to interface the system with an audio equalizer to filter and enhance the usable signal. Again, experiment and experience is the best solution to quickly set up and obtain optimum performance.

A setup using the scattered reflection utilizes the light detected off the optical axis. The signal will be weaker over a given distance and will require more careful alignment.

You now have a choice of using the individual modules mounted on tripods for a demonstration of the concept and experiment. You may also choose to retrofit the modules within a sturdy housing similar to what is shown in the following data, providing a usable, field-worthy system of medium performance.

The objective is to allow an optical alignment between the laser impact point on the window and the return signal being coincident (coming back on same axis) to the optical axis of the receiver. Once this initial alignment is accomplished, it is only necessary to "tweak" the micro adjust screws that secure the laser optimizing the signal from any reasonable distance. It is assumed some signal is always detectable once initially aligned.

A test-tone signal modulates the laser at 1 to 2 kHz. This scheme provides easy access to aligning the optical axis. You now carefully search for the test tone, which is clearly detected with the optical receiver. Night vision equipment also can be an aid in initial alignment.

Mounting the laser in our test prototype involved floating it inside a stable housing, allowing several degrees of both vertical and horizontal adjustment for final "tweaking" using vernier adjust screws. The lens used on our prototype was from a low-cost video camera and screwed into a mating adapter plate firmly attached to the housing.

Obviously, a certain amount of mechanical ingenuity will be required in finalizing the system. The suggested assembly of this integrated system is shown in Figures 13-15, 13-16, and 13-17.

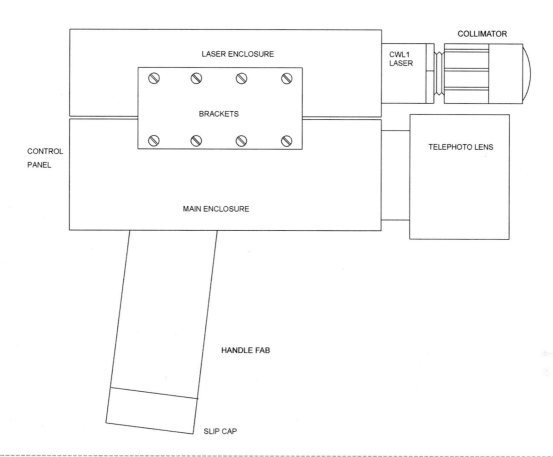

Figure 13-15 *Suggested system assembly—side view*

Exact dimensions are not given, as this might limit you due to the access of materials and the use of your own ingenuity. We show our approach to be used only as a guide that may be closely or partially followed.

Assembly of a Field-Worthy, Integrated System

The laser receiver innards are placed inside a sturdy housing and secured in place. An adapter fitting is secured to the exit end, allowing the fitting of your choice of telephoto lens. The laser transmitter is housed with a collimator and is secured to the laser receiver by two aluminum brackets.

This assembly requires good accuracy for maintaining close optical alignment, as any adjustment is via two micro-head adjusting screws for azimuth and elevation adjustments for fine-tuning.

You will note that the laser module is supported at its midsection by two rubber O-rings that provide a flexing motion both fore and aft. Compression springs are placed directly opposite the contact points of the micro-head adjusting screws. These springs load the position of the laser module beyond the optical alignment point and now require contact pressure of the micro head adjusting screws to adjust back into alignment against the spring pressure. The front points provide pivoting for elevation adjustments, whereas the rear points provide for the azimuth adjust.

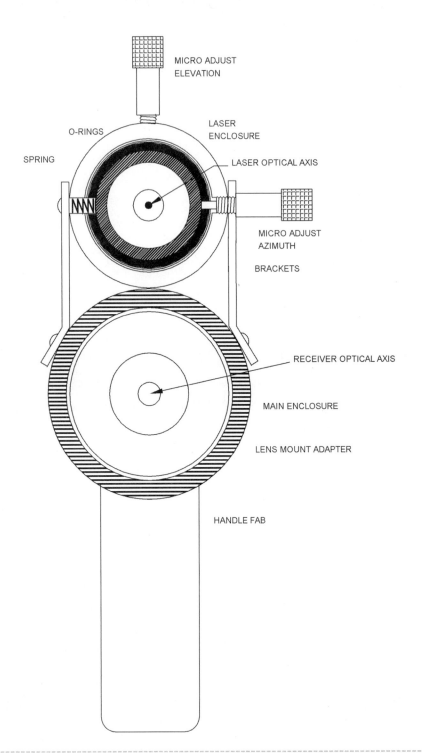

Figure 13-16   *Suggested system assembly—front view minus optics*

Parts and Fabrication Description

You must supply the following parts for final enclosure and assembly. Note that the recommended housing is aluminum tubing with at least a 1/4-inch wall

thickness. Plastic PVC may be used for easier fabricating at the cost of mechanical stability.

- **Main enclosure** $2^3/_8$ OD $\times$ 1/4-inch wall aluminum tubing. Fabricate a hole for the press fit of the handle section. Note 1/4–20 holes for mounting to tripods.

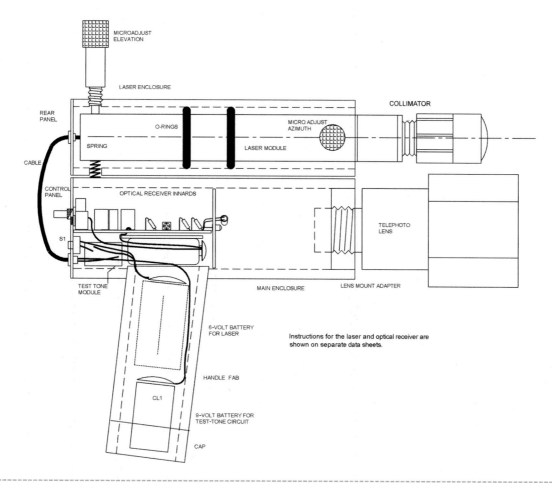

Figure 13-17   *Suggested system assembly—x-ray view of side*

- **Laser enclosure** $1^3/_4$ OD $\times$ $^1/_4$-inch wall aluminum tubing. Fabricate as shown for spring-retaining cutouts and holes for micro-adjusting verniers.

- **Handle** $1^5/_8$-inch aluminum tubing for handle and battery enclosure.

- **Brackets**   Fabricate two brackets for securing the laser and main enclosure together. It is important that this step be as precise as possible, as the two-enclosure axis must be parallel.

- **Control panel**   You may use the front panel of the laser receiver module. Fabricate additional holes for the test-tone switch and feed-through bushing for the cable going to the laser section.

- **Rear panel**   For the rear section of the laser enclosure with a hole for cable bushing.

- **Lens mount adapter**   Fabricate for a $^1/_2$-inch-thick piece of PVC to fit into the *innerdiameter* (ID) of the main enclosure tube. Cut out

the center to mate with the telephoto lens of choice.

- **O-rings**   For sleeving over the laser module and sandwiching with the laser enclosure tube. The position along the length may be adjusted for best movement using vernier adjusters. Experiment for best results.

- **Springs**   Compression springs for maintaining a positive pressure against micro-adjusting vernier screws through their adjusting movement.

- **Micro-head adjusting screws**   For precise elevation and azimuth positioning.

- **Cable**   A multiconductor for 6-volt battery power and test-tone input to the laser module.

- **Switches**   For power and test-tone control to the laser module.

- **6-volt battery**   Uses a four AA battery pack for powering the laser module and is controlled by one of the switches.

- **9-volt battery** For powering the test-tone circuit.

- **Telephoto lens** An optional choice will greatly determine the potential range and performance of the system. Mates to the mount adapter.

- **Cap** Slip-on plastic cap for the retaining end of the handle.

- **Cushioned headset with individual volume controls**

Table 13-1 Laser receiver parts list

Ref. #	Qty.	Description	DB#
R1, 4	2	10-ohm, $^1/_4$-watt resistor (brn-blk-blk)	
R2, 3, 8, 12, 13, 15, 16, 17, 18	9	100K, $^1/_4$-watt resistor (brn-blk-yel)	
R5, 20	2	1K, $^1/_4$-watt resistor (brn-blk-red)	
R6		10K, $^1/_4$-watt resistor (brn-blk-or)	
R7		2.2K, $^1/_4$-watt resistor (red-red-red)	
R9, 10, 11, 19	4	150K, $^1/_4$-watt resistor (br-grn-yel)	
R14	1	3 meg, $^1/_4$-watt resistor (or-blk-grn)	
R21		5K trimpot resistor	
R22		10K pot and switch	
C1, 7	2	100 mfd/25 vertical electrolytic capacitor	
C2, 4, 5	3	.1 mfd, 25-volt disc capacitor (104)	
C3, 14	2	47 mfd/25-volt vertical. electrolytic capacitor	
C8, 10, 13	3	01/25-volt disc capacitor (103)	
C9		470 pfd, 25-volt disc capacitor (471)	
C11, 12	2	.001 mfd, 25-volt disc capacitor (102)	
IC1		LM074 or 324 14-pin DIP note	
IC2		LM386 B8 pin DIP	
J1		Stereo jack for P1	
CL1		9-volt battery clip	
Q1		L14G3 phototransistor	
LED1		IR light-emitter diode	
PC1		HGAPC PCB	
CAP1		2$^3/_8$-inch plastic cap	
CAP2		2-inch plastic cap	
EN1		6$^3/_4$-inch $\times$ 2$^3/_8$-inch sked 40 PVC tube	
HS3		High-quality headsets	
EXTUBE		3- $\times$ 2-inch schedule 40 PVC (see Figure 13-13)	
LEN1		45- $\times$ 89-millimeter convex lens	
LENRET1		2-inch plastic cap #A2 (see Figure 13-13)	

Table 13-2   Laser transmitter parts list

Ref. #	Qty	Description	DB #
R1, 5	2	100-ohm, $1/4$-watt (br-blk-br) resistor	
R2		470-ohm, $1/4$-watt (yel-pur-br) resistor	
R3, 4	2	27-ohm, $1/4$ watt (red-pur-blk) resistor	
R6		5,000-ohm trimpot resistor 502	
R9	1	1K, $1/4$-watt (br-blk-red) resistor	
R10		5.6 m, $1/4$-watt (gr-bl-gr) resistor	
C1, 7	2	.01 mfd/50-volt disc capacitor	
C2		10 mfd/25-volt vertical electrolytic capacitor	
C3	1	1 mfd/25-volt vertical electrolytic capacitor	
Z1		3-volt zener diode 1N5221B	
Q1, 4	2	PN2907 PNP TO92 transistor	
Q2, 3	2	PN2222 NPN TO92 transistor	
LED1		Any high-brightness LED	
SOCK1		Four-pin transistor socket	
BH1		Four AA cell holder	
CL1		Battery clip	
S1		Small toggle switch; key or push button may be used	
BUSS24		#24 bus wire for extension leads of SOCK1	
WR6		12-inch #22 vinyl hookup wire	
PC1		CWL1 PCB	PCCWL1
LASMOD1		Laser diode with integrated optics	LASMOD
LD1		10-milliwatt, 880 nm laser diode; see text	IRLD1
HS1NK		Special fabricated aluminum heatsink, lens holder, and hardware	HSINK
LEN1		Basic lens in threaded fitting for 9-millimeter diode	LENS13
LEN2		15- × -25-millimeter double concave DCV negative glass lens	LE15-25
LEN3		24- × 75-millimeter double convex DCX glass lens	LE2475
EN1		1- × $1/16$- wall × 7-inch length of clear polycarbonate tubing	
LTUB1		.835- × .6- × 2-inch length of schedule $40 1/2$ inches PVC tubing	
ADJMALE		$1/2$-inch sked 40 female slip to male thread GENOVA 30405	
ADJFEMALE		$1/2$-inch sked 40 female slip to female thread GENOVA 30305	
CAP1		1-inch plastic cap	
SW1		6-32 × $1/2$-inch nylon screw	
LABEL CLASSIIIBSET		Labels cert, class, and aperture	

Laser Window Bounce Listening Device

Part Three

Chapter Fourteen

Twelve-Inch-Spark Tesla Coil

Use Caution Around Sensitive Electronic Equipment

This fascinating visual and audible display project never fails to attract attention as fiery bolts of lightning jump into the open air. The project, when properly built and tuned, will generate some 12-inch spark discharges (see Figure 14-1). Operation is done from a standard 3-wire 115 *volts-alternating current* (VAC) outlet and requires caution in assembling and operation. The transformer secondary output is 23 milliamps and is not considered lethal but can provide a nasty shock and burn that the victim will remember. The system output, while quite spectacular, can be contacted by firmly holding a metal object and bringing it near the output terminal. You will experience a very mild shock.

Expect to spend around $100 for this very rewarding project. Assembly is more mechanical than electrical and will require basic skills in the use of hand tools. The winding of the secondary coil may require

building a simple jig and fixture to allow turning for applying the fine wire windings.

Tesla Coil

A Tesla coil is one of the most fascinating electrical display devices to see in operation; a large unit can produce a continuous spark exceeding the height of the coil. Electric discharges that simulate lightning bolts will produce cracks of noise louder than a rifle shot. These sparks, as well as being very impressive and attention getting, can also can produce bizarre effects in most common materials. For example, wood may explode into splinters or be made to glow with an eerie, reddish light from within. Insulating materials seem to be useless against this energy. Lights energize without wires, and sparks and corona in the form of St. Elmo's fire occur within proximity of the device. High-energy electric and magnetic fields render electronic equipment useless. Phenomena not normally associated with standard high-voltage electricity become apparent in the form of many weird and bizarre effects.

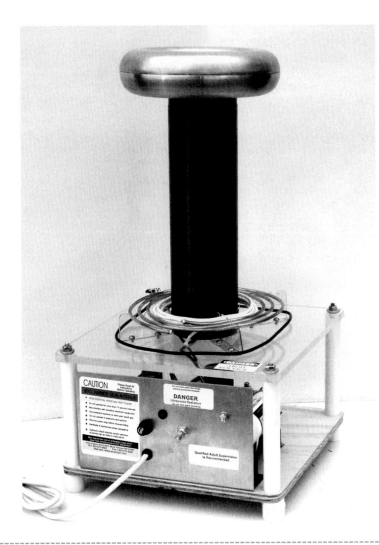

Figure 14-1   *Twelve-inch spark Tesla coil*

Brief Theory

A Tesla coil is a high-frequency resonant transformer. It differs from a conventional transformer in that the voltage and current relationships between the primary winding and secondary winding are independent of the winding turn's ratios. A working apparatus basically consists of a secondary (LS1) and a primary (LP1) coil. It is obvious that the primary circuit is capacitance dominant, and tuning the primary circuit via taps along the primary coil alters the frequency accordingly. However, this relative fine-tuning of the primary circuit to the secondary is mandatory for proper operation. Force driving the secondary coil will produce hot spots and an interwinding breakdown along with other negative results.

Introduction to Your Easy-to-Build Tesla Coil

The circuit schematic in Figure 14-2 shows a step-up transformer producing 6,500 volts at 23 milliamps from the 115 AC line. This voltage-current combination can produce a painful shock. The builder must use adequate caution, just as being around any live 115 VAC circuit. When in doubt, consult someone experienced with this equipment. Safety rules should be followed at all times.

The device also produces ozone; therefore, use it in a well-ventilated area. Do not use it for prolonged periods of time. Thirty seconds at any one time is ample for any demonstration. Avoid eye exposure to the spark gap and use ample protection, such as

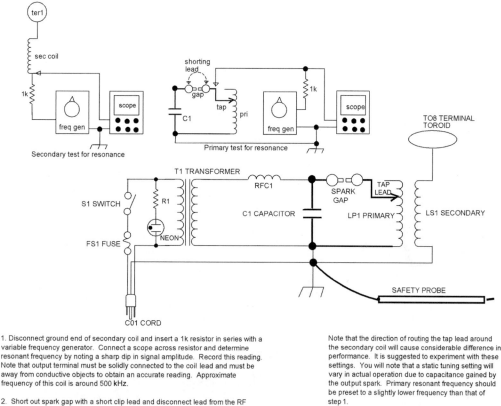

1. Disconnect ground end of secondary coil and insert a 1k resistor in series with a variable frequency generator. Connect a scope across resistor and determine resonant frequency by noting a sharp dip in signal amplitude. Record this reading. Note that output terminal must be solidly connected to the coil lead and must be away from conductive objects to obtain an accurate reading. Approximate frequency of this coil is around 500 kHz.

2. Short out spark gap with a short clip lead and disconnect lead from the RF choke. Connect scope and generator combination to tap of primary coil. Start at maximum turns and note a sharp rise in the voltage wave form at some frequency. Record and repeat at various tap points.

Note that the direction of routing the tap lead around the secondary coil will cause considerable difference in performance. It is suggested to experiment with these settings. You will note that a static tuning setting will vary in actual operation due to capacitance gained by the output spark. Primary resonant frequency should be preset to a slightly lower frequency than that of step 1.

Figure 14-2   *Circuit schematic*

safety glasses or shielding, as dangerous ultraviolet light is emitted.

The unit, when constructed, can develop voltages up to and in excess of 250,000 volts. (This is the DC value of voltage that would be necessary to produce the arc lengths obtainable with this Tesla coil model.) It will cause a gas discharge lamp, such as a regular household fluorescent lamp, to glow up to a distance of several feet from the unit. The high-voltage output coil terminal can actually be touched with a piece of metal held securely in the demonstrator's hand, creating quite a conversation piece.

Circuit Theory

The device, as shown in Figure 14-3, consists of a secondary coil, LS1, containing approximately 500 turns wound on a *polyvinyl chloride* (PVC) form 13 inches long. This coil possesses an inherent resonant fre-

quency determined by its inductance and capacity, usually around 500 kHz. A primary circuit consisting of a drive coil (LP1) and a capacitor (C1) are impulse driven by a spark gap (SGAP1). This primary circuit should also have a resonant frequency equal to that of the secondary coil for maximum performance. It is possible to "force drive" the secondary coil with fewer results. The output voltage of the device is dependent on the ratio of Q between these two coils.

The primary coil has an adjustable tap that allows for fine-tuning. It should be noted that it does not take much in the way of added capacitance to the secondary to alter its resonance point. Even a change in the output terminal will require a readjustment of the tap.

Transformer T1 supplies the necessary high voltage. It is rated at 6,500 VAC at 20 milliamperes. (A larger-capacity transformer will produce more output but may stress the other circuit components.) This voltage charges the primary resonating "tank" capacitor (C1) to a voltage where it fires the spark gap

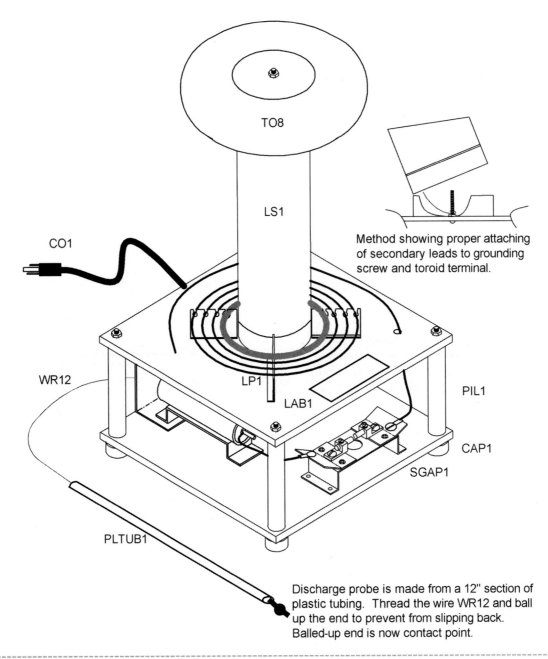

Method showing proper attaching
of secondary leads to grounding
screw and toroid terminal.

Discharge probe is made from a 12" section of
plastic tubing. Thread the wire WR12 and ball
up the end to prevent from slipping back.
Balled-up end is now contact point.

Figure 14-3   *Completed Tesla coil rear view*

(SGAP1), producing an impulse of current through the primary inductance (LP1) where oscillations take place. The frequency is determined by the inductance and capacity values of the primary circuit. The voltage output of the secondary coil, LS1, is usually approximately related to the primary voltage Vp multiplied by C1/C2 where C1 equals the primary capacity, VP equals the spark gap discharge voltage, and C2 equals the secondary coil capacitance (usually relatively small). Another way of expressing this relation is that the output volts are dependent on the

input drive voltage times the ratio of primary Q to secondary Q. Information describing some of Tesla's works are available on our site at www. amazing1.com.

Construction Steps

Note that the layout must be closely followed only where actual dimensions are shown. Otherwise, trial-

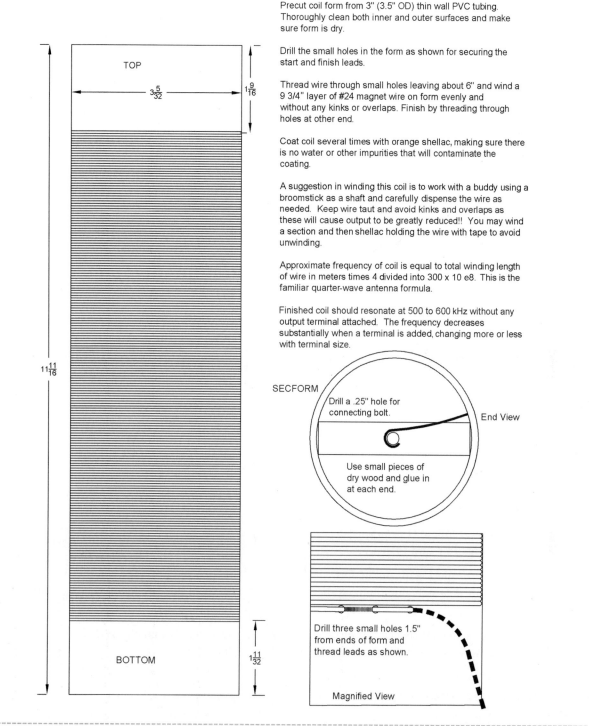

Precut coil form from 3" (3.5" OD) thin wall PVC tubing. Thoroughly clean both inner and outer surfaces and make sure form is dry.

Drill the small holes in the form as shown for securing the start and finish leads.

Thread wire through small holes leaving about 6" and wind a 9 3/4" layer of #24 magnet wire on form evenly and without any kinks or overlaps. Finish by threading through holes at other end.

Coat coil several times with orange shellac, making sure there is no water or other impurities that will contaminate the coating.

A suggestion in winding this coil is to work with a buddy using a broomstick as a shaft and carefully dispense the wire as needed. Keep wire taut and avoid kinks and overlaps as these will cause output to be greatly reduced!! You may wind a section and then shellac holding the wire with tape to avoid unwinding.

Approximate frequency of coil is equal to total winding length of wire in meters times 4 divided into 300 x 10 e8. This is the familiar quarter-wave antenna formula.

Finished coil should resonate at 500 to 600 kHz without any output terminal attached. The frequency decreases substantially when a terminal is added, changing more or less with terminal size.

TOP

$3\frac{5}{32}$

$1\frac{9}{16}$

$11\frac{11}{16}$

BOTTOM

$1\frac{11}{32}$

SECFORM

Drill a .25" hole for connecting bolt.

End View

Use small pieces of dry wood and glue in at each end.

Drill three small holes 1.5" from ends of form and thread leads as shown.

Magnified View

Figure 14-4 *Secondary coil*

position the components as shown in the figures and use the "eyeball" approach to locate them. Please note that you may make substitutions that may increase or decrease performance.

1. Assemble subassembly secondary coil (LS1), as shown in Figure 14-4.

2. Assemble the RFC1 subassembly and fabricate the CBKT1 capacitor bracket, as shown in Figure 14-5. Drill two small holes for threading the 4-inch connection leads. Evenly and tightly wind 40 to 50 turns of #26 magnet wire. Note the mounting scheme of sleeving into a plastic cap (CAP15) screwed

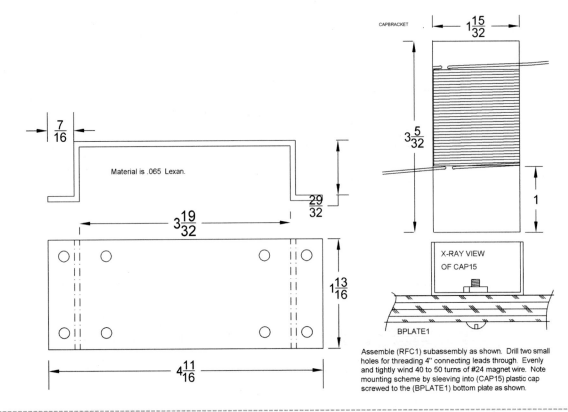

CAPBRACKET

$1\frac{15}{32}$

$3\frac{5}{32}$

1

$\frac{7}{16}$

Material is .065 Lexan.

$\frac{29}{32}$

$3\frac{19}{32}$

$1\frac{13}{16}$

$4\frac{11}{16}$

X-RAY VIEW
OF CAP15

BPLATE1

Assemble (RFC1) subassembly as shown. Drill two small
holes for threading 4" connecting leads through. Evenly
and tightly wind 40 to 50 turns of #24 magnet wire. Note
mounting scheme by sleeving into (CAP15) plastic cap
screwed to the (BPLATE1) bottom plate as shown.

Figure 14-5   *CBKT1 cap bracket and RFC1 assembly*

to a base plate (BPLATE1) shown later in Figure 14-16.

3. Fabricate the SGBKT spark gap bracket and four primary coil-holding brackets (PCBKT1) from .065 polycarbonate (Lexan) plastic, as shown in Figures 14-6 and 14-7.

4. Fabricate the two spark gap cooling fins (SGFIN1) from .065-inch aluminum, as shown in Figure 14-6.

5. Fabricate front panel (PANEL1) from .065-inch aluminum, as shown in Figure 14-8.

6. Fabricate TPLATE1 from $\frac{3}{8}$ - to $\frac{1}{2}$-inch thickness plywood, as shown in Figure 14-9.

7. Fabricate BPLATE1 from $\frac{3}{8}$ - to $\frac{1}{2}$-inch plywood, as shown in Figure 14-10. Drill holes when the layout is verified in step 12. More serious craftsman may want to replace the plywood with clear Lexan or Plexiglas. This approach creates a professional-looking project.

8. Fabricate four pillars (PIL1) at $5\frac{3}{4}$-inch lengths of $\frac{1}{2}$-inch PVC tubing, as shown in Figure 14-3.

9. Attach capacitor C1 to the bracket, as shown in Figure 14-11.

10. Assemble the spark gap assembly (SGA-PASSY) as shown in Figure 14-12.

11. Add several $\frac{1}{4}$-20 hex nuts to a length of threaded rod (TROD). Cut four pieces to $7\frac{1}{4}$ inches and dress the cut ends with a small file. Reform the threaded ends using the nuts as a threading die.

12. Lay out bottom plate (BPLATE1) and verify the components' locations. Drill the mounting holes for securing the components, as shown in Figure 14-13.

13. Assemble the four PVC caps with pillars at the corners using the threaded rods and hardware. Assemble the components using the 6-32 × 1-inch screws with washers and nuts. Note the front panel is secured by the sandwiching action of T1 and the capacitor bracket. Attach the solder lugs as shown in Figure 14-13.

14. Wire up the high-voltage section, as shown in Figure 14-14. Note all leads to the capacitor

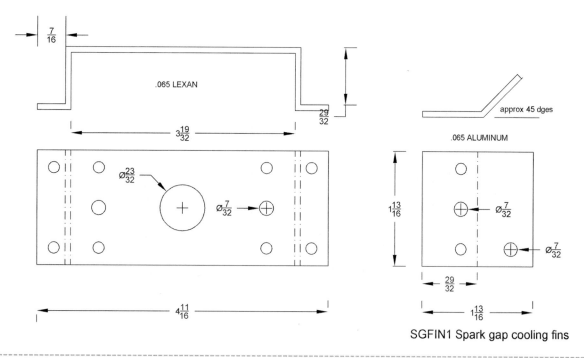

.065 LEXAN

approx 45 dges

.065 ALUMINUM

SGFIN1 Spark gap cooling fins

Figure 14-6 *SGPK1 spark gap brackets*

and the spark gap must be direct. Avoid loops and use short pieces of stripped WR12 soldered to the respective lugs. The T1 transformer lead to the *radio frequency* (RFC) choke is soldered as shown. The secondary ground lead is soldered to LUG2. You should end up with the grounding lead to the secondary, the primary coil center connection with the lug, a lead for the primary tap, and a lead to the safety discharge probe.

15. Assemble PANEL1 and wire the primary 115 VAC input section, as shown in Figure 14-15. Note that it is important to verify a proper ground of the green wire of the cord (CO1) being electrically connected to the frame of T1, the PANEL section, and the secondary return lead.

16. Attach a 3½-inch plastic cap (CAP35) to the center via a ¼-20 × 1-inch brass screw and nut. This point is the earth-grounded common return of both the primary and secondary coils. Attach the four coil brackets using #6 ⅜-inch brass, flathead wood screws. Verify that the bracket holes line up with the pilot holes so that everything fits as shown in Figure 14-16. If not, redo the pilot holes and assemble as shown.

17. Measure an 8-foot length of wire (WR12) and thread one end through the clearance hole as

shown and wind six full turns, placing them into the larger clearance in the four brackets as you go. Shape for a neat circular shape, as shown in Figure 14-17. Strip off the end as it must be inserted and soldered into the ³/₁₆-inch copper tubing (COTUB316) for the remaining part of the primary coil.

18. Continue by winding the attached 7-foot section of the ³/₁₆-inch copper tubing as shown, snapping the cutouts into the brackets. Attempt a neat circular look, as shown in Figure 14-17.

19. Assemble and check the integrity of all the wiring and mechanical assembly. Label the unit as shown in Figures 14-3 and 14-18.

20. Open up the spark gap or place a piece of insulating material between the electrodes to prevent firing. Verify the proper assembly and plug the unit into a 115 VAC, three-wire, grounded outlet. Check the action of S1 and note the lamp NE1 igniting. Quickly short out the spark gap electrode attached to C1 to the chassis ground with the safety probe. Note a loud, bright discharge occurring. Only perform this momentarily as it subjects components to unusual stress and only serves to verify proper operation at this point. Also remove the power cord and material between the spark gap electrodes.

Material is .065 clear polycarbonate (Lexan) plastic. Three pieces required.

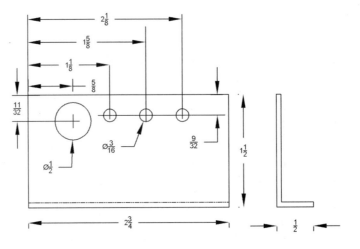

Dimensions shown must be followed on this drawing to obtain proper coupling of the primary and secondary coils.

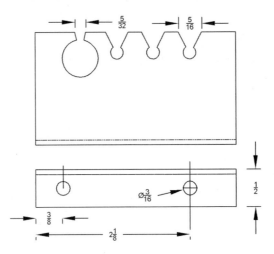

Follow this drawing to get the correct slotting. Use as a pattern or template. Slots must retain wire and tubing but allow snapping into position.

Figure 14-7 *PCBK1 primary coil holding brackets (four required)*

21. Feed the tap wire through the appropriate hole and attach a crocodile clip (CLIP1). Insert the secondary coil assembly into cap CAP350, attaching the ground return wire to the screw. This lead should be as short as possible, as breakdown will occur inside of the coil. Attach the output terminal to the top of the coil using the same procedure. See Figure 14-3.

    Before proceeding to the next step, you must always remove the power cord before making adjustments. The tap lead is a shock hazard,

producing 23 milliamperes to earth ground. The 115 VAC power lines must be avoided, as they can be lethal.

22. When firing up your coil, start with the spark gap set at $1/8$ inches and the adjustment tap at the outermost winding tap to provide maximum inductance. Note that this lead also contributes to total inductance when routed as shown (the same direction as turns of the primary coil). It should be possible with this coil to easily obtain 10- to 12-inch streamers when properly adjusted. Careful adjustment of the

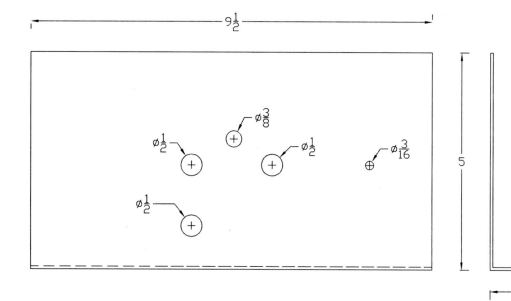

Figure 14-8   *Panel made of .065-inch aluminum*

tap location and gap adjustment will greatly enhance performance. When operating for a prolonged period, always note the potential breakdown points, indicated by heavy corona or premature sparking. These points must be corrected or a burning tracking condition will result.

Special Notes

The tap lead is shown being 3 feet long, enough to wrap itself around the primary coil assembly, providing an additional turn. You may shorten this lead to a shorter length, bearing in mind that it will require readjustment, as it is part of the primary coil.

If you cannot obtain open-air discharges of 10 to 12 inches, it will be necessary to experiment with the tap position along with the spark gap spacing. Open-air discharges are where the sparks emit from the output terminal into the open air. These will be longer than if they were direct to a discharge probe. Point-to-point discharge lengths of 7 to 10 inches are possible and will be more intense but shorter than the open-air discharges.

The spark gap should be set to a maximum length where reliable operation occurs without extinguishing. The coil should not be operated for periods over

30 seconds due to *radio frequency interface* (RFI) potentials and overheating. You may have to place the system in a Faraday cage as shown on our plans as #FARA1. This will be necessary around sensitive electronic and communications equipment.

The secondary coil leads must never be allowed to route inside the coil form, as breakdown and irreversible damage will occur (see Figure 14-3). Figure 14-4 shows a wooden block with leads secured.

Note the coil is referenced TOP and BOTTOM, where one end is slightly spaced shorter than the other. The shop unit seems to work best with the short end at the bottom. Coupling between the coils is critical and you may experiment by placing the secondary coil on wooden blocks with various thicknesses and check the results. A toroid terminal is recommended for optimum performance and you may use metal bowls.

Application and Operation

Please note this Tesla coil device produces electromagnetic radiation that can damage or interfere with certain types of equipment. Operation inside a screen room or Faraday cage may be required for FCC compliance.

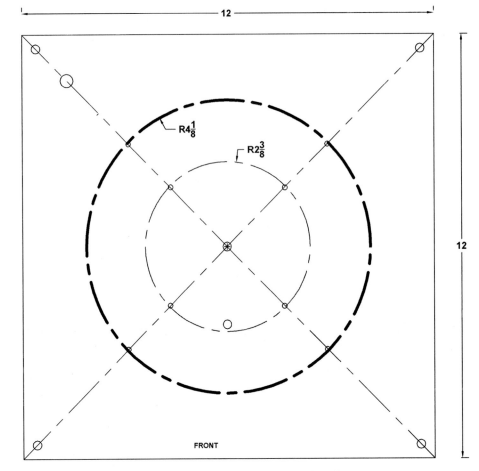

12

12

R4⅛

R2⅜

FRONT

Material for this piece may be 3/8" to 1/2" finished plywood. Plexiglas for this section will add greatly to the finished aesthetics of the project.

1. Note that only holes shown dimensioned must be precise. Others may be eyeballed.

2. Corner holes are 1/4" and are located 1/2" from edges.

3. Use small pilot holes for #6 wood screws for securing the four BKPC1 brackets.

4. Dimensioned holes are referenced from exact center.

Figure 14-9 *TPLATE1 fabrication*

As stated previously, remove the power cord before making any of the following adjustments. The tap lead when ungrounded from the primary is a shock hazard and can produce 23 milliamperes to ground.

Optimum output requires the proper selection of the correct tap position on the primary coil. Start at the end of the outermost turn and reconnect the tap to the exposed sections, changing a quarter of a turn at a time while noting the increasing or decreasing output. You may secure it via soldering when the proper position is determined. Caution: Keep this

lead flat, as breakdown to the secondary will permanently damage the windings. Note the tap lead clip position and mark it when using different terminals. Note Figure 14-17 for proper routing of the tap lead, which we found tuned reasonably close with this terminal.

The capacitance of the output terminal will greatly affect the tuning, requiring more turns the larger the terminal size and vice versa. To operate the coil, you need to perform the following steps:

1. Place the coil on a table away from sensitive electronic equipment. Computers must be

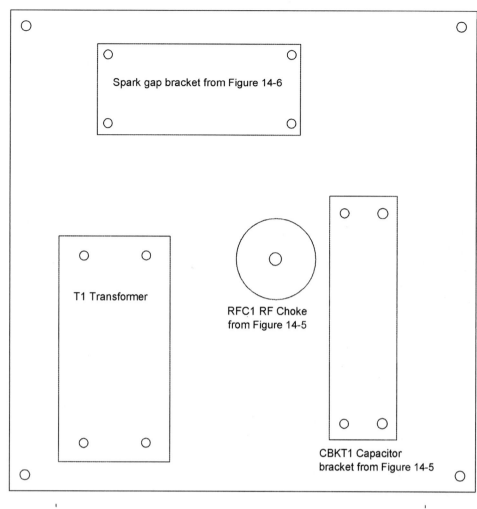

FRONT

Use 3/8" to 1/2" plywood

1. Corner holes are 1/4" and mate to those on TPLATE1 in Figure 14-9.

2. Trial-position referenced subassemblies as shown and mark hole locations for drilling 1/8" clearance holes.

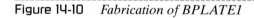

Figure 14-10   *Fabrication of BPLATE1*

removed from the area and disconnected from power if on the same line. Observe, read, and heed all warning labels.

2. Make sure the switch is "off" on the front panel before plugging the coil into a three-wire grounded outlet. Defeating the ground on the plug will result in system failure, potential fire, and an extreme shock hazard.

3. Attach the secondary coil as shown in Figure 14-3. Make sure that leads from the coil are as short as possible, allowing only what is necessary to make the connections to the bottom

grounding screw and repeating for the top terminal. Failure to do this will result in irreversible damage to the coil.

4. The spark gap can be increased just before reliable firing becomes a problem. Further spacing may produce intermittently longer discharges but may overstress the capacitor and transformer. Always set for a reliable, steady firing of continued operation. It is a good idea to space the gap at $1/8$ of an inch when first firing up, and increase as described previously.

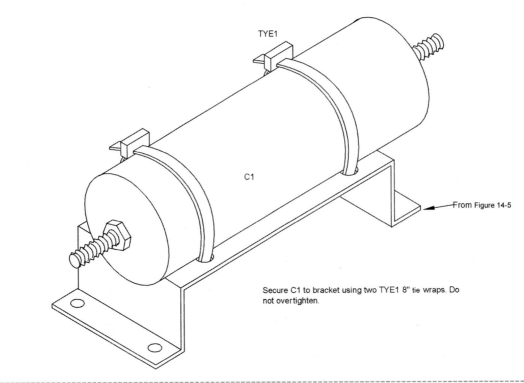

TYE1

C1

From Figure 14-5

Secure C1 to bracket using two TYE1 8" tie wraps. Do not overtighten.

Figure 14-11 *Isometric view of CBKTASSY capacitor bracket assembly*

Assemble as shown and verify proper alignment of tungsten electrodes.

Preset gap to 1/4" and secure with lock nuts.

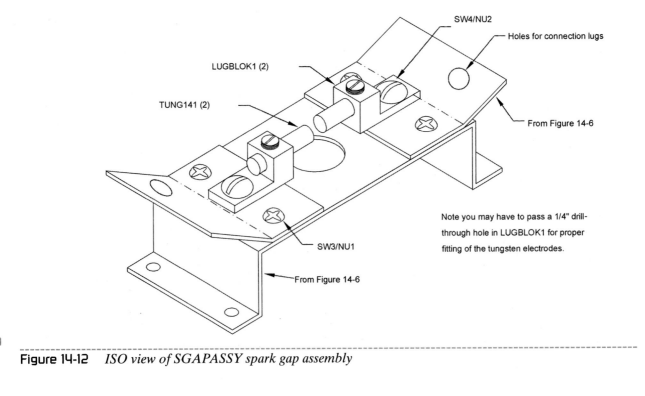

SW4/NU2

Holes for connection lugs

LUGBLOK1 (2)

TUNG141 (2)

From Figure 14-6

Note you may have to pass a 1/4" drill-through hole in LUGBLOK1 for proper fitting of the tungsten electrodes.

SW3/NU1

From Figure 14-6

Figure 14-12 *ISO view of SGAPASSY spark gap assembly*

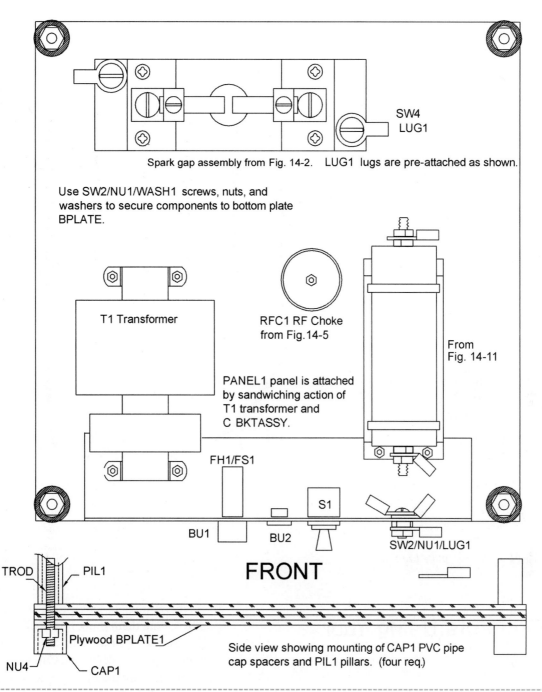

Spark gap assembly from Fig. 14-2. LUG1 lugs are pre-attached as shown.

Use SW2/NU1/WASH1 screws, nuts, and washers to secure components to bottom plate BPLATE.

T1 Transformer

RFC1 RF Choke from Fig. 14-5

From Fig. 14-11

PANEL1 panel is attached by sandwiching action of T1 transformer and C BKTASSY.

SW4
LUG1

FH1/FS1

S1

BU1

BU2

SW2/NU1/LUG1

FRONT

TROD   PIL1

Plywood BPLATE1

NU4   CAP1

Side view showing mounting of CAP1 PVC pipe cap spacers and PIL1 pillars. (four req.)

Figure 14-13 *Bottom plate assembly*

The secondary coil should be terminated into a conductive object such as a sphere or torus. A metal 13- to 16-ounce coffee can or salad bowl may also be used with reasonable results.

5. The trick now is to adjust the tap lead for the maximum spark output from the particular terminal used. Note the lead position along

with the spark output of the secondary coil; the discharge length increases or decreases as adjustments are made. Note that a variance of several inches along the primary coil winding can make a noticeable difference in output. The open-air discharges will be longer than the point-to-point ones.

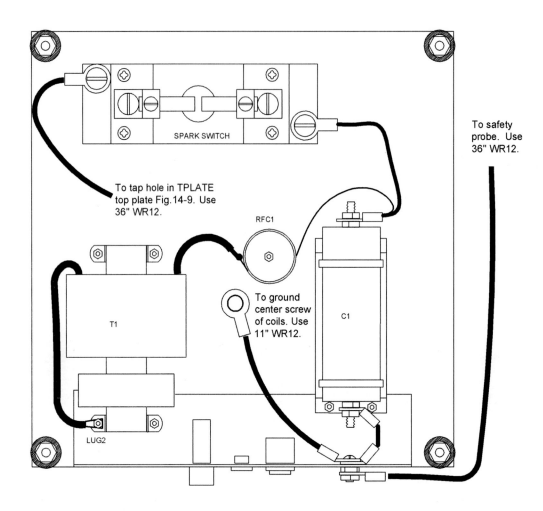

SPARK SWITCH

To safety probe. Use 36" WR12.

To tap hole in TPLATE top plate Fig.14-9. Use 36" WR12.

RFC1

To ground center screw of coils. Use 11" WR12.

T1

C1

LUG2

LUGS may be replaced by forming circular loops at the lead ends and securing under the screw and nuts using flat washers. If using stranded wire, always pre-tin before forming to keep intact.

Figure 14-14   *Bottom plate high-voltage wiring*

Experiments Using Your Coil

You may perform the following experiments with your coil:

- **Adjustment**   Suspend a grounded metallic object above the device. Start at about 3 inches of separation and make adjustments, increasing the separation until the optimum spark length is obtained. (Note grounding means connection to the metal base.)

- **Effect on human body**   Use caution as this may cause a reflex secondary reaction even though it is painless. Hold a metal object tightly and advance to the coil terminal. Note the painless, tingling sensation. Fake out your pals by letting them think it is really painful. This demonstrates the "skin" or surface effect of high-frequency electricity. Caution! Make sure you are standing on a dry, nonconductive surface.

- **Effect on insulators**   Place various objects on top of the coil and note the effects of the high-frequency electricity. Glass or other usual insulators do not stop the sparks. Experiment using objects such as lightbulbs, bottles, glass, and so on.

- **Effect on partial insulators**   Use some wood pieces about $12 \times 1 \times 3$-inches and notice red streaks and other weird phenomena occurring from within the piece. Try other materials as well.

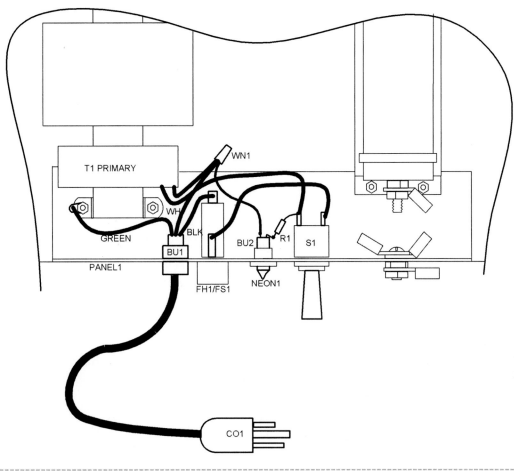

Figure 14-15 *Low-voltage wiring*

- **Ionization of gases** Obtain a fluorescent lamp or plasma globe and allow it to come within several feet of the device. It will glow and produce light without a direct connection, clearly demonstrating the effects of the electric and magnetic fields on the gas. Note the distance from the coil that the lamp will glow. Experiment using a neon lamp in place. Obtain other lamps and note the colors, distances, and other phenomena.

- **Induction fields** Obtain a small filament-type lamp, such as a flashlight bulb, and connect it between a large 1½- to 2-inch-diameter metal or wire loop. The lamp will light due to energy coupled by induction. You will note that a current is required to light this type of lamp and is entirely different than the radia-

tion field that ionizes and causes the gas lamps to glow.

- **Ion motors** Create and carefully balance the rotor as shown. Use a piece of thin, spring beryllium copper. Pinprick the center with a center punch for the bearing point. The rotor will spin at high speeds if it is carefully balanced, demonstrating ion propulsion.

Note that your Tesla coil output is hungry for "capacitance." It is suggested that you obtain an 8-inch toroidal terminal such as our TO8 at www.amazing1.com. You may substitute two metal bowls and attach them together to produce a spherical-shaped object. The torus, however, provides electrostatic shielding of the coil, eliminating a discharge at these points.

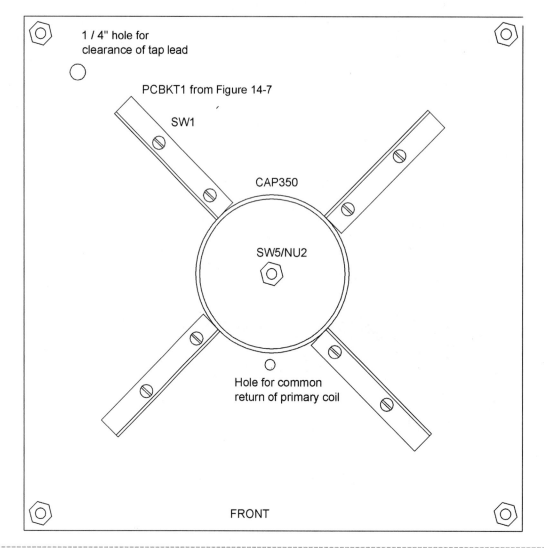

1 / 4" hole for
clearance of tap lead

PCBKT1 from Figure 14-7

SW1

CAP350

SW5/NU2

Hole for common
return of primary coil

FRONT

Figure 14-16   *Top-plate layout showing component mounting*

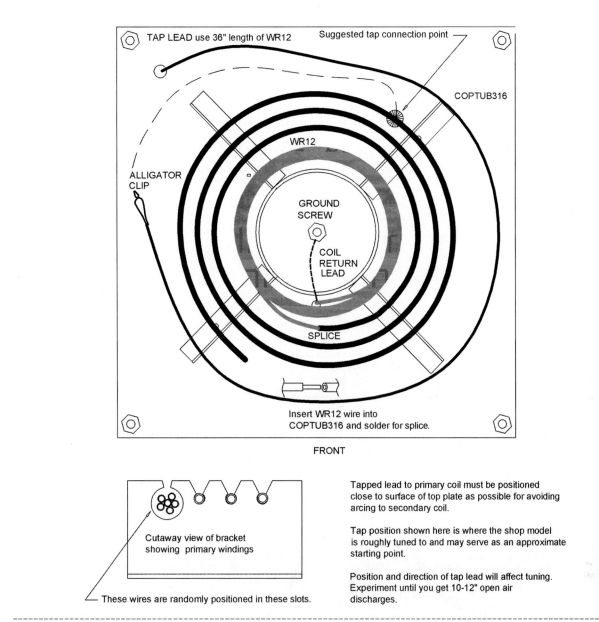

TAP LEAD use 36" length of WR12   Suggested tap connection point

COPTUB316

WR12

ALLIGATOR CLIP

GROUND SCREW

COIL RETURN LEAD

SPLICE

Insert WR12 wire into
COPTUB316 and solder for splice.

FRONT

Cutaway view of bracket
showing primary windings

These wires are randomly positioned in these slots.

Tapped lead to primary coil must be positioned
close to surface of top plate as possible for avoiding
arcing to secondary coil.

Tap position shown here is where the shop model
is roughly tuned to and may serve as an approximate
starting point.

Position and direction of tap lead will affect tuning.
Experiment until you get 10-12" open air
discharges.

Figure 14-17   *Top-plate layout showing primary coils*

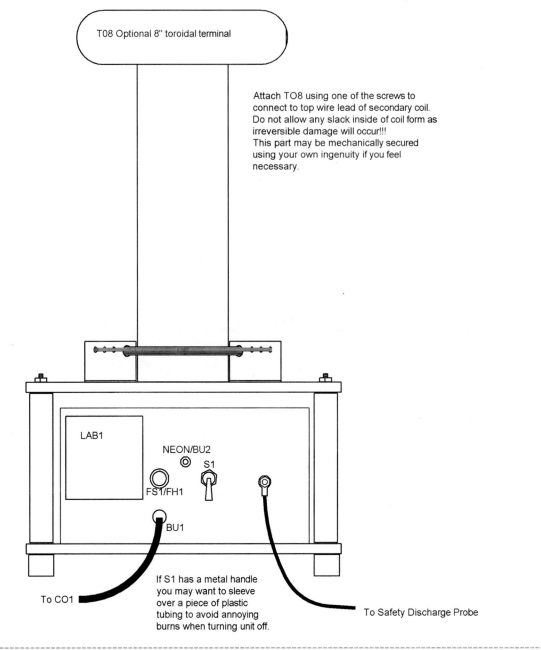

T08 Optional 8" toroidal terminal

Attach TO8 using one of the screws to connect to top wire lead of secondary coil. Do not allow any slack inside of coil form as irreversible damage will occur!!!
This part may be mechanically secured using your own ingenuity if you feel necessary.

LAB1

NEON/BU2

S1

FS1/FH1

BU1

To CO1

If S1 has a metal handle you may want to sleeve over a piece of plastic tubing to avoid annoying burns when turning unit off.

To Safety Discharge Probe

Figure 14-18   *Front view showing the control panel*

Table 14-1 Twelve-inch-spark Tesla coil parts

Ref. #	Qty.	Description	DB#
TPLATE1	1	Top plate, fabricated as shown in Figure 14-9	
BPLATE1	1	Bottom plate, fabricated as shown in Figure 14-10	
CBKT1	1	Capacitor bracket, fabricated as shown in Figure 14-5	
SGBKT1	1	Spark gap bracket, fabricated as shown in Figure 14-6	
SGFIN1	2	Heat fins, fabricated as shown in Figure 14-6	
PANEL1	1	Control panel, fabricated as shown in Figure 14-8	
PCBKT1	4	Primary coil brackets, fabricated as shown in Figure 14-7	
LS1	1	Secondary coil form, fabricated as shown in Figure 14-4	
RFC1	1	RFC choke coil form, fabricated as shown in Figure 14-5	
PIL1	4	5 $^{1}/_{2}$ - × $^{1}/_{2}$ -inch PVC tubing for pillar spacers, as shown in Figure 14-3	
TRAN1	4	6500-volt/.02-milliampere current limited core and coil transformer	6KV/02A
C1		.005 mfd, 10 Kv AC special polypropylene capacitor	.005 M/10 Kv
TUNG141	2	$^{1}/_{4}$ - × 1-inch pure tungsten rod, cut and faced, as shown in Figure 14-12	TUNG141
S1		*Single pole single throw* (SPST) 6-amp toggle switch (see Figure 14-15)	
FH1		Panel-mount fuse holder (see Figure 14-15)	
FS1		3-amp, slow-blow, 3 ag size fuse (see Figure 14-15)	
NEON		Neon indicator lamp (see Figure 14-15)	
CO1		3-wire #18 power cord (see Figure 14-15)	
BU1		Cord clamp bushing (see Figure 14-15)	
BU2		Neon lamp bushing (see Figure 14-15)	
R1		39K, $^{1}/_{4}$-watt resistor (see Figure 14-15)	
WN1		Small wire nut (see Figure 14-15)	
CAP158		1 $^{5}/_{8}$-inch plastic cap (see Figure 14-5)	
CAP350		3 $^{1}/_{2}$-inch plastic cap (see Figure 14-16)	
CAP	4	$^{3}/_{4}$-inch PVC cap used for feet (see Figure 14-13)	
PLTUB1		10 to 12 inches of $^{3}/_{8}$ -inch OD plastic tubing (see Figure 14-3)	
TYE1	2	8-inch tie wraps (see Figure 14-11)	
CLIP1		Small crocodile clip (see Figure 14-17)	
WR12	15 feet	#12 AWG stranded, vinyl-covered wire	
SW1	8	#6 × $^{3}/_{8}$ brass wood screws	
SW2	14	#6-32 × 1 screws, Phillips	
SW3	4	#6-32 × $^{1}/_{4}$ screws, Phillips	
NU1	20	#6-32 hex nuts	
WASH1	28	#6 × $^{1}/_{2}$ flat washers; use under #6 screws and nuts	
SW4	2	$^{1}/_{4}$-20 × $^{1}/_{2}$ -inch brass screw	

Table 14-1 Continued

Ref. #	Qty.	Description	DB#
SW5	1	$^1/_4$-20 × 1-inch brass screw	
NU2	4	$^1/_4$-20 brass hex nut	
NU3	4	#8-32 hex nut for threads of C1	
COTUBE	7 feet	$^3/_{16}$ coiled copper tubing	
ROD1	4	$^1/_4$-20 × 7 $^1/_4$ -inch threaded rod (see Figure 14-13)	
NU4	12	$^1/_4$-20 steel hex nut	
LUG1	8	$^1/_4$-20 × #12 solder ring lug (see Figure 14-13 and 14-14)	
LUG2	2	#6 solder lug for green lead of CO1 and T1 ground (see Figure 14-14)	
LUGBLOK	2	$^1/_4$-20 block lug #YLA6 (see Figure 14-12)	
LAB1		General label (see Figure 14-18)	
LAB2		Spark gap danger label (see Figure 14-3)	
TO8		8- × 2-inch spun aluminum toroidal terminal (optional)	TO8

Chapter Fifteen

Two-Inch-Spark Tesla Coil with Timer

This low-cost and interesting project generates a continuous spark discharge with a variable rate of current. It operates from a 12-volt wall adapter or a battery for portable use or a science project where outlets may not be available. The device (see Figure 15-1) has a built-in timer, allowing you to set the off and on times. This property provides excellent display for desk or bar conversation pieces where it will automatically turn on for several seconds and recycle, surprising those in the area as the unit generates this noisy visual display.

Construction involves minimal electronic experience. Expect to spend around $25 for this rewarding and interesting conversation piece with most parts readily available. Those that are special, including a *printed circuit board* (PCB), may be obtained through www.amazing1.com. Table 15-1 outlines the parts needed for this project.

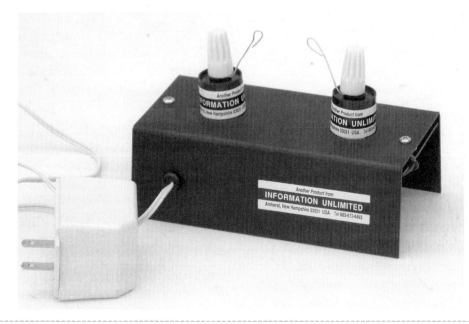

Figure 15-1 *Spark Tesla coil with timer*

Circuit Theory

Figure 15-2 shows a high DC voltage being produced by a blocking oscillator circuit consisting of a transformer (T1) being switched on and off by a transistor (Q1). The current through the primary winding (I) rises as a function of Et/L (when Q1 is on), where E is the applied voltage, in this case 12 *volts-direct current* (VDC), and L is the primary inductance of T1. This rise in current induces a voltage in the feedback winding further holding Q1 on, due to supplying the base current through resistor R2 and the speed-up capacitor C2. When the core of T1 saturates because of a high primary DC current, the induced base voltage goes to zero, turning off Q1. This results in a reverse voltage induced in the secondary forward-biasing diode (D1) and charging capacitors (C3 and C4). When the charging capacitor reaches the trigger voltage of the SIDAC silicon switch (around 300 volts), it now turns on, dumping the energy stored in the capacitor into the primaries of pulse transformers (T2 and T3). This energy causes a rapid rise in the current "forward," inducing the high-voltage output pulses required for the lightning display.

You will note that two pulse transformers are connected out of phase (reversed connected) relative to one another. This approach generates twice the output of what is possible from one transformer, now being in excess of 50,000 volts. You will note switch S1 controls the primary power. Switch S2 selects the display texture. The base of Q1 is the control port for the output of current sink Q2, which is controlled by switch S3.

Construction Steps

To begin the project, follow these steps:

1. Identify all parts and pieces and verify them with the bill of materials.

2. Insert the components, starting from one end of the perforated circuit board, and follow the locations shown in Figure 15-3, using the individual holes as a guide. Use the leads of the actual components as the connection runs,

which are indicated by the dashed lines. It is a good idea to trial-fit the larger parts before actually starting to solder.

Always avoid bare wire bridges, messy solder joints, and potential solder shorts. Also check for cold or loose solder joints.

Pay attention to the polarity of the capacitors with polarity signs and all the semiconductors. The transformer position is determined using an ohmmeter as instructed in Figure 15-3 . Note the SIDAC may have two or three pins. Only the outer ones are used and may connect in either way, as the part is not polarized.

3. Cut, strip, and tin the wire leads for connecting to S1, S2, and S3 and solder them. These leads should be 4 to 5 inches long.

4. Fabricate the plate section from a $4^3/4 \times 2^1/4 \times .06$ piece of plastic. This is the base plate for mounting and gluing the T2 and T3 pulse coils/transformers.

5. Prewire T2 and T3, as shown in Figure 15-3, at the separation distance of 2 inches. Use short pieces of vinyl wire to extend these leads. Splice in 5-inch leads for interconnecting to the board.

6. Carefully position the wired pulse coil assembly to the plastic plate and secure it with silicon rubber cement (a *room temperature vulcanizing* [RTV] adhesive). Clamp it in place to hold it in position as the cement sets. It is important to keep these coils straight for aesthetics.

7. Attach the discharge electrode wires using the wire nuts. You may solder these wires but must use care as excessive heat can internally damage T2 and T3.

8. Preconnect the pulse coil assembly to the board as shown. Connect the wall adapter using the wire nuts, observing proper polarity.

Electrical Pretest

To test the project's wiring, follow these steps:

1. Separate the ends of the discharge wires to approximately 2 inches. Preset the trimmer

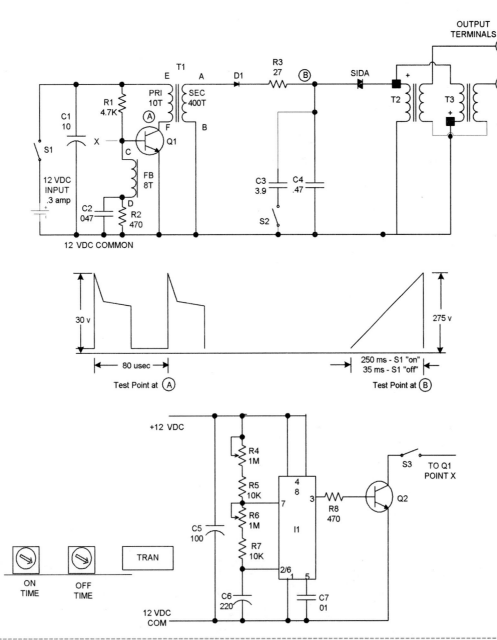

Figure 15-2 *Tesla lightning generator schematic*

pots to midrange and the slider switches S1 and S3 to "off."

2. Turn on S1 and note a discharge occurring between the pulse coils. Change the position of the toggle switch S2 and note the discharge texture changing. Identify the switch position for heavy or light spark display.

3. Turn on S3 and note the display cycling on and off at an approximate rate of 100 seconds on and 100 seconds off. These times are independently variable over a wide range. Our suggested setting is 100 seconds on and 100 seconds off. The device can be left on continuously with these low-duty cycle settings.

Final Assembly

For the final assembly, follow these steps:

1. Fabricate the enclosure from a $6^{1}/_{4}$ - $\times$ $5^{1}/_{4}$ - $\times$ .06-inch piece of plastic, as shown in Figure 15-4.

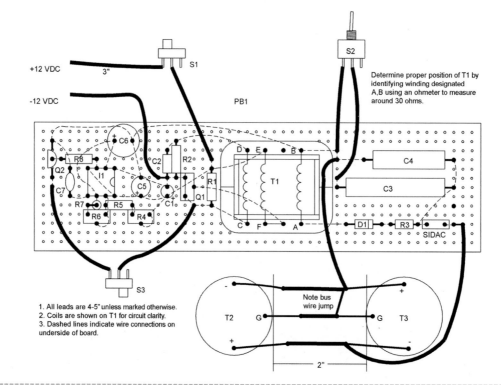

+12 VDC

3"

-12 VDC

S1

S2

PB1

Determine proper position of T1 by
identifying winding designated
A,B using an ohmeter to measure
around 30 ohms.

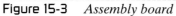

1. All leads are 4-5" unless marked otherwise.
2. Coils are shown on T1 for circuit clarity.
3. Dashed lines indicate wire connections on
   underside of board.

Note bus
wire jump

Figure 15-3 *Assembly board*

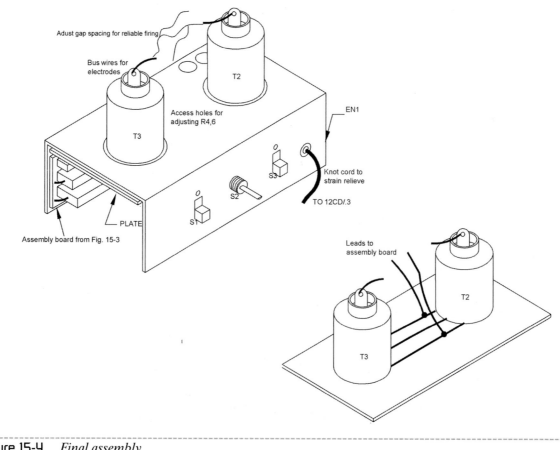

Adust gap spacing for reliable firing

Bus wires for
electrodes

Access holes for
adjusting R4,6

T3

EN1

Knot cord to
strain relieve

TO 12CD/.3

PLATE

Assembly board from Fig. 15-3

Leads to
assembly board

T2

T3

Figure 15-4 *Final assembly*

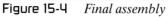

You can use a clear or colored piece for this part. Fabricate holes for T2 and T3, observing the proper alignment with the glued plate assembly as in the prior step. Fabricate the remaining holes for switches, power leads, and trimpot access holes.

2. Glue the pulse coil assembly, followed by the assembly board. Mount the controls and dress the wire leads for a neat-appearing assembly.

3. Verify the operation and preset controls for the desired spark display and cyclic timing.

Table 15-1 Tesla 2-inch spark coil parts

Ref. #	Qty.	Description	DB#
R1		4.7K, $\frac{1}{4}$-watt carbon film resistor (yel-pur-red)	
R2, 8	2	470-ohm, $\frac{1}{4}$-watt carbon film resistor (yel-pur-br)	
R3		27-ohm, $\frac{1}{4}$-watt carbon film resistor (red-pur-blk)	
R4, 6	2	1M trimmer resistor vertical mount	
R5, 7	2	10K, $\frac{1}{4}$-watt carbon film resistor (br-blk-or)	
C5		100 mfd, 25-volt electrolytic capacitor vertical mount	
C6		220 mfd, 25-volt electrolytic capacitor, vertical mount	
C7		.01 mfd, 50-volt disk ceramic capacitor	
C1		10 mfd, 25-volt electrolytic capacitor, vertical mount	
C2		.047 mfd, 50-volt polyester capacitor, marked 2A473 on green body	
C3		3.94 mfd, 350-volt polyester capacitor	
C4		.47 mfd, 250-volt polyester capacitor	
Q1		MJE3055 NPN transistor TO220	
Q2		NPN PN2222 GP transistor	
I1		555-dual in-line package DIP timer	
D1		IN4007 1 Kv rectifier diode	
SIDAC		300-volt SIDAC switch marked K3000; see text	SIDAC
T1		Switching square-wave transformer 400V	TYPE1PC
T2, 3	2	25 Kv pulse transformers	CD25B
S1, 2, 3	3	SPST 3-amp toggle switch or equivalent	
PB1		5 × 1.5 .1 grid perforated circuit board	
PCTLITE		Optional PCB replaces PB1	PCTLITE
WR20B	36 inches	#20 vinyl stranded hookup wire, black	
WR20R	36 inches	#20 vinyl stranded hookup wire, red	
WN1	2	Small wire nuts, #71B	
EN1		Enclosure 6 $\frac{1}{4}$ × 5 $\frac{1}{4}$ × .062 plastic, see Figure 15-4	
PLATE		2 $\frac{1}{4}$ × 4 $\frac{3}{4}$ × .062 plastic partition	
TAPE		6 × 1 × .125 double-sided sticky tape	
12DC/.3		12-volt DC .3-amp wall adapter	12DC/.3

Chapter Sixteen

Tesla Plasma and Ion Projects

This multifunctional electrical project is very suitable for science fairs, as the device is powered by a 12 *volts-direct current* (VDC) battery or a low-voltage wall adapter transformer (see Figure 16-1). Many interesting action and display experiments are shown in detail with safety stressed throughout, thus suitable for the younger hobbyist. Construction involves basic wiring and soldering with some basic mechanical assembly.

Most parts are readily available. Those that are special, including a *printed circuit board* (PCB), may be obtained through www.amazing1.com. Expect to spend $25 to $50 to complete the project as shown. Table 16-1 outlines the parts needed for the project.

Experiments and Applications Using Plasma

Circuit Theory

Figure 16- 2 shows transistor Q1 connected as a free-running flyback oscillator with a resonant frequency determined by capacitor CX and the short-circuit primary inductance (LP) of transformer T1. The leakage inductance of T1 must be high as it is where the inductive energy is stored when Q1 is biased on. The current now ramps up as a function of $i = ET/L$. Energy is equal to $W = Li^2/2$.

Circuit operation commences when 12 VDC is applied and resistor R2 biases Q1 on, and the current now flows through the primary of T1, inducing a high voltage into the secondary winding. Oscillation is maintained by the base of Q1 being controlled by the feedback voltage from the feedback winding on T1. The feedback current is limited by resistor R2. Capacitor CX aids in speeding up the turnoff time of Q1. Resistor R3 and capacitor C5 form a filter to prevent self-oscillation at the resonant frequency of the secondary of T1.

Construction

To begin the project, follow these steps:

1. Lay out and identify all the parts and pieces. Verify them with the parts list, and separate the resistors as they have a color code to determine their value. Colors are noted on the parts list.

2. Obtain the available PCB as shown in Figure 16-3 or fabricate a piece of perforated circuit board (perf) to match the PCB (PC) as laid

Figure 16-1   *Tesla plasma and ion generator*

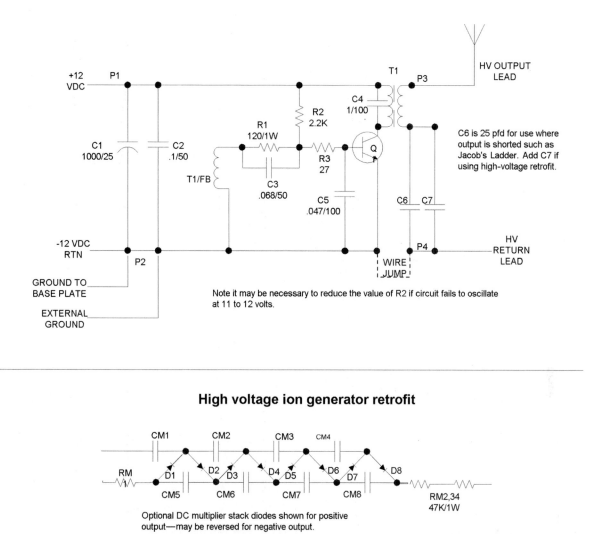

High voltage ion generator retrofit

C6 is 25 pfd for use where output is shorted such as Jacob's Ladder. Add C7 if using high-voltage retrofit.

Note it may be necessary to reduce the value of R2 if circuit fails to oscillate at 11 to 12 volts.

Optional DC multiplier stack diodes shown for positive output—may be reversed for negative output.

Figure 16- 2 *Tesla plasma and ion schematic*

out. Note that the size of the PC board is 3½ - × 1⅜-inches and contains the silk screening that shows the positioning of the mounted parts. Note that the Q1 transistor must be mounted so that it is flush with its mounting surface at a right angle. This step is important for proper heat sinking and mechanically stability.

3. If you are building from a perforated board, insert components starting in the lower left-hand corner. Pay attention to the polarity of the capacitors with polarity signs, as well as all the semiconductors.

   Route the leads of the components as shown and solder as you go, cutting away unused wires. Attempt to use certain leads as the wire runs. Follow the dashed lines on the assembly

drawing as these indicate connection runs on the underside of the assembly board.

4. Attach three 6-inch #22-20 leads as shown for input power (P1 and P2) and external grounding. Attach a short piece of bus wire for grounding of the base section. Leads are routed through the BU1 bushing and are twisted in the final assembly steps. Attach pieces of bus wire to the output connections (P3 and P4) and shape them to fit around the screws as shown.

5. Double-check the accuracy of the wiring and the quality of the solder joints. Avoid wire bridges, shorts, and any close proximity to other circuit components. If a wire bridge is necessary, sleeve some insulation onto the lead to avoid any potential shorts.

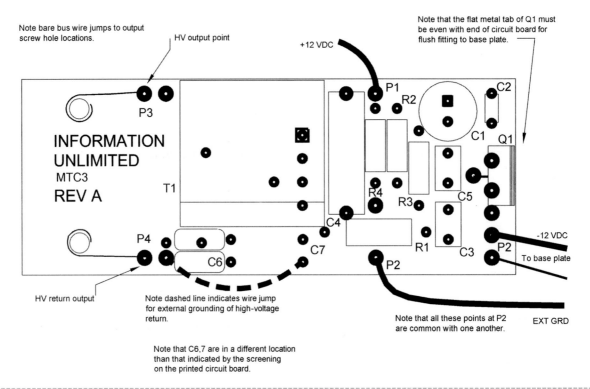

Figure 16-3 *Assembly board showing part identification*

Pretest

To test the circuit out, temporarily and quickly connect a 12 VDC source of 1 amp to the input power leads, and note a current draw of around .3 to .5 amps. Touch the output lead with an insulated metal tool and note a small arc of ¼ to ½ inches. Do not operate for more than several seconds as Q1 will overheat because it requires mounting to the base section for heat sinking. You may want to verify the wave shape on the collector of Q1, using a scope. It should resemble that shown in Figure 16-2.

Mechanical Assembly

Assembling the parts is done by following these steps:

1. Fabricate the base section, BASE, from a 3 × 3 × .063 aluminum plate and drill a hole to mount the heat sink of Q1, a small hole for SW2 to mount the BK1 bracket, and a ³/₈-inch clearance hole for the BU1 bushing. Trial-fit everything before actually drilling. Fold down the corner sections of the base to form the legs as shown in Figure 16-4.

2. Fabricate the bracket BK1 from a piece of formable plastic that is 3 × 1 ¼ × .063 inches. Fold a ½-inch lip for mounting to the base section. Note the clearance hole for SW2. Position as shown, abutting to the rear of the assembly board. This will provide a stable mounting scheme.

3. Finally, assemble as shown in Figure 16-4. Secure the assembly board to the bracket using a plastic tie wrap. Ground the base section with a short piece of bus wire under SW2 and attach the wire to the P2 line on the board.

4. Fabricate the shroud tube from a piece of clear 10- × 1³/₄-inch OD plastic tubing. This piece slips over the assembly and provides enclosure protection while allowing visibility of the circuitry. This is an obvious asset when making a science project presentation. Note you will have to file slots or grooves at the bottom to allow airflow and to clear screw SW2.

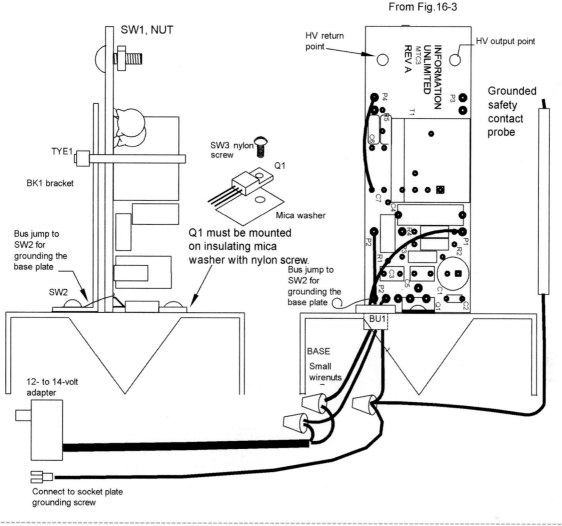

Figure 16-4   *Mechanical assembly*

Test and Function Selection

Connect a 12 to 14 VDC source of 1 to 2 amps, and test it by touching the output with the end of a fluorescent tube, noting a moderate bright glowing. Study the following data to decide the experiments you wish to pursue. The unit can be operated from any 12- to 14-volt battery pack that can supply a current of 1 to 2 amps or more.

It is important, where noted, that the unit be earth grounded with a separate grounding connection to a known, grounded object. A wall adapter, when used, will require a separate grounding wire for proper operation as the output circuitry is usually isolated from the primary plug side, as there is no third wire grounding pin. The screw of a receptacle plate can be used for this external connection, as is noted on the drawings. When using a converter, automatic grounding is usually via the third prong on the power plug, which is the same as the negative lead output (see Figure 16-4). Experimenting with ions will require assembly of the magnifier retrofit as shown in Figure 16-10.

Experiments and Applications Using Plasma

- **Radiated power**   Energize a light without any connections (see Figure 16-5).

Radiated Power Project

Obtain a gas discharge lamp such as a piece of neon or use the test neon lamp in the kit and, holding one of the leads, bring it near the radiator, noting it glowing up to a foot or so from the unit. Try with a larger fluorescent tube.

Gas Tube Energy Supply

Touch one end of a household fluorescent tube to the output radiator and note lamp glowing with only the single connection. This demonstrates the high-frequency ground current flowing via the electrical capacity of the discharge tube.

Pyrotech Display

Obtain some steel wool and attach a small tuft to the output lead. Draw an arc with a grounded object and note strands brightly burning. Experienced personnel may wish to add certain oxidizing chemicals to greatly enhance this effect. *USE CAUTION*.

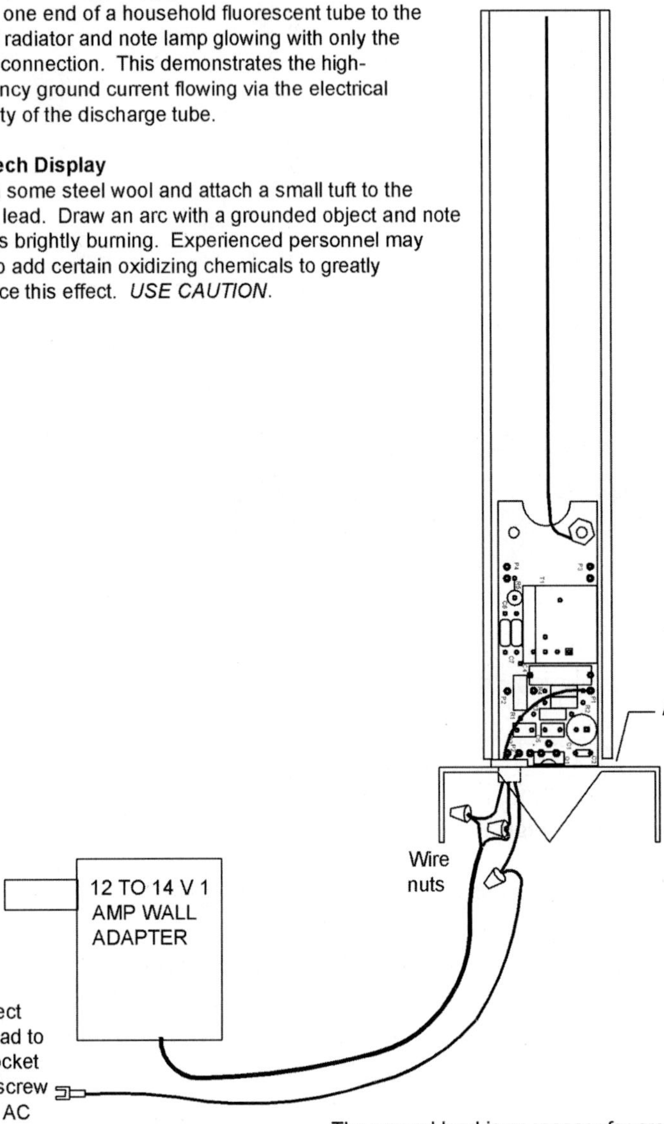

Clear plastic should be removed for access to output. It is necessary to prevent accidental contact that can cause annoying but not dangerous shocks.

Shape a 10" length of #14 bus lead or 1/16" brass rod and shape as shown with a loop to fit around the screw for securing in place. This serves as a small energy radiator.

Air circulation gap

Copyright 10/97

12 TO 14 V 1 AMP WALL ADAPTER

Wire nuts

Connect this lead to the socket plate screw of the AC receptacle.

The ground lead is necessary for proper operation of this project as capacitive ground currents are produced.

Figure 16-5 *Set up for radiated power*

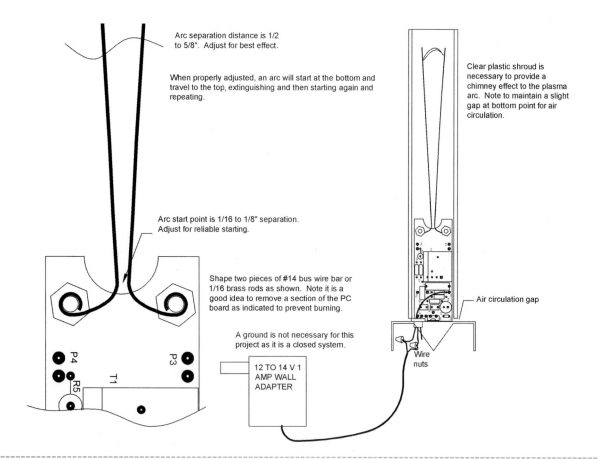

Arc separation distance is 1/2 to 5/8". Adjust for best effect.

When properly adjusted, an arc will start at the bottom and travel to the top, extinguishing and then starting again and repeating.

Clear plastic shroud is necessary to provide a chimney effect to the plasma arc. Note to maintain a slight gap at bottom point for air circulation.

Arc start point is 1/16 to 1/8" separation. Adjust for reliable starting.

Shape two pieces of #14 bus wire bar or 1/16 brass rods as shown. Note it is a good idea to remove a section of the PC board as indicated to prevent burning.

Air circulation gap

A ground is not necessary for this project as it is a closed system.

Wire nuts

12 TO 14 V 1 AMP WALL ADAPTER

P4   P3

R5   T1

Figure 16- 6   *Jacob's ladder project*

- **Gas discharge power supply**   Energize colorful neon and other gas-filled tubes (see Figure 16-5).

- **Electrical pyrotechnic display**   Burn up a piece of steel wool in a shower of sparks (see Figure 16-5).

- **Jacob's ladder**   Build the popular traveling plasma arc machine seen in Frankenstein movies (see Figure 16-6).

- **Build an effective deodorizing machine**   It actually produces fresh air (see Figure 16-7).

- **Magic electric man**   Energize a fluorescent tube by the touch of your hand (see Figure 16-8).

- **Plasma etching**   Use an electrical arc to create intricate designs in wood and plastic (see Figure 16-9).

Experiments and Applications Using Ions

- **Ion high speed motor**   Spin a metal rotor into the thousands of RPM (see Figure 16-1 and 16-10).

- **Force field**   Forces an object on to a surface with considerable pressure (see Figure 16-10).

- **Motion freezing**   Capture high-speed periodic motion as a moment in time (see Figure 16-10).

- **Charge transmission and capacitor charging**   Charge objects to high voltages without contact (see Figure 16-10).

- **Negative ion generator and ion wind**   Produce a high flux of beneficial negative ions. Easily detectable (see Figure 16-10).

This project as shown requires our 30 mg ozone cell #SS01 and is available at www.amazing1.com.

Input voltage will play a major role in the performance of this project. If the cell constantly sparks over it will be necessary to reduce the input voltage below 12 VDC. Proper operation is evidenced by bluish coronal fire emanating from the sharp points of the cell. You should notice a fresh air smell similar to that after a good thunderstorm. This smell is not ozone but is nitrous oxide formed from ozone and nitrogen, the main ingredient to nitric acid!! It is OK in small amounts for deodorizing but must be controlled.

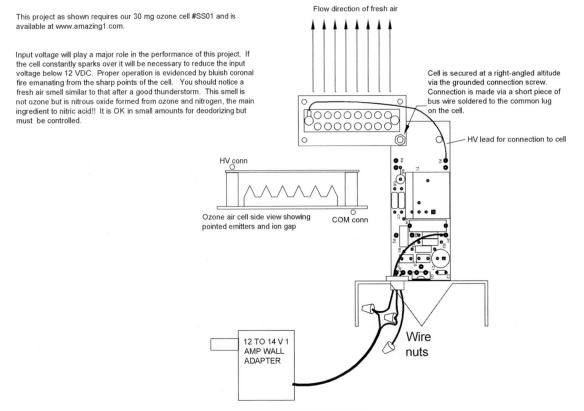

Flow direction of fresh air

Cell is secured at a right-angled altitude via the grounded connection screw. Connection is made via a short piece of bus wire soldered to the common lug on the cell.

HV lead for connection to cell

HV conn

Ozone air cell side view showing pointed emitters and ion gap

COM conn

12 TO 14 V 1 AMP WALL ADAPTER

Wire nuts

The ground lead is not necessary for proper operation of this project.

Figure 16-7   *Set up for deodorizing machine*

Caution This project can produce painful and annoying shocks if improperly used. It should only be attempted by those experienced in use of HV for magic shows, props, etc.

Instructions - Grab a fluorescent or neon tube in one hand and with the other **tightly hold** a metal object at least 1" in diameter by 6 to 8 in. in length. (touching as much surface area as possible with your hand). Turn the unit on and contact the HV output with the metal object, noting the lamp in your other hand energizing, indicating an electric current flow through your body without any shock!! Bring the lamp near a large metal object or another person and note it getting brighter. This experiment affects the skin where high-frequency currents flow on the surface of the person's body, lighting the lamp and the capacitive properties of conductive objects where they seemingly attract electrical energy.

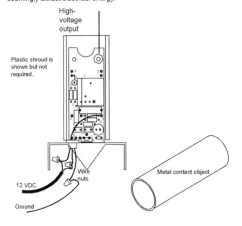

High-voltage output

Plastic shroud is shown but not required.

Wire nuts

Metal contact object

12 VDC

Ground

Power to this unit can be any 12 to 14 VDC source capable of supplying 1 amp. The ground lead must be grounded for proper operation.

Figure 16-8   *Magic electric man*

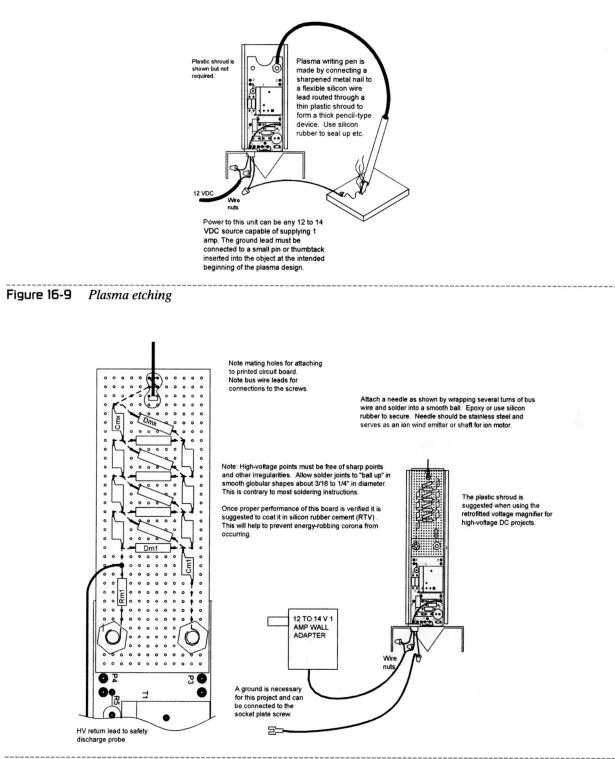

Use a plasma arc to burn in etches and designs into nonconductive materials such as wood, plastic, etc. (THIN WOOD WORKS EXCELLENT). This effect is very similar to the old wood-burning sets but can produce much finer and precise designs once mastered.

Burn must be prestarted with an existing arc drawn from a grounded point to start the carbon channel. We attach the ground return wire to a thumbtack. Once the arc trace is started, it can be continued as desired. Our lab samples were done on a 1/4" piece of mason slightly dampened.

This setup can also be used for **Kirlian Photography projects.** Experimenters may wish to consider our available #KIRL1 plans.

Plastic shroud is shown but not required.

Plasma writing pen is made by connecting a sharpened metal nail to a flexible silicon wire lead routed through a thin plastic shroud to form a thick pencil-type device. Use silicon rubber to seal up etc.

12 VDC

Wire nuts

Power to this unit can be any 12 to 14 VDC source capable of supplying 1 amp. The ground lead must be connected to a small pin or thumbtack inserted into the object at the intended beginning of the plasma design.

Figure 16-9   *Plasma etching*

Note mating holes for attaching to printed circuit board. Note bus wire leads for connections to the screws.

Attach a needle as shown by wrapping several turns of bus wire and solder into a smooth ball. Epoxy or use silicon rubber to secure. Needle should be stainless steel and serves as an ion wind emitter or shaft for ion motor.

Note: High-voltage points must be free of sharp points and other irregularities. Allow solder joints to "ball up" in smooth globular shapes about 3/16 to 1/4" in diameter. This is contrary to most soldering instructions.

Once proper performance of this board is verified it is suggested to coat it in silicon rubber cement (RTV). This will help to prevent energy-robbing corona from occurring.

The plastic shroud is suggested when using the retrofitted voltage magnifier for high-voltage DC projects.

12 TO 14 V 1 AMP WALL ADAPTER

Wire nuts

A ground is necessary for this project and can be connected to the socket plate screw.

HV return lead to safety discharge probe

Figure 16-10   *Magnifier retrofit*

Cmx   Dmx   Dm1   Cm1   Rm1   P4   R5   T1   P3

Ref. #	Qty.	Description	DB#
R1		120 ohms, 1 watt (br-red-br)	
R2		1.5K, 1/4 watt (br-grn-red); see note on schematic	
C1		1000 mfd/25-volts electrolytic vertical capacitor	
C2		.1 mfd/50-volts plastic capacitor	
C3		.068 mfd/50-volts plastic capacitor	
C4		1 mfd/250-volts metallized plastic capacitor	
C5		.047 mfd/100-volts plastic capacitor	
C6, 7	2	25 pfd at 6 Kv ceramic capacitor	
Q1		MJE3055t TO220 power tab NPN transistor	
T1		Special high-voltage, high-frequency transformer	28K089
WR20	48 inches	#20 vinyl wire	
WRBUSS	16 inches	#14 straight US wire or 1/16-inch brass rod for ladder elements in Jacob's ladder	
PCMTC		PCB or use a piece of 3 1/2 - × 1 1/2-inch .1 grid perf board	PCMTC
BASE		3 × 3 .063 Al plate with folded corners fabricated as shown	
BU1		3/8-inch plastic bushing	
PLTUBE		3/8- × 6-inch length of plastic tubing for safety probe and plasma pen	
BK1		3- × 1 1/4-inch .063 plastic bracket	
SW1	2	6-32 3/4 Phillips screw	
SW2		#6 × 3/8-inch Phillips sheet metal screw	
SW3		6-32 × 1/2-inch nylon screw	
NUT	3	6-32 kep nut	
MICA		Mica washer for TO 220 transistor	
TUBE		10 × 1 3/4 OD × 1 1/16 wall, clear CAB plastic tubing	
TYE1		4-inch tie wrap	
WN1		Small wire nuts	
14/1G		14 volt, 1 amp with internal ground	14DC/1
NE1		Small neon lamp with leads	
NE15		15-inch length of colorful neon in green, red, blue, purple, white	NE15COL
WRHV	12 inches	Flexible silicon 20 Kv wire	
CM1-8	8	500 Pfd/10 Kv ceramic disk capacitors	500/10 Kv
DM1-8	8	10 Kv, 5-milliampere, 100 ns high-voltage rectifiers	VG12
RM1		220 1/2-watt resistor (red-red-br)	
RM2, 3		47K, 1-watt resistor (yel-pur-or)	
PB1		4 1/4 × 1 1/4 .1 x .1 grid perforated circuit vector board	

Table 16-1 Continued

Ref. #	Qty.	Description	DB#
ROTOR		Ion-propelled, four-point aluminum rotor	ROTOR
BALL38		Small brass ball with threaded hole	
		Miscellaneous hardware	
		Small needle for ion motors	
		Piece of masonite for plasma writer	
		6 to 8- × 1 ½-inch piece of metal tube or rod for electric man experiment	
		Miscellaneous wire, hardware, and so on	

Tesla Plasma and Ion Projects

Chapter Seventeen

Solid-State Tesla Coil

This versatile project shows how to convert a readily available television flyback transformer into a high-frequency, high-voltage generator operating from 12 volts or a battery source (see Figure 17-1). The assembled device is excellent for powering all kinds of gas display pieces from plasma balls to an everyday lightbulb. Chapter 20, "Amazing Plasma Tornado Generator," shows how to actually build a plasma tornado with some fascinating and unique properties. The solid-state Tesla coil can be used to create ozone, corona and brush discharges, and electric pyrotechnics, including a small Jacob's ladder.

Figure 17-1 *Solid-state Tesla coil*

The construction here includes an easy-to-build circuit costing under $50 with parts readily available. Those that are special, such as a *printed circuit board* (PCB), may be obtained through www.amazing1.com. Flyback will require some re-work involving disassembly and the addition of a 10-turn coil. Some small hand and soldering tools are required. The necessary parts are listed in Table 17-1.

Circuit Theory

The output is a result of a resonating action of the secondary coil occurring around 50 to 70 kHz. Under these "tuned" conditions, the transformer requires considerable power and produces high-voltage discharges that would quickly break down the insulation if left on for any period of time. Immersion in an oil bath can help limit overheating and eventual break-down, but is not necessary if operated as directed.

Figure 17-2 shows the primary transformer T1 being driven by two *metal-oxide-semiconductor field effect transistors* (MOSFETs), Q1 and Q2, in a push-pull configuration. This approach utilizes the full core potential of the flyback and reduces the electrical strain on the MOSFETs as they run very cool, even at 5 to 7 amps input. A driver circuit (I1) generates complementary outputs 180 degrees out of phase

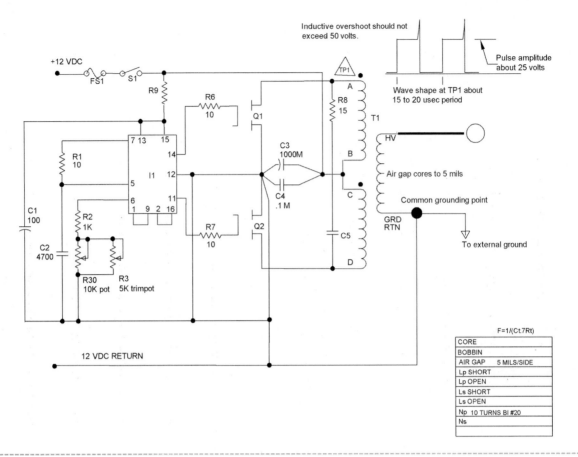

Figure 17-2   *Circuit schematic*

with built-in dead time. The frequency is made variable by control pot R30 and range-adjust trimpot R3. This adjustment allows a wide range of frequency, driving T1 out of the resonant mode where it can power a voltage multiplier and make a variable high-voltage DC source.

Capacitor C2 and R3 determine the operating frequency. Resistor R2 sets the upper frequency limit, and resistor R1 sets the switching dead time for reliable operation. Resistor R9 and capacitor C1 provide decoupling between the MOSFETs and driver I1. Resistors R6 and R7 eliminate parasitic oscillation from occurring on the gates of Q1 and Q2. Resistors R8 and C5 form a snubber network to clamp the voltage spike generated by the leakage inductance of T1. The capacitance of the snubber absorbs the current charge of the spike, limiting the voltage peak to a safe level. The MOSFETs would quickly break down as a result of high-voltage spikes if it were not for this network. C3 and C4 bypass any of the high frequencies, appearing at the primary center point of T1.

The power requirement for the unit is 12 to 14 *volts-direct current* (VDC) at a maximum of 5 to 7 amps when tuned to the resonant frequency of T1. It is suggested that the entire T1 flyback assembly be immersed in transformer oil if full output operation is anticipated over a period of time. The oil bath both cools and provides added insulation to the high voltages developed, but is not necessary for normal operation.

Assembly

To begin the project, follow these steps:

1. Lay out and identify all the parts and pieces, and check them against the parts list.

2. Assemble T1, the ferrite transformer, in Figure 17-3. Measure the inductance if you have an LCR (inductance, capacity, resistance) meter for verifying the values given.

Due to the difficulty in reworking certain flyback transformers this ready-to-use part is available on www.amazing1.com Fortunately some units are easy to disassemble while others are not.

1. Remove the "U" bolt and one of the core halves. Some units may require chipping away the binding material with a pointed object until the cores will come apart using moderate force.

2. Form a bobbin from a piece of plastic or cardboard tubing, as shown, of a length that allows core pieces to touch one another.

3. Bifilarly (parallel) wind two different color #18 magnet wires for 10 turns of these double windings, leaving 8" leads. Note the different colors will help identify the lead ends.

4. Scotch tape the face of each core half so that when reassembled there will be two pieces separating each side of the cores. These should produce a 5 mil gap at each junction.

5. Place the coil wound in step 3 onto the cores as shown and tape tightly into place.

6. Identify the secondary lead return that will be the leads attached to the base. Usually any one will work but it is suggested to use an ohmmeter and use the highest resistive combination. Carefully attach an external wire to this point and strain relieve with some silicon rubber. Verify other leads are not shorted together.

You may verify the inductance as follows secondary open:
A to B&C are around 15 microhenries.
D to B&C are around 15 microhenries

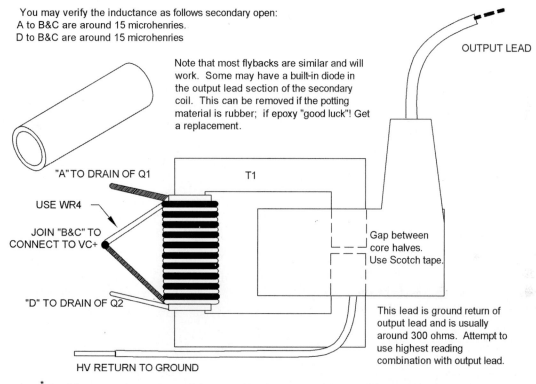

OUTPUT LEAD

Note that most flybacks are similar and will work. Some may have a built-in diode in the output lead section of the secondary coil. This can be removed if the potting material is rubber; if epoxy "good luck"! Get a replacement.

"A" TO DRAIN OF Q1

T1

USE WR4

JOIN "B&C" TO CONNECT TO VC+

Gap between core halves. Use Scotch tape.

"D" TO DRAIN OF Q2

This lead is ground return of output lead and is usually around 300 ohms. Attempt to use highest reading combination with output lead.

HV RETURN TO GROUND

*#18 solid magnet wire can be used; however, high-frequency LITZ wire will give a slight improvement. You can make this wire by obtaining 6 pieces of #26 magnet wire and twist together as a single wire.

Figure 17-3 *Flyback rework for T1*

3. Fabricate the MTGBKT mounting bracket (see Figure 17-4) and the channel/cover combination as shown in Figure 17-5.

4. Assemble the board as shown in Figure 17-6.

   a. Insert capacitors C1 and C3, and note the polarity, as these are electrolytic.

   b. Insert C2, C4, and C5.

   c. Insert R1, R2, R5, R6, R7, and R8. Note that the exposed lead of R8 should attach to Q1 for an easy access test point.

   d. Insert Q1, Q2, and I1, noting the polarity.

   e. Solder in the remaining components and clip off the excess leads.

5. Mount the components to the chassis bracket MTGBKT. Wire the assembled board to the

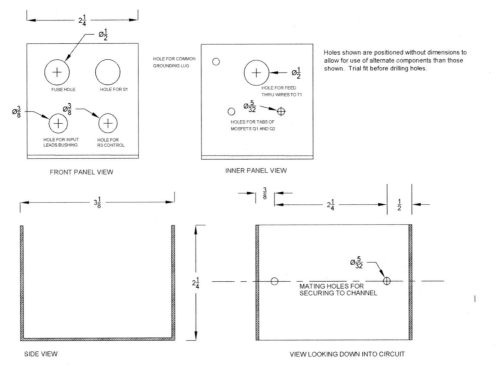

Figure 17-4   *Mounting bracket fabrication*

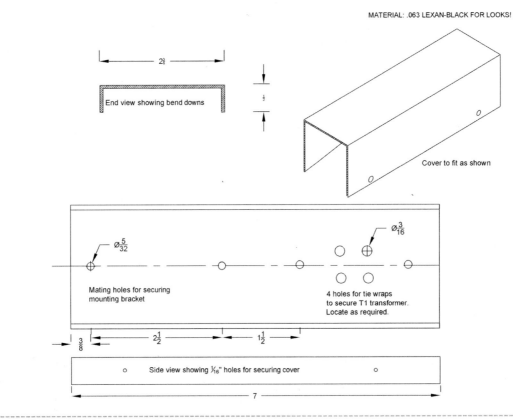

Figure 17-5   *Fabrication of plastic channel and cover*

Chapter Seventeen

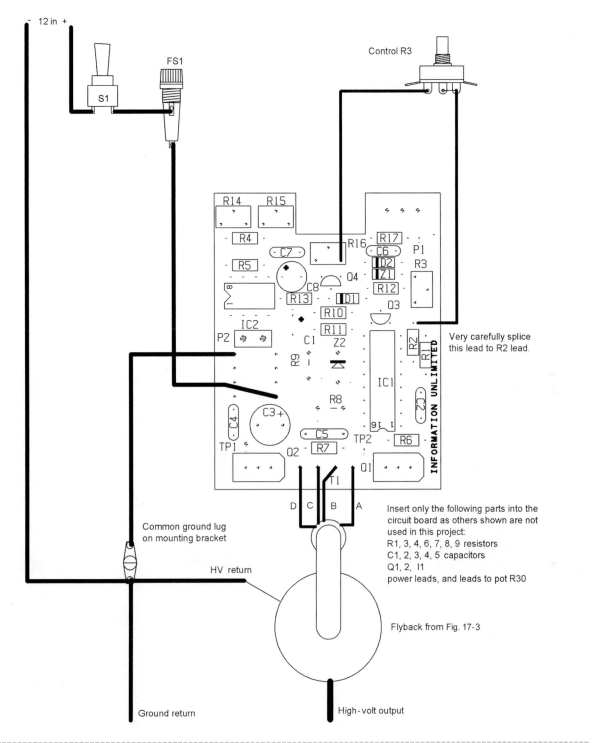

Figure 17-6   *Assembly board and external wiring*

mounted parts and the primary leads of T1, as shown in Figure 17-6 and Figure 17-7, using the appropriate lengths of the #20 vinyl hookup wire. Note the mounting scheme of Q1 and Q2, as they must be insulated from the chassis bracket. Assemble this to the channel via two 6-32 × ½ screws and nuts

SW2/NU1. Note the thin piece of plastic insulating material for the PC board.

6.   Mount a 3.5-inch plastic cap, CAP2, to the channel using two screws, nuts, and small, flat washers SW2/NU1/WA1. Note the four holes for the two tie wraps to secure T1 in place via

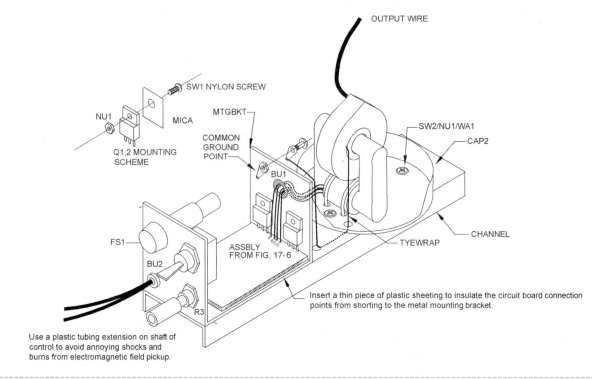

OUTPUT WIRE

SW1 NYLON SCREW

NU1

MICA

MTGBKT

COMMON
GROUND
POINT

SW2/NU1/WA1

CAP2

Q1,2 MOUNTING
SCHEME

BU1

FS1

CHANNEL

BU2

TYEWRAP

ASSBLY
FROM FIG. 17-6

R3

Insert a thin piece of plastic sheeting to insulate the circuit board connection
points from shorting to the metal mounting bracket.

Use a plastic tubing extension on shaft of
control to avoid annoying shocks and
burns from electromagnetic field pickup.

Figure 17-7   *Isometric view of the total assembly*

the primary section shown in Figure 17-7. Note the ground return lead of T1 being routed through a small hole beneath the channel and returning through another hole to connect to common ground point.

7. Finally, assemble everything as shown and insert a 4- × 3½-inch *polyvinyl chloride* (PVC) tube (EN1), as shown in Figure 17-8.

8. Verify all wiring with the schematic. Use an ohmmeter to check that the drain tabs of Q1 and Q2 are insulated from the metal chassis. Check for solder joints, shorts, and so on.

9. If you decide you require oil filling, CAP2 will need to be replaced with an actual PVC flat cap fitting, intended for a 3-inch (actually 3.5-inch OD) schedule 40 PVC tube. These parts are available through most local hardware stores. Add a sealant using PVC cement and primer as directed on the container. Note that the oil filling is only needed for light loads where the output voltage will be excessively high.

Test Steps

To test the project, follow these steps:

1. Turn R3 to full *counterclockwise* (CCW) and S1 off. Insert a temporary 10-amp fuse into the holder.

2. Connect a test lead at the chassis ground and place the other end about one inch from the output lead of T1. This is a crucial step.

3. Connect to a suitable 12-volt converter or a high-current battery. It is a good idea to monitor the input amps for circuit performance verification.

4. Turn the power switch S1 on and note a quiescent current of 1 amp. Slowly adjust R3 clockwise and note a rise in the current of about 2 amps along with some corona at the output terminal. This is the low-output mode and can be used for continuous operation without overheating. Preset R4 to midrange.

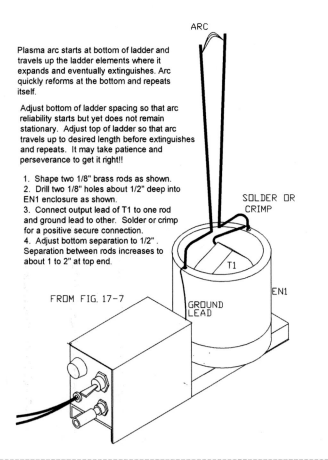

Plasma arc starts at bottom of ladder and travels up the ladder elements where it expands and eventually extinguishes. Arc quickly reforms at the bottom and repeats itself.

Adjust bottom of ladder spacing so that arc reliability starts but yet does not remain stationary. Adjust top of ladder so that arc travels up to desired length before extinguishes and repeats. It may take patience and perseverance to get it right!!

1. Shape two 1/8" brass rods as shown.
2. Drill two 1/8" holes about 1/2" deep into EN1 enclosure as shown.
3. Connect output lead of T1 to one rod and ground lead to other. Solder or crimp for a positive secure connection.
4. Adjust bottom separation to 1/2". Separation between rods increases to about 1 to 2" at top end.

ARC

SOLDER OR CRIMP

T1

EN1

FROM FIG. 17-7

GROUND LEAD

Figure 17-8   *Final view of Jacob's ladder project*

5. Continue tuning R3 and note a very sharp rise in the current (about 7 amps) with the output coming to life. Immediately shut it down as the coil can be damaged in this mode without a suitable load such as connection to a neon or fluorescent tube.

   If you have a scope, you may short the output of T1 to ground and note the test patterns included in Figure 17-2. This verifies the operation. Note that the transformer is designed to allow proper switching of the MOSFETs even with shorted output.

6. Replace the 10-amp fuse with a 5- to 7-amp value.

7. Insert a four-inch enclosure tube (EN1) into the bottom cap (CAP1). Use PVC plumber cement to seal these pieces as they must not leak the transformer oil. Do this only after correct circuit performance has been verified as you cannot service T1 without hacksawing the enclosure apart.

8. Fill the enclosure with oil to the top of the T1. It is not necessary to seal the top cap if the unit is always operated in its upright position. Note this step is not necessary for use with the enclosed experiments when operated as shown.

9. Proceed to conduct the experiments as shown in Figures 17-8 and 17-9. Experiment using pieces of steel wool, needles, and fluorescent and gas-filled lamps, and observe how different materials react to the high-frequency energy. Caution: Obtain some potassium nitrate and sprinkle it onto some steel wool. Note the pyrotechnic display. Danger: Do not use chlorates or perchlorates.

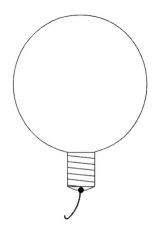

1. Carefully solder a piece of thin bus wire to the center terminal of lightbulb. This is usually lead and solders easily. Caution to not overheat.

2. Connect output lead from T1 to this lead. Bulb should be secured to top of EN1 via a small bracket or other suitable means. Use a nonconductive material.

3. Use a 5", clear 100-watt "DECOR" lamp, preferably one made by Sylvania or GE as these seem to work the best. Experiment with other bulbs as you may get some interesting results.

4. Do not leave this display on for long periods as high-frequency energy may quickly puncture the thin glass envelope of these bulbs.

Figure 17-9   *A poor man's plasma globe*

Table 17-1   Solid-state Tesla coil

Ref. #	Qty.	Description	DB#
R1, 6, 7	3	10-ohm, ¼-watt resistor (br-blk-blk)	
R2		1-kilo-ohm, ¼-watt resistor (br-blk-red)	
R3		5 kilo-ohm trimpot	
R9		10-ohm, ½-watt resistor (br-blk-blk)	
R8		15-ohm, 3-watt resistor (br-grn-blk)	
R30		10K, 17-millimeter potentiometer	
C1		100 mfd, 25-volt vertical electrolytic capacitor	
C2		4700 pfd, 50-volt polyester capacitor	
C3		1000 mfd, 25-volt vertical electrolytic capacitor	
C4		.1 mfd, 100-volt polyester capacitor	
C5		.0033 mfd, 250-volt polypropylene capacitor	
I1		Integrated circuit driver 3525	
Q1, 2	2	MOSFETs IRF540	
T1		Reworked flyback transformer; see Figure 17-3	FLYGRA
WR1BLK	6 feet	#20 vinyl hookup wire, black	

Table 17-1 Continued

Ref. #	Qty.	Description	DB#
WR1REDK	6 feet	#20 vinyl hookup wire, red	
WR1GRN	6 feet	#20 vinyl hookup wire, green	
WR2	6 inches	#20 bus wire	
PCTCL		PCB or perforated circuit board .1 × .1 grid	PCVARG
SW1	2	6-32 × ½-inch nylon screws	
SW2	5	6-32 × ½-inch steel screws	
NU1	7	6-32 nuts	
MICA	2	Mica washers for Q1 and Q2	
TYEWRAPS	2	6-inch nylon tie wraps	
FH1/FS1		Fuse holder and 5-7 amp fuse	
S1		SPST 5-amp switch	
LUGS	2	#6 solder lugs	
BU1		⅜-inch plastic bushing	
BU2		Strain relief plastic bushing	
MTGBK1		Mounting bracket, as shown in Figure 17-4	
CHANNEL		Plastic channel base, as shown in Figure 17-5	
PLASTIC		2- × 2-inch thin plastic insulating sheet	
CAP2		3 ½-inch plastic cap	
EN1		4- × 3 ½-inch schedule 40 PVC tube	
COVER		Plastic cover, as shown in Figure 17-5	
SLEEVE		1- × ¼-inch vinyl tube for control pot, which prevents annoying burns	
12DC/7		12-volt, 5- to 7-amp converter for main power	

Chapter Eighteen

Thirty-Inch-Spark Tesla Lightning Generator

An advanced electrical project is an excellent display for museums or can be a fascinating and rewarding project for the serious and experienced hobbyist. Highly energetic audible and visual bolts of lightning jump into empty space, providing a spectacular effect. This can be an excellent advertising and attention-getting display when properly set up (see Figure 18-1).

The project is shown using basic materials but will require certain specialized parts readily available through our web site at www.amazing1.com (see Table 18-1). Expect to spend around $400 for the special parts with the others available at a local hardware store. The unit utilizes dangerous high voltages and is not recommended for inexperienced personnel. Safety is fully emphasized throughout these construction plans.

Basic Points to Consider Before Building This Project

Your Tesla coil produces large amounts of electromagnetic energy. It may damage computer systems and cause destructive interference to communication and sensitive electrical equipment. The system should be operated within a shielded enclosure, such as a Faraday cage, if near such sensitive equipment.

The primary circuitry, consisting of transformers T1 through T4, produces lethal currents. Human contact with these points, when the system is connected to a power source, can result in a dangerous and fatal shock or serious burns.

Never stand on a conductive surface such as cement or wet ground when operating this equipment. The proper grounding of the system is very important for safe and optimum operation. Omission of the line bypass capacitors C1 and C2 can create an unsafe condition with a voltage breakdown in the primary feed lines.

Never operate the device in a flammable atmosphere as sparks can cause ignition. Low, overhead wooden structures are also prone to fire hazards. Also, always provide adequate ventilation as the discharge produces ozone in large amounts.

It is often a merit of Tesla coil operation to make physical contact with the secondary spark discharge for demonstration purposes. An experienced and qualified person should only attempt to do such a demonstration. The secondary return of the output coil must be directly grounded to earth.

Never leave the system unattended where children or other unqualified personnel may turn it on. Also, there is no need to energize the coil for longer than 10 to 20 seconds at a time.

This coil can produce up to 36-inch discharges. Therefore, it is suggested to position the main power

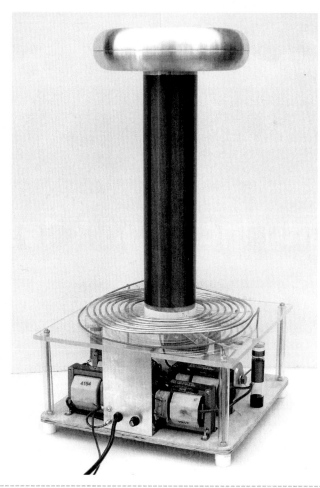

Figure 18-1   *The 30-inch-spark Tesla lightning generator*

switch from a remote point as the sparks may jump toward you if you are standing too close to the output coil. All metal controls should be insulated because contact may cause irritating burns.

Do not use near pacemakers or other similar devices. Always warn spectators as to the possible danger of being near this device if wearing or using sensitive equipment. Also do not operate this device near computers.

Brief Theory of Operation

Transformers T1 through T4 step up the household 115 to 8,000 *volts-alternating current* (VAC) and charges the "tank" capacitor C3. This capacitor now discharges through the *spark gap switch* (SPKGAP) on voltage peaks and steps a pulse of current into the

primary coil (LP1). This sets up a resonant voltage of a frequency determined by the inductance of LP1 and the capacity of C3. Energy is now coupled into the secondary coil (LS1), also tuned to the resonant frequency of the primary circuit. The secondary energy now "rings down" with an exponential decaying waveform. The high voltage output produced in the secondary is now a function of the ratios of primary LP1 to secondary LS1 Q factors or capacities.

It is important to note that voltage is not dependent on turn ratios. The spark gap switch must turn off in order to not allow the "secondary ring down" energy to couple back into the primary circuit. The spark gap uses multiple gaps to enhance the positive turnoff and to prevent ionization from excessive heating. The spark gap electrodes can be brass or steel for limited use but should be tungsten for prolonged operation. Input may be wired for 110- or 220-volt operation as shown in Figure 18-2.

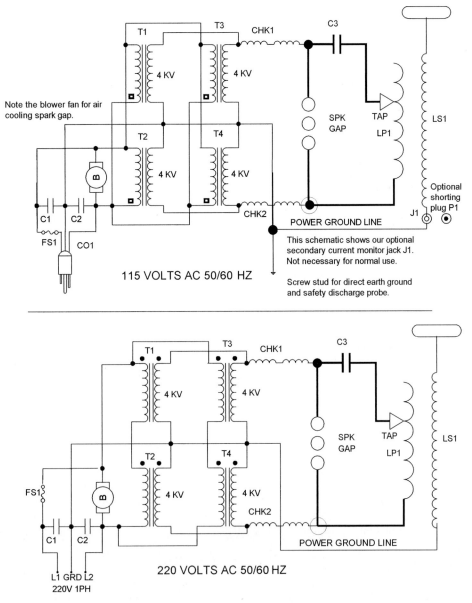

Figure 18-2 *Schematic for both 115- and 220-volt operation*

Circuit Description

Figure 18-2 shows transformer T1, T2, T3, and T4 connected in a parallel series, parallel combination to produce 8,000 volts at 60-*milliampere* (ma) output. The secondary coils are connected to a common neutral point to provide a midpoint ground. This scheme provides 8 Kv between the transformer endpoints but only 4 Kv from any end point to ground. The primaries are wired using standard 115 VAC wiring techniques.

It is important to note the phasing dots shown adjacent to the windings. These must be followed in both primary and secondary circuits. Use Figure 18-2 for wiring points.

Caution: Note that the primary wiring must be isolated from the secondary. The secondary coil is connected directly to the ground plate bracket (GRDPLT).

The fuse (FS1) is rated for 10 amps and is necessary to prevent catastrophic damage from faults, breakdown, and so on. A heavy three-wire line cord (CO1) is strain relieved to the ground plate bracket. This bracket also serves as a connection point for all grounds of the system. It should be connected directly to a dedicated, solid earth ground. You will note capacitors C1 and C2 connected across the AC lines to the ground plate bracket. These capacitors bypass any "kickback" pulses from entering the house wiring where damage to sensitive electronic equipment is minimized (not eliminated). The system might be connected to a remote variable transformer for output adjustment.

The output of the unit can cause annoying burns and shocks from contact to the metal controls. Operation can be controlled by simple removal and insertion of the power plug or a remotely located switch.

Description of Major Components

The role main components perform in your Tesla coil.

LS1 Secondary Coil

This is where the high voltage is produced. The coil form must be an excellent insulator and have a low dissipation factor to the high-frequency currents. It preferably should be of a material that will not readily "carbon track" in the event of spark breakover. Turns must be even and properly spaced. Turn crossovers or overlaps will always cause serious performance problems and must be avoided like the plague. The resonant frequency of the secondary coil can approximately be calculated by considering it to be a quarter-wave section of the length equal to the actual physical length of the wire used. A reduction in this figure can be fudged due to extra capacitance as a result of ionization at the top of the coil when a discharge occurs.

Output Terminal

The output terminal of your system is shown as a 12-inch toroidal terminal. These are expensive and hard to find. You may use stove pipe elbows as a substitute at the cost of overall appearance.

The purpose of the terminal is twofold. First, it electrostatically shields the top winding of the secondary coil from arcing into the open air. This burns the coil and can result in performance degradation. Second, the addition of electrical capacity to the top of a quarter-wave system will enhance current flow through the coil. This property will increase the spark energy at the cost of fewer discharges per unit time. Mathematically speaking, this capacity is unlimited, with the exception of the resonant frequency decreasing to an unworkable value. We are currently designing a computer program on this important property when it is used for voltage magnification and other nonmagnetically coupled resonant systems.

LP1 Primary Coil

This coil is combined with capacitor C3 and must form a resonant circuit equal in frequency to that of the LS1 secondary coil with its associated output terminal. It is made tunable via a tap that enables connection anywhere along its spiraling turns. The wire used is heavy, bare copper or copper tubing and should be $3/16$ of an inch or thicker to accommodate the high-flowing primary tank currents as a result of its high Q factor.

Coupling

The secondary coil LS1 is coupled to the LP1 primary coil and must be inherently tuned to the same frequency for efficient operation. The coupling of these circuits must not be too tight as beat frequencies can cause hot spots along the secondary coil. Coupling that is too loose, however, will not allow a proper energy transfer between the circuits. You may want to experiment by changing the position of LS1 by placing it on wooden blocks.

SPKGAP Spark Gap Switch

This is where energy stored in capacitor C3 is switched into the primary inductor LP1. The spark gap electrodes must allow for clean "makes" and "breaks." Adjustment is usually critical to allow C3 to charge sufficiently before a breakdown or switching occurs. Remember, system energy is a function of the square of the charging voltage across the primary capacitor. It is important that the gap cleanly shuts down before the secondary current reaches it maximum value. The energy in the secondary current must not couple back into the primary one, as it will cause erratic spark gap operation, destructive voltage nodes, hot spots, and so on. The use of tungsten for the spark switch electrodes is recommended if the coil is to be frequently used.

C3 Primary Capacitor

This is where the energy is stored that is exchanged with the primary inductor at a rate equal to the resonant frequency. It must be capable of handling high currents and have a low dissipation factor for efficient operation. Note a special capacitor must be used in this circuit and attention must be given to its dissipation factor and reverse current-handling capability.

CHK1—RF Choke

This part is necessary to block the high resonant frequency as well as the harmonic voltages and currents from feeding back into the transformer. These currents can create destructive voltages that will most certainly cause premature breakdown of this part.

Assembly Steps

The following steps are those used by our laboratory when assembling this device. You may implement your own ideas, but we cannot guarantee the performance claimed if circuit parameters and values are changed. The dimensions shown may also vary with different mechanical parts.

1. Study all of the the plans, schematics, and figures.

2. Identify the parts and pieces if you purchase the kit. Important: The hardware located near the primary coil must be brass.

3. Assemble the secondary coil designated as LS1, as shown in Figure 18-3.

4. Assemble the chokes, designated CHK1, as shown in Figure 18-4 (two are required).

5. Add the top and bottom base sections, designated BASTOP and BASBOT, as shown in Figure 18-5. Note that not all holes are prefabricated because they are best determined by the actual placement of the components.

6. Fabricate the plastic parts (PRIBKT) per Figure 18-6 (four are required). Then, fabricate the parts for a spark switch, as shown in Figure 18-7. Note that the hole location on these pieces must be accurate, as shown. Also note the top and front view of the spark switch assembly, as shown in Figures 18-6 and 18-8.

7. Fabricate the metal parts designed GRDPLT (see Figure 18-9). Note the position of the holes for mounting T1, T2, T3, and T4.

8. Assemble the primary coil section designated BTCPRI (see Figure 18-10). This can easily be done because the 3/16-inch copper tubing is usually neatly coiled up in its shipping container. Mount the PRIBKT brackets in place with small #6 brass wood screws. Carefully fit the coil turns of the copper tubing in place as shown. Connect a 12-inch piece of insulated #12 wire to the copper tube by inserting the striped end into the tube and soldering it with a heavy iron or propane torch. Note the end of the outside copper tube winding adjacent to the hole for the tap lead. Use some silicon rubber (*room temperature vulcanizing* [RTV] adhesive) and apply at slots in PRIBKTS to hold the copper tubing in place. Allow 24 hours to set. Cut a 3-foot piece of the #12 wire for the coil tap lead and attach the crocodile clip for connection to the copper tube.

9. Assemble the parts to the bracket, designated GRDPLT, as shown in Figure 18-9. Note the fuse holder, strain relief, grounding screw (GRDSCR), and current monitor jack, as shown in Figure 18-11. Positioning is not critical but should allow adequate clearance of the connection points.

The transformers are mounted to the bottom base (BASBOT) via suitable hardware. Use Figures 18-12a and 18-12b and attempt to follow the layout as closely as shown. Note that

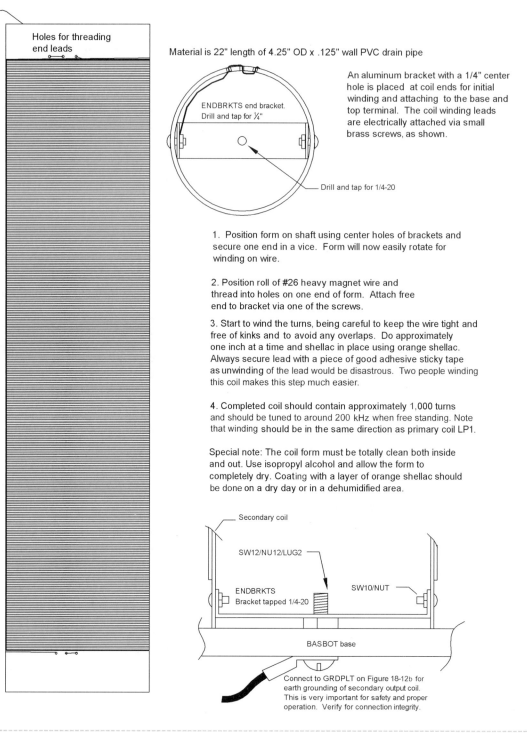

Holes for threading
end leads

Material is 22" length of 4.25" OD x .125" wall PVC drain pipe

ENDBRKTS end bracket.
Drill and tap for ¼"

An aluminum bracket with a 1/4" center
hole is placed at coil ends for initial
winding and attaching to the base and
top terminal. The coil winding leads
are electrically attached via small
brass screws, as shown.

Drill and tap for 1/4-20

1. Position form on shaft using center holes of brackets and
secure one end in a vice. Form will now easily rotate for
winding on wire.

2. Position roll of #26 heavy magnet wire and
thread into holes on one end of form. Attach free
end to bracket via one of the screws.

3. Start to wind the turns, being careful to keep the wire tight and
free of kinks and to avoid any overlaps. Do approximately
one inch at a time and shellac in place using orange shellac.
Always secure lead with a piece of good adhesive sticky tape
as unwinding of the lead would be disastrous. Two people winding
this coil makes this step much easier.

4. Completed coil should contain approximately 1,000 turns
and should be tuned to around 200 kHz when free standing. Note
that winding should be in the same direction as primary coil LP1.

Special note: The coil form must be totally clean both inside
and out. Use isopropyl alcohol and allow the form to
completely dry. Coating with a layer of orange shellac should
be done on a dry day or in a dehumidified area.

Secondary coil

SW12/NU12/LUG2

ENDBRKTS
Bracket tapped 1/4-20

SW10/NUT

BASBOT base

Connect to GRDPLT on Figure 18-12b for
earth grounding of secondary output coil.
This is very important for safety and proper
operation. Verify for connection integrity.

Figure 18-3   *Secondary coil assembly*

the inner mounting legs of the transformer are
electrically contacted to the grounding
bracket GRDPLT, providing a positive earth
ground for the transformer frames.

Finally, assemble everything as shown in the
remaining figures. Note the individual

schematic showing both 115 and 230 VAC
operation. Note the wire nuts are shown connecting to primary wires. You may use terminal strips or another more suitable means.
Remember these points are at 115/220 VAC
and must support at least 5 amps.

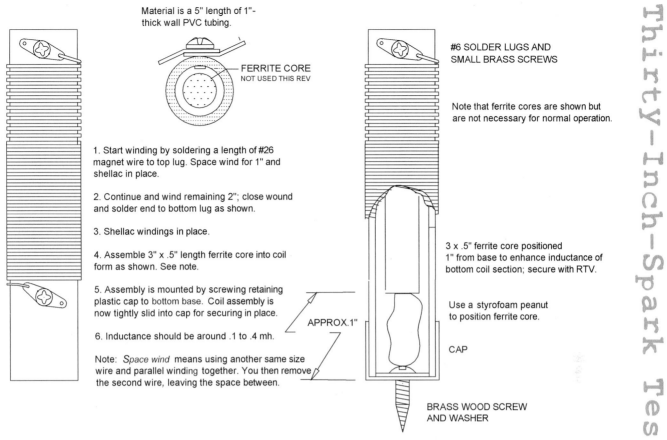

Material is a 5" length of 1"-thick wall PVC tubing.

FERRITE CORE
NOT USED THIS REV

#6 SOLDER LUGS AND
SMALL BRASS SCREWS

Note that ferrite cores are shown but
are not necessary for normal operation.

1. Start winding by soldering a length of #26
magnet wire to top lug. Space wind for 1" and
shellac in place.

2. Continue and wind remaining 2"; close wound
and solder end to bottom lug as shown.

3. Shellac windings in place.

4. Assemble 3" x .5" length ferrite core into coil
form as shown. See note.

5. Assembly is mounted by screwing retaining
plastic cap to bottom base. Coil assembly is
now tightly slid into cap for securing in place.

6. Inductance should be around .1 to .4 mh.

Note: *Space wind* means using another same size
wire and parallel winding together. You then remove
the second wire, leaving the space between.

3 x .5" ferrite core positioned
1" from base to enhance inductance of
bottom coil section; secure with RTV.

Use a styrofoam peanut
to position ferrite core.

CAP

APPROX.1"

BRASS WOOD SCREW
AND WASHER

Figure 18-4   *CHK1 choke protection coils*

The wire in spark gap assembly is as shown in Figure 18-13. Note the mounting of C3 via two heavy tie wraps.

Testing Steps

If you live in a congested area, close to an airport, or near computers and other sensitive electronic equipment, it will be necessary to test this system within a Faraday cage (FAR1). It is also strongly advised to test this system in an area with wooden floors, as dangerous ground currents resulting from accidental contact will be reduced. To conduct the test, follow these steps:

1. Carefully position the secondary coil bottom bracket over the center screw and rotate until it secures in place.

2. Verify the correct wiring, the proper clearance of high-voltage points, and a proper grounding to the GRDPLT ground plate.

3. Set the two spark gaps to about $1/16$ of an inch.

4. Connect the tap lead to the outermost primary turn. You will note that the flexible primary tap lead is used as a full turn of the primary winding and adds another turn depending on the position of the tap connection. This is important, as the system may not properly tune with the toroidal terminal without this added turn.

5. Connect a meter to J1 or short (not necessary for a basic setup).

6. Attach a toroid or another similar terminal. Note that a metal bowl will suffice in a pinch. Screw on to the top secondary bracket and secure with a nut.

7. You may use a 2- to 3-inch screw as an output terminal if not using the toroid. Attach and

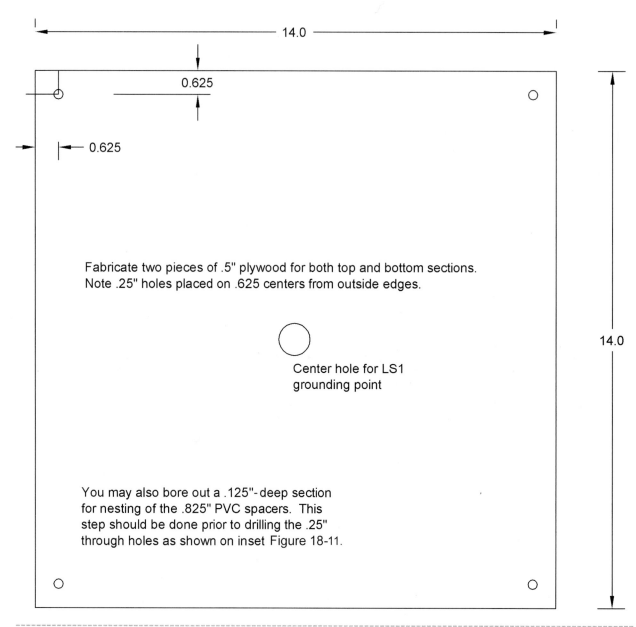

14.0

0.625

0.625

14.0

Fabricate two pieces of .5" plywood for both top and bottom sections. Note .25" holes placed on .625 centers from outside edges.

Center hole for LS1 grounding point

You may also bore out a .125"-deep section for nesting of the .825" PVC spacers. This step should be done prior to drilling the .25" through holes as shown on inset Figure 18-11.

Figure 18-5   *Fabrication of both the top and bottom pieces BASTOP and BASBOT*

screw into the center hole of the top bracket from the bottom. Note that maximum output will require the tap lead now being moved inwards several turns. Do not allow discharges to occur from the actual top winding as you may cause burning of the PVC coil form.

8. Place a grounded contact approximately 8 inches from the terminal and secure it in place.

9. Quickly plug the unit in and note a discharge occurring at the test contact. If not, recheck the system.

10. Separate the test contact from the point where the spark is erratic and connect the tap to the next inner winding on LP1. Reapply power.

11. Repeat the previous steps in an attempt to find the exact point on the primary tap for maximum output. Slightly open the spark gaps and repeat. Note that discharges into open air will be considerably longer than those from point to point. You may tune for this effect if the unit is for display purposes.

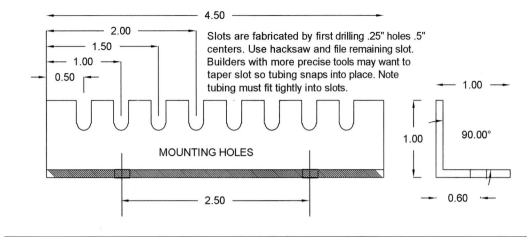

Material is .06" Lexan (polycarbonate) sheet   PRIBKT Primary coil brackets (4 required)

Slots are fabricated by first drilling .25" holes .5" centers. Use hacksaw and file remaining slot. Builders with more precise tools may want to taper slot so tubing snaps into place. Note tubing must fit tightly into slots.

MOUNTING HOLES

Spark gap top view

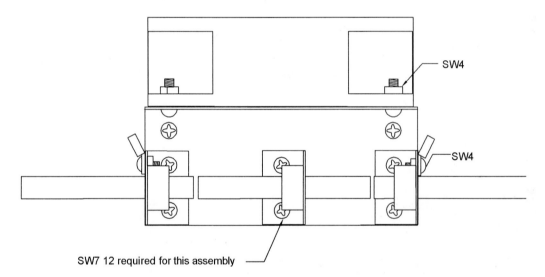

SW4

SW4

SW7 12 required for this assembly

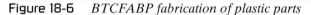

Figure 18-6   *BTCFABP fabrication of plastic parts*

Note that there will be a second-order differential effect when the gap is opened due to changes in dynamic capacity resulting from space volume ionization. This effect will manifest itself as requiring slightly more inductance in the primary. We are currently developing an advanced program in MATH-CAD for those into the complex higher math and electrical physics of resonant high-voltage systems.

12. Experiment by using different values of coupling by raising the secondary coil on wooden blocks and readjusting all parameters. A round or square wooden block that fits inside the secondary coil will position and secure it in place. Always use brass screws for fastening. You may have to remove the bottom bracket. Use your own ingenuity in performing this experiment.

It is also a good idea to keep notes as you experiment.

Material is #22-gauge galvanized sheet metal.

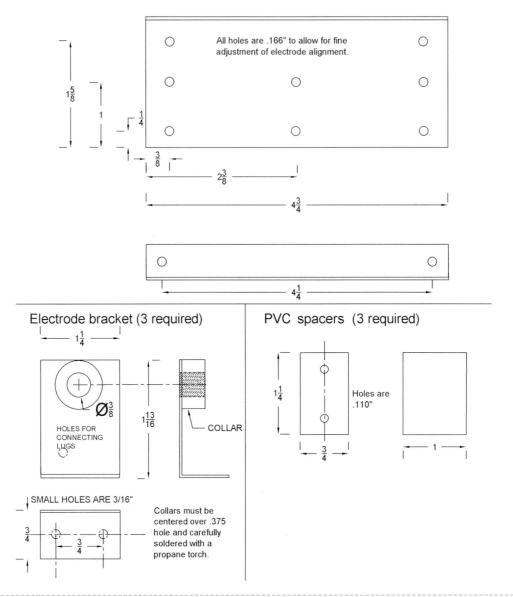

Figure 18-7   *Spark switch fabrication*

Special Notes

Note the optional current meter (M1) can be used for an indication of fine-tuning by adjusting to a maximum value. An initial operation where tuning may be way off can result in erroneous current peaks, especially if coupling is too tight. Use a thermo couple device if available.

The secondary coil in the system acts similar to a quarter-wave antenna with top loading capacity. This means a current node at the base in amps. If you depend on the green wire of the line cord for grounding, you force your electrical wiring to become a part of this system. This is not a good idea as a voltage gradient can result along the wiring run. This voltage value is determined by many complex factors such as frequency, harmonic content, and, of course, a varying

1. Fabricate and verify pieces as per Fig. 18-7.

2. Steel collars must be properly positioned and soldered to brackets. Use a propane torch and ingenuity in this step. Holes must allow proper alignment of electrode faces.

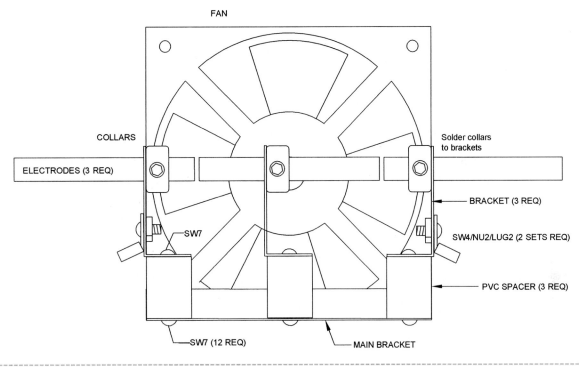

Figure 18-8 *Spark switch assembly and fan*

amount of parallel and series complex impedances along this run. What this means in simple language is that you should ground the coil to earth with the shortest direct lead possible. This can also eliminate a good part of interference coupled back to the power line.

The coil form used for the secondary coil should be an excellent insulator, plus be a relatively lossless dielectric at the operating frequency. A ruby mica coil form would be ideal if it existed. A material with this property can be expensive. Thin-wall PVC tubing, while not the best, is a good compromise between cost and performance. Unfortunately, PVC is hygroscopic (it absorbs moisture), but it can be treated by driving out the moisture with a heat gun and then sealing in orange shellac or an equivalent (do this on a dry day).

Caution: PVC tubing may give off undesirable gases under high-voltage stress. Lexan polycarbonate

tubing is preferred if you are planning to use your coil for continual use.

Not all the holes on the fabrication points are dimensioned. It is suggested that you trial-fit and fabricate as you go.

The CHK1 and CHK2 chokes are designed to provide a relatively high impedance to the resonant operating frequency and any of its associated harmonies. This is necessary to choke off these high frequencies from reaching the power transformer, T1 to T4, where damage to the secondary winding inevitably results. You may use 1K 50-watt wirewound resistors in place of these chokes.

The spark gap is connected across the transformer to limit voltage stress on their secondary windings. The gaps are fabricated from three pieces of 2-inch $\times$ $3/8$-inch tool steel or tungsten rod. A muffin fan keeps the assembly cool (see Figure 18-13). Be cautious of any ultraviolet hazards. Spark gap discharges emit

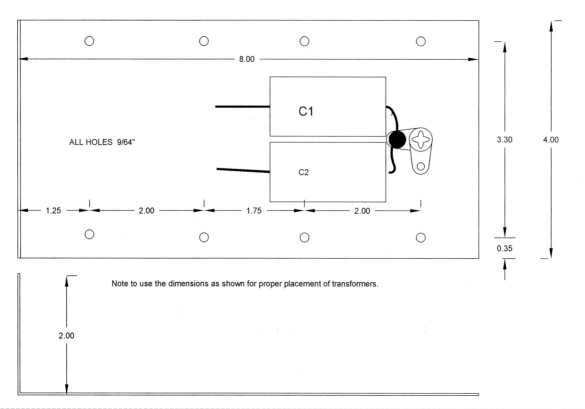

Material is .032 aluminum.

8.00

C1

ALL HOLES 9/64"

C2

3.30

4.00

1.25   2.00   1.75   2.00

0.35

Note to use the dimensions as shown for proper placement of transformers.

2.00

Figure 18-9   *BTCFABM fabrication and assembly of GRDPLT ground plate*

hazardous ultraviolet emissions and must not be viewed directly. Observe them through a clear piece of plastic or protective eyewear.

Tungsten is recommended for long-term reliable coil operation. However, tungsten is expensive and those requiring short-term operation may use drill rod/tool steel. It will be necessary to reface electrodes more often than if they were tungsten.

Use the remaining four feet of #14 wire to make a ground discharge probe that must connect directly to the ground plate (see Figure 18-11). This can be used for drawing sparks from the output of the coil.

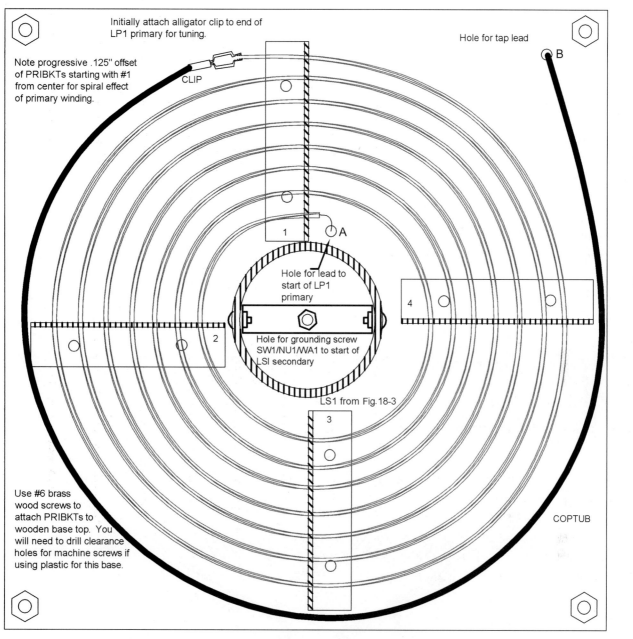

Dashed lines show possible routing of tap lead when used with optional toroidal terminal. This is a rough adjustment of tap setting and will require repositioning to obtain maximum output. It does, however, suffice as a starting point and should produce some impressive streamers into open air. You will note the routing of this tap lead is in the direction as the primary winding and actually adds a turn. Going the other way would effectively remove a turn. The routing and positioning shown will vary dependent on assembly.

Figure 18-10 *Assembly of primary coil*

Suggested labels if you intend to use your coil for demonstrations

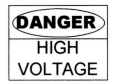

SCREW TO OUTPUT
TERMINAL

X-RAY OF SPACER ASSEMBLY

PLYWOOD or ACRYLIC TOP
6" X .875" PVC SPACERS AND 8"
1/4-20 THREADED BRASS ROD

SPCR

Note cutout for nesting of
the PVC spacers. Not
necessary for assembly.

PLYWOOD BOTTOM

PVC CAPS USED FOR FEET

Do not view spark gap
without protective eyewear.

From Fig. 18-3

CAUTION Please read:

Faraday shield may be required.

Operate in short 10-second intervals.

Keep clear of sensitive electronic equipment.

Do not operate in explosive atmosphere.

Remove power plug before servicing.

Ventilate if ozone becomes excessive.

Output requires careful selection of tap lead.

Blowup of tap clip or
use a crocodile clip

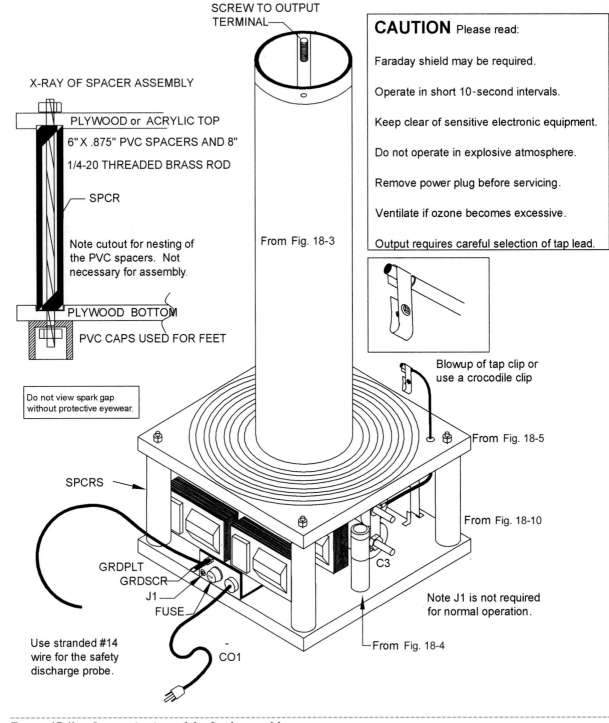

From Fig. 18-5

From Fig. 18-10

C3

SPCRS

GRDPLT
GRDSCR
J1
FUSE

CO1

Use stranded #14
wire for the safety
discharge probe.

Note J1 is not required
for normal operation.

From Fig. 18-4

Figure 18-11 *Isometric view of the final assembly*

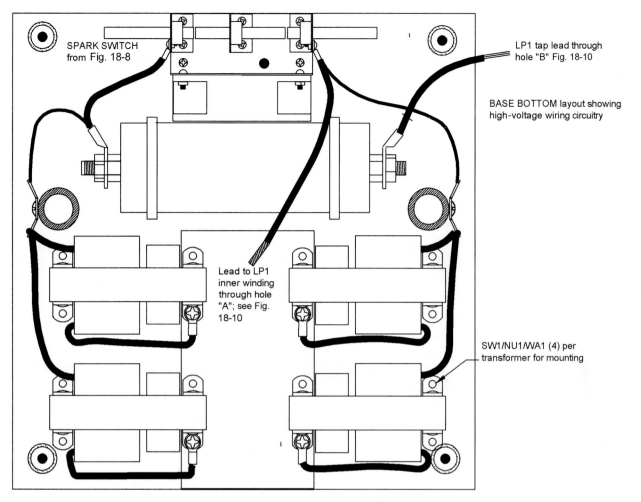

SPARK SWITCH
from Fig. 18-8

LP1 tap lead through
hole "B" Fig. 18-10

BASE BOTTOM layout showing
high-voltage wiring circuitry

Lead to LP1
inner winding
through hole
"A"; see Fig.
18-10

SW1/NU1/WA1 (4) per
transformer for mounting

Figure 18-12a   *BTCAS assembly and first-level wiring*

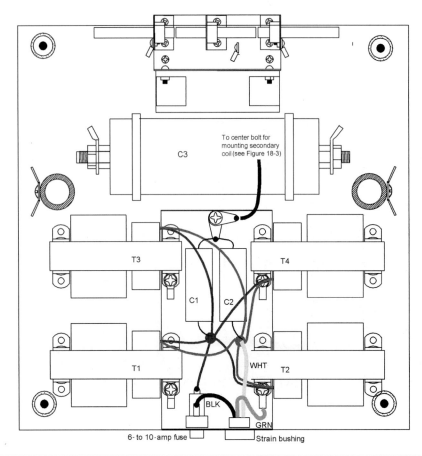

BASE BOTTOM layout showing low-voltage wiring. Note shown for 115 VAC.

Special note on phasing: The first step used in our laboratory is to initially phase T1 and T2 to obtain 8,000 volts with a center point ground. Repeat with T3 and T4. The second step is to now phase the above combinations for 8,000 volts at 60 ma.

DANGER as the above parameters can be lethal if improperly contacted.

Figure 18-12b   *BTCBASP assembly aid for second-level wiring*

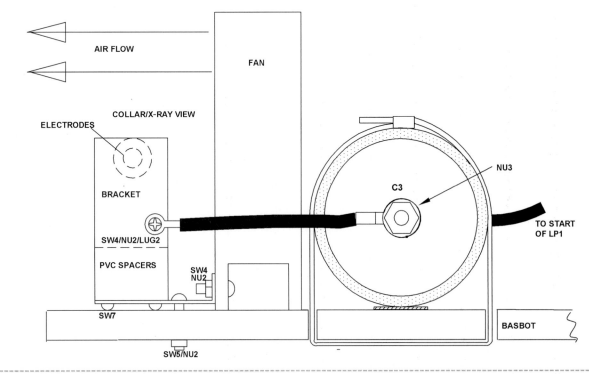

Figure 18-13   *Side view of spark gap and capacitor*

Table 18-1 Thirty-inch lightning bolt generator parts

Ref. #	Qty.	Description	DB#
T1, 2, 3, 4	4	4000-volt, 30-milliampere current-limited transformer	#4 Kv/.03 amp
C1, 2	2	.1 mfd @ 630-volt capacitor	#.1M/630
C3		.022 mfd, 10 KVAC pulse capacitor	#.02 M/10 Kv
FS1		Fuse holder and 10-amp fuse	
CO1		Three-wire, heavy-duty line cord	
BU1		Line cord bushing	
CLIP		Crocodile clip for primary coil tap	
WIRE		Eight-foot section of #14 PVC stranded hookup wire	
COPTUB	50 feet	$3/16$-inch roll of copper tubing	
TOROID		(Optional) 12- × 3-feet or use stove pipe elbows	
WIRENUTS	2	Large wirenuts	
FAN		$4^1/2$-inch 115 AC muffin fan (high output)	
TYEWRAP	3	15-inch heavy nylon tie wraps for attaching C3	
LS10	1	Assembled secondary coil as per Figure 18-3	
CHK10	2	Assembled chokes as per Figure 18-4	
FORMLS1		22- × 4.25-inch OD $1/8$ wall PVC tube for LS10	
FORMCHK	2	5- × 1-inch OD sked 80 PVC tube for CHK1 form	
PRIBKT	4	$4^1/2$ × 2 × .0625 lexan for primary bracket	
GRDPLT		4 × 10 × .063 aluminum for ground plate	
BASTOP/BOT	2	14 × 14 × $1/2$ finished plywood (see Figure 18-5)	
SPCR	4	6- × 1-inch OD PVC tube (see Figure 18-11)	
ELECTRODES	3	2- × $3/8$-inch tungsten or tool steel	#TUNG38
MAIN BRACKET		Fabricate as shown in Figure 18-7	
PVC SPACERS	3	Fabricate as shown in Figure 18-7	
ELTRODE BRKETS	3	Fabricate as shown in Figure 18-7	
END BRKTS	2	Fabricate for top and bottom of secondary coil as shown in Figure 18-3	
SW1	16	6-32 × 1-inch screws	
SW3	4	6-32 × $1/4$-inch screws	
SW4	6	8-32 × $1/2$-inch screws	
SW5	2	8-32 × 1-inch screws	
SW6	4	$1/4$-20 × 8-inch threaded rod (see Figure 18-11)	
SW7	16	#6 × $1/4$-inch sheet metal	
	1	8-32 × 2-inch brass screw	
NU1	16	6-32 nuts	
NU2	8	8-32 nuts	

Thirty-Inch-Spark Tesla Lightning Generator

Table 18-1 Continued

Ref. #	Qty.	Description	DB#
NU3	12	$1/4$-20 nuts	
WA1	17	$1/2$-inch flat washers	
LUG1	9	#6 solder lugs	
LUG2	5	$1/4$-20 lugs	
PVCCAP	4	PVC end caps for bottom feet (see Figure 8-11)	
COLLARS	3	$3/8$ ID $\times$ $3/4$ OD steel collars with set screws	
SW2	10	$6 \times 1/2$ brass wood screws	
SW10/NUT	4	6-32 $\times$ $3/8$-inch brass machine screws and nuts	
SW12	2	$1/4$-20 $\times$ $1 1/2$-inch brass machine screw	
NU12	4	$1/4$-20 brass hex nuts for previous screws	

Part Four

Traveling-Plasma Jacob's Ladder

This popular electrical display saw great fame in the old *Frankenstein* horror movies of the 1930s. A continuous traveling arc of electrical plasma climbs a metal ladder, expanding and eventually evaporating into space (see Figure 19-1). At that moment, the arc reinitiates at the bottom of the ladder and now continually repeats the cycle. It must be noted that these older displays required dangerous and lethal amounts of electricity to obtain the desired results. This property often discouraged their use in public places.

This project generates a plasma arc over 3 inches long that travels a ladder height of nearly 24 inches. It rapidly recycles and contains an arc power adjustment.

Not only is this a rewarding and energetic display, any shock hazard is greatly reduced by the use of our patented safety shutdown circuitry. Any contact to the ladder elements results in immediate circuit shutdown, avoiding any shock. Even if the shutdown were disabled, contact would only cause a minor burn as the output energy is at a high frequency.

Most parts are readily available and are listed in Table 19-1 at the end of the chapter. Those that are specialized, including a *printed circuit board* (PCB), may be obtained through www.amazing1.com. Expect to spend $50 to $75 to complete the project as shown.

Origin

The Bible tells the story of Jacob's dream about a ladder that extended from Earth to heaven. Jacob, the son of Isaac, was the father of the founders of the 12 tribes of Israel. Among sailors, however, a Jacob's ladder is a long rope ladder that is hung over the side of a ship so the harbor pilot can climb aboard.

Basic Description

The power supply for this project forms electric arcs across two diverging stainless steel strips (LADDER). The 16-inch-long strips are mounted on insulating blocks to eliminate possible leakage. The stainless steel strips are separated by about a quarter-inch at their bases and diverge to a distance of about 3 to 4 inches at their upper ends.

The strips form a gap across the secondary winding of the output transformer. After power is turned on, the air dielectric breaks down due to the almost short-circuit state across the lower end of the gap, and an electric arc is formed. As the arc heats up, thermal convection causes the arc to rise up the V-shaped ladder. As the plasma arc ascends the ladder,

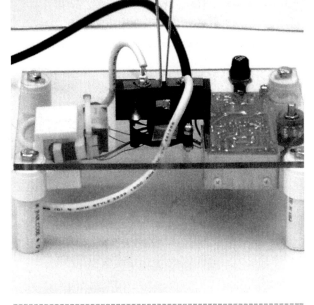

Figure 19-1 *Jacob's ladder plasma machine*

its length increases, thereby increasing the arc's dynamic resistance and thus increasing power consumption and heat. This causes the arc to stretch as it rises, and it extinguishes when it reaches the top of the ladder. When this happens, the transformer output momentarily exists in an open circuit state until the breakdown of the air dielectric produces another arc at the base of the ladder and the sequence repeats.

Circuit Descriptions

We suggest you refer to the circuit description sections in Chapter 8, "Handheld Burning CO_2 Gas Laser," as this project is very similiar, with the exception of the added safety shutdown circuit shown in Figure 19-2. This circuit detects any abnormal ground currents that would be produced by accidental contact to one of the ladder elements. Current now flows through the resistor divider (R5 and R9) and biases diode (D6) into conduction. A rectified voltage is now developed across capacitor (C8) and trimpot (R4). The trimpot sets the trip point of the silicon-controlled crowbar switch, shorting out the supply voltage to IC1 and turning off the power circuit.

Assembly Steps

To begin the project, follow these steps:

1. Study Figures 19-3 and 19-4 showing the power board. Assemble as shown, adding the safety shutdown components shown in Figure 19-2. The added parts are R1, R4, R5, R9, C8, C12, D6, and SCR and are included on the parts list in this chapter.

2. Assemble the heat sink bracket as shown in Figure 8-8 (Chapter 8).

3. Assemble the main power board shown in Figure 19-3.

4. Finally, assemble everything as shown in Figures 19-4 and 19-5, and read all special notes and options.

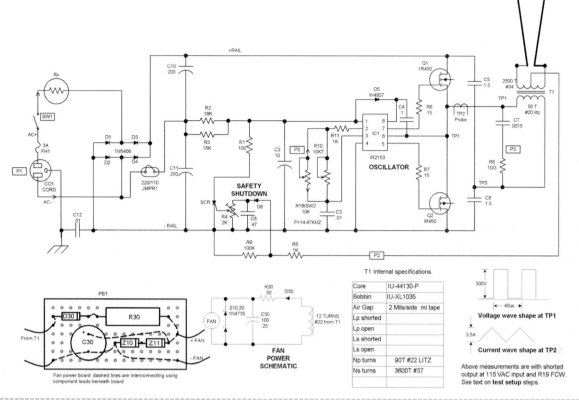

Figure 19-2 *Jacob's ladder circuit schematic*

Testing the Circuit

To test the circuit, follow these steps (see Figure 19-2):

1. Set the bottom of the ladder elements to no more than $1/8$ of an inch separation and temporarily short circuit the ladder elements with a clip lead. Preset the trimpots R4 and R10 to midrange (12 o'clock).

2. Connect an oscilliscope to TP1 and TP2. If you do not have a scope and your assembly is correct, you may assume the circuit is correctly functioning if the following measurements are read.

3. Obtain a 40-watt lightbulb used as a ballast and an isolated source of 115 *volts-alternating*

5. Verify all the wiring for the solder shorts, the correct components, the polarity of parts where noted, and the general overall quality of the assembly.

current (VAC), preferably with a adjustable voltage transformer (variac) and a current-reading meter. A suggested circuit is described in Chapter 8 (see Figure 8-3).

Rotate R19 full *clockwise* (CW) and quickly adjust R4 to read a period of 40 *microseconds* (usec) (25 kHz). Turn R9 back full *counterclockwise* (CCW).

4. Turn off the power and remove the short across the ladder elements, as directed in step 1. Turn on the power and advance R19 in a CW direction, noting an arc that forms and attempts to rise up the ladder. Switch out the ballast, noting the full action of the plasma arc rising and breaking at around 3 inches and repeating. Check that the fan is fully on and measure 12 *volts-direct current* (VDC) across the leads.

5. Observe the line current, noting that it rises to around 1 to 2 amps as the arc rises and breaks at the ladder top.

6. Firmly grab a metal object and contact one of the ladder elements. The unit should quickly

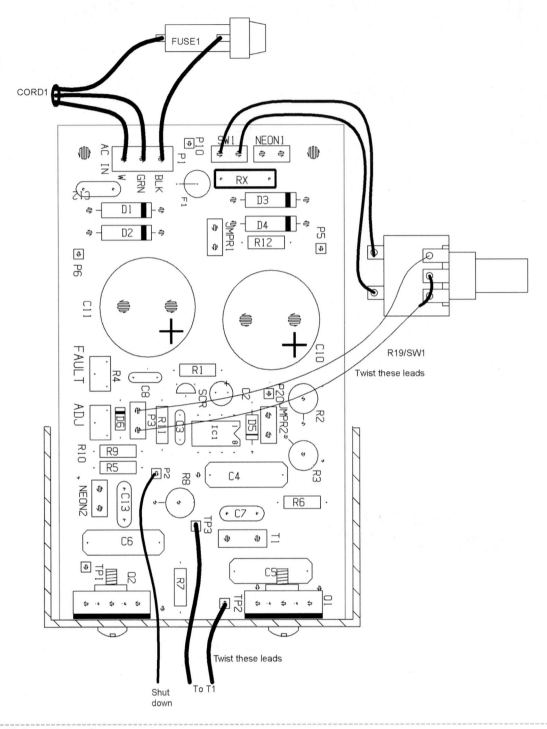

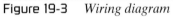

Figure 19-3 *Wiring diagram*

shut down. Note that R4 may be adjusted CW to increase shutdown sensitivity or to full CCW to disable. The normal setting is at midrange.

7. Even though the circuit is intended for continuous use, the project as described should not be left on unattended.

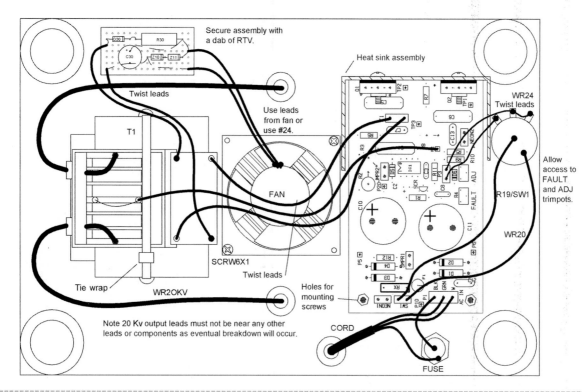

Figure 19-4   *Bottom view of parts layout*

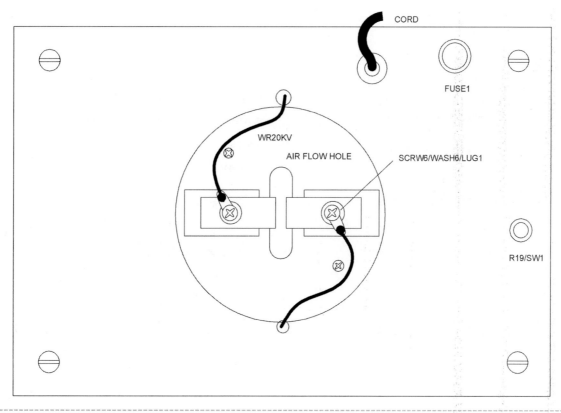

Figure 19-5   *Top view of parts layout*

Table 19-1 Jacob's ladder parts

Ref. #	Qty.	Description
R1		100-ohm, $1/4$-watt resistor (br-blk-br)
R2, 3	2	18K, 3-watt *metal oxide* (mox) resistor
R4		2K vertical trimpot
R5, 11	2	1K, $1/4$-watt resistor (br-blk-red)
R6, 7	2	15-ohms, $1/4$-watt resistor (br-grn-blk)
R8		10-ohm, 3-watt mox resistor
R10		10K vertical trimpot
R30		30-ohm, 3-watt mox resistor
R9		100K, $1/4$-watt resistor (br-blk-yel)
R19/SW1		10K potentiometer and 115 VAC switch
Rx		In-rush current limiter #CL190
C2		10 mfd/25-volt vertical electrolytic capacitor
C3		.01 mfd/100-volt plastic capacitor
C4		.1 mfd/600-volt metalized polypropylene capacitor
C5, 6	2	1.5 mfd/250-volt metalized polypropylene capacitor
C7		.0015 mfd/600-volt metalized polypropylene capacitor
C8		.47 mfd/50-volt plastic capacitor
C10, 11	2	220 to 330 mfd/200-volt vertical electrolytic capacitor
C12		.01 mfd/1 Kv disc capacitor
C30		100 mfd/25-volt vertical electrolytic capacitor
D1, 2, 3, 4	4	IN5408 1 Kv, 3-amp rectifier
D5		1N4937 1 Kv fast-switching diode
D6		IN914 silicon general purpose diode
D30		IN4001 1-amp rectifier diode

Ref. #	Qty.	Description
IC1		IR2153 dual in-line package driver
Q1, 2	2	IRF450 *metal-oxide-semiconductor field effect transistors* (MOSFETs)
SCR		Sensitive gate silicon-controlled rectifier #EC103D
Z1, 2	2	6-volt 1-watt zener diodes #IN4735
PC1		PCB and wire as shown in Figure 19-3 #PCLINE
THERMO1	2	Insulating thermo pads for Q1 and Q2
SWNYLON	2	6-32 $\times$ $1/2$-inch nylon screws and metal nuts
FH$1/3$ A		Panel-mount fuse holder and 3-amp slow-blow fuse
CO1		Three-wire #18 power cord
T1		High-voltage ferrite switching transformer, #JACKT1
WR24	3'	#24 red and black pieces of hookup wire
WR20	6'	#20 red and black pieces of hookup wire
WR20KV	5 feet	20 Kv silicon high-voltage wire
BASE		6 $\times$ 9 $\times$ $1/4$ finished plywood or Lexan, fabricated as required
LADDERS	2	18 $\times$ $1/2$ $\times$ .05 stainless steel, fabricated as shown
BLOCKS	2	1 $1/4$ $\times$ 1 $\times$ $3/4$ PVC or Teflon blocks
FEET	4	1 $\times$ 2 wooden dowels or plastic
TYE12		12-inch tie wrap for securing T1
SCRW6	8	#6 $\times$ $3/8$-inch blunt sheet metal screws
WASH6	2	#6 $\times$ $1/2$-inch washers
SCRW4$\times$1	2	#4-44 $\times$ $1/2$-inch screw and nuts for attaching the assembly board
SCRW8$\times$.5	4	#8 $\times$ $1/2$-inch wood screws for attaching feet pieces
LUG	1	#6 solder lugs

Chapter Twenty

Amazing Plasma Tornado Generator

This project provides a simple yet quite spectacular display of various forms of electrical plasma (see Figure 20-1). The medium used is safe, ordinary, rarefied air pumped down to a rough vacuum of approximately $1/2$ millimeters (which is measured in Torr, which is a measure of pressure equal to 1 millimeter of mercury). The plasma display takes on various forms from a well-defined, multivortex swirling tornado to a column of orange saucer-shaped disks. Hand proximity to the container produces an interactive mechanism where the tornado can be controlled both in position, movement, and intensity.

The low-cost construction includes the use of a 1-gallon glass pickle jar as the plasma display vessel. A wide-mouthed jar is preferred with a brass cap that will allow the soldering of the necessary fittings. The electrical input is supplied by the high-frequency, high-voltage Tesla project described in Chapter 17, "Solid-State Tesla Coil." You will need access to a vacuum pump usually found in most high school science labs that will pump down the air. A pumped-down jar—one that is processed with a vacuum—is available from amazing1.com. Properly processed, the display will last up to a year.

Theory of Operation

An evacuated glass container is sealed and pumped to .5 to 2 Torrs of pressure. A metal cap seals the con-

Figure 20-1   *Plasma tornado enclosure jar*

tainer and serves as an electrode for charging the remaining thin gas mixture. The voltage applied to the cap is at a potential of 10 to 20 thousand volts and is at a high frequency of approximately 25 kHz. The capacitive effect of the thin gas causes a current to flow, creating the plasma discharges. One may visualize the device in the following manner: A capacitor is formed by the conductive gas inside the container forming one plate, the glass envelope being the insulating dielectric and the outer air serving as the other plate. Any conducting object brought near the container now only enhances this effect and appears to draw the plasma arc to the point of contact.

The vacuum will vary along with the physical parameters of the container and can be adjusted to enhance the type of discharge desired. A pressure where the plasma discharges are most defined may be critical. Increases will create a broken, wispy effect, whereas a further decrease will broaden the discharge and eventually form striations as a series of weird, orange disks. Any weird effects are possible from changing pressure, power frequencies, and voltage.

The effect of where the conduction of a gas peaks at a certain pressure is known as the *Townsend effect* and becomes an important factor in the design of vacuum systems where medium- to high-voltages are encountered. The device as described does not use any gas other than the existing atmosphere rarefied by evacuation. Other colors and effects are limitless when the builder chooses to charge the unit with other gases or combinations of pressures.

Project Description

The described device is intended for display purposes, novelty decorations, and special effects, as well as an educational science fair project demonstrating plasma that is controlled electrically and magnetically. Special materials treated by a controlled plasma beam can also be realized.

Plasma is often considered to be the fourth state of matter. It consists of atoms that are ionized and demonstrates peculiar effects unlike the other three forms of matter.

Columns of pinkish and purplish plasma are attracted to external influences such as fingers and other objects when placed on or near the display container. These columns of plasma light span the entire length of the display container, dancing and writhing with a tornado-like effect. Balls of plasma and fingers are created and controlled by simply touching the container. This effect cannot be effectively or justifiably described in words and can only be appreciated when actually observed.

The device is low powered, high frequency, and high voltage, producing the necessary parameters for obtaining the described plasma effect. This generator utilizes conventional electrified circuitry, consisting of a transistor switching the ferrite core of a high-voltage resonant transformer (similar to a TV flyback). Power for the transistors is obtained from a simple step-down transformer and rectifier combination.

Assembly Instructions

This project will require a properly working Tesla driver, as described in Chapter 17.

Obtain a 1-gallon pickle jar or the equivalent with a brass- or tin-plated metal cap for the display container, as shown in Figures 20-2 and 20-3. The display container must have a provision for depressurizing and then being permanently sealed. Again, this metal cover of a pickle jar makes an excellent choice because a piece of copper capillary can be directly soldered to it, forming a good, vacuum-tight seal and allowing pinching off for sealing. Should the display container require repumping, the pinched capillary may be easily reopened for reconnection to the vacuum system.

The display container may be mounted on a suitable stand that houses the generator beneath, resembling a water cooler. Refer to the figures for final assembly and pump down instructions.

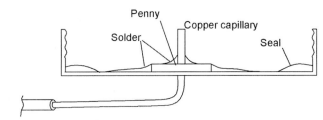

Note that the penny stiffens up the thin metal cover to provide a sturdy mounting point for the capillary tubing.

Automotive vacuum hose

Cover preparation:

1. Drill a .125" hole in the center of a copper penny and in the jar cover.
2. Throughly clean the penny and cover to a bare shiny surface.
3. Obtain a propane torch and solder the tube to the penny and to the jar cover as shown. Try not to burn the jar seal. Verify that the solder completely seals this point as any leaks will prevent operation.
4. Throughly clean the jar and let it dry. Put in microwave and "nuke" for 30 seconds.
5. Apply vacuum grease to the jar seal and tightly screw on the cap.
6. Unit is ready to pump down.

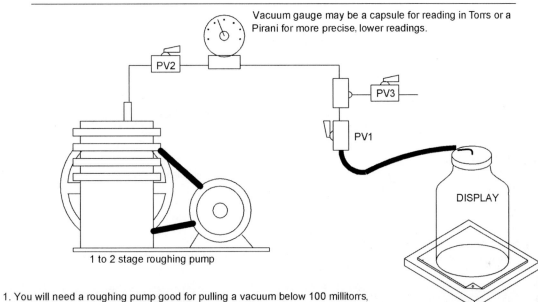

Vacuum gauge may be a capsule for reading in Torrs or a Pirani for more precise, lower readings.

1 to 2 stage roughing pump

1. You will need a roughing pump good for pulling a vacuum below 100 millitorrs, a Pirani gauge, and a bleed-off valve PV3.
2. Connect output of power supply to metal base section. Verify proper isolation from any conducting objects as this point must support the high voltage and high-frequency energy for this setup.
3. Slide copper capillary into appropriate section of vacuum hose.
4. Allow to pump down to limit of system—should be below 100 millitorrs.
5. Bleed off to .5 Torr and apply power, noting rarefied air glowing a purplish pink.
6. Disconnect from vacuum system and pinch off section of rubber hose as shown.
7. Connect power supply output now to copper tube on cap and reapply power, noting a well-defined tornado-shaped discharge extending the full length of jar. Display is visible under normal lighting but should be quite spectacular in the dark. See note below.
NOTE: Display may vary with temperature, proximity to other objects, grounding, and, of course, pressure of air inside. Many display variations are possible and experimentation is suggested.

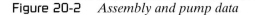

Figure 20-2   *Assembly and pump data*

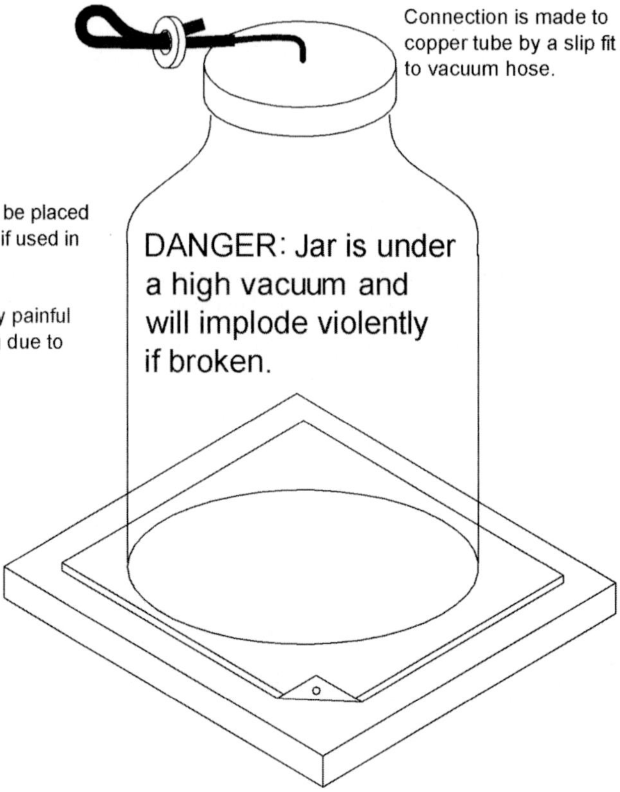

Jar is sealed via a small washer pinching the folded rubber hose. This method provides an excellent temporary seal that is easily removed for repumping etc.

Caution: Jar is under a high vacuum and should be placed in a meshed bag to contain flying shards of glass if used in public.

Output from power supply can cause a moderately painful shock or burn. It is not injurious or life threatening due to low current and high frequency.

Insulating plate–use glass or plastic. Dry wood will work but might leech off energy. You may also support it on a glass mixing bowl.

Use a metal plate by grounding, ungrounding, or attaching various lengths of wire etc. for best results.

Connection is made to copper tube by a slip fit to vacuum hose.

DANGER: Jar is under a high vacuum and will implode violently if broken.

Operation

1. Place an approx. 6 x 6" piece of metal on a larger piece of glass or other insulating material.
2. Place jar as shown on metal plate. Do not ground for now.
3. Verify proper operation of power supply as shown on instructions.
4. Connect up green grounding lead to a positive earth ground. Failure to do this will result in improper operation.
5. Connect output lead to copper capillary tube exiting the jar cover.
6. Rotate power control full CCW and turn on power switch.
7. Rotate power control until you get desired effect that should be a pink column of energized plasma resembling a tornado.

Bring hand near jar and note attraction of display.
This demonstrates the capacitive effect produced by proximity of two conductive objects. The high-frequency current now wants to flow between these points that form a capacitive reactance.

Hold a fluorescent lamp near the jar and note it lighting! This demonstrates the radiative effects of the energized plasma and provides an interesting science project.

If you are fortunate enough to have access to a vacuum pump, experiments with different pressures can yield some real interesting results.

Adjust power control knob for maximum display. Note a peak in adjustment.
Do not leave on for extended periods of time until you check for heating of supply and jar.

Figure 20-3   *Plasma tornado jar setup*

Application and Operation

The display is inside a glass enclosure and resembles a tornado shape of glowing and swirling plasma. It dances and jumps to anything brought near it and is highly visible, even in normal, fluorescent lighting. This sensitivity to any external capacity creates many bizarre and seemingly striking effects. The plasma also can light up a fluorescent lamp when brought near the glass enclosure without any wires or connections of any kind. This feature demonstrates the highly radiative properties of the plasma field and serves as an excellent science fair project or a unique conversation piece.

Special Note

Your display will vary over time, starting out as a wide, undefined glow with purplish and orange disks and eventually forming a defined purple tornado-like vortex extending to the length of the container. This change is due to the pressure increasing and is a function of leakage and internal impurities. The system, properly processed, should be active for up to months before repumping is required.

Chapter Twenty-One

Plasma Lightsaber

This chapter shows how to construct a novelty product that provides a bizarre special-effect light display. It is fashioned after the *Star Wars* lightsaber (see Figure 21-1) and uses a recently patented phenomenon involving "traveling plasma" (our patent #5,089,745). Total control of this display effect is accomplished by simple grip contract on the saber's handle. No switch is used in any way. Energized plasma (an electrically ignited gas giving off visible light) travels up the saber, illuminating its length as it moves. The effect is greatly enhanced in darkness and is quite dramatic when properly controlled by the user. This gives the effect of a controllable length of visible light appearing to emanate from the handle of the device and continues out into space. A striking visual display now results when operated at low light levels.

Expect to spend $25 to $35 for this awesome, attention-getting display project. Instructions are detailed with all the specialized parts, the *printed circuit board* (PCB), and the actual plasma tube available from www.amazing1.com. The parts list is included in Table 21-1 at the end of the chapter.

Circuit Theory

The system utilizes a high-frequency, high-voltage plasma power source that requires only one electrode or an external capacitive electrode for input to the plasma display discharge tube (see Figure 21-2). The external capacitive effect greatly reduces the cost of producing this plasma tube, as no internal electrode or glass-to-metal seals are required. Also eliminated are any grounds or electrical returns required in conventional systems.

Ignition of the plasma discharge appears to occur, extending outwardly into space without a return connection. In actuality, high-frequency electrical currents flow through the capacitive reactance of the plasma tube with the surroundings where the glass enclosure acts as the dielectric between the two. The user, by hand contact with the control pads, forms the other plate of this virtual capacitor.

The circuit consists of transistor Q3 connected as a Hartly-type oscillator where its collector is in a series

Figure 21-1   *Lightsaber*

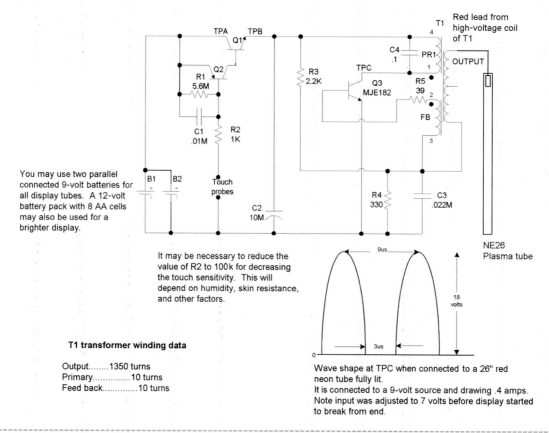

You may use two parallel connected 9-volt batteries for all display tubes. A 12-volt battery pack with 8 AA cells may also be used for a brighter display.

It may be necessary to reduce the value of R2 to 100k for decreasing the touch sensitivity. This will depend on humidity, skin resistance, and other factors.

T1 transformer winding data

Output........1350 turns
Primary...............10 turns
Feed back..............10 turns

Wave shape at TPC when connected to a 26" red neon tube fully lit.
It is connected to a 9-volt source and drawing .4 amps. Note input was adjusted to 7 volts before display started to break from end.

Figure 21-2   *Plasma lightsaber schematic*

with the primary winding, PR1, of transformer T1 and is energized by batteries B1 and B2. The drive signal to its base is obtained by a feedback winding (FB) properly phased to allow oscillation to take place. The base current is limited by resistor R4 and biased into conduction by resistor R3. Capacitor C3 speeds up the switching times. The oscillations produced are at a frequency of approximately 100 kHz. This is usually determined by the resonant frequency of T1 and tank capacitor C4.

The output of Q3 is controlled via the conductance of pass transistor Q2 by biasing its base with a ramp signal from transistor Q1. C2 bypasses any high-frequency switching currents to the common line of the circuit. This approach provides a positively defined state between the energized and denergized plasma, hence its lit display length.

The current through Q2 and therefore the power to Q3 is controlled by the DC ramp amplifier Q1. The Q1 transistor is now controlled when a base current flows through resistor R2. This occurs when the user's fingers simultaneously touch the two external pad

contacts biasing Q1 to a point dependent on the user's contact resistance. This effect produces the variable current ramp that controls the current through pass transistor Q2, hence controlling the output of Q3. No off/on switch is necessary since total power is controlled by the user's finger contact, a capacitor C1 bypasses any external signals that may cause premature operation, and R1 controls the sensitivity range of the necessary contact resistance for full ignition as well as linearity.

Construction

The device can be built in two parts, consisting of the display and power sections. These are easily separated for convenience should the plasma display discharge tube become broken or damaged. Also, we must consider the option of using display tubes with other gases, producing different colorful effects.

The display section of the device can consist of a 12- to 36-inch length of small-diameter neon or another gas tube. Only one electrode is necessary on the display tube, thus eliminating unnecessary costs. This internal gas tube is centered in a clear or colored plastic tube that serves for protection from breakage and provides a more enhanced visual effect due to its diffusive, refractive, and diffractive optical properties.

Construction Steps

Figure 21-3 shows the foil traces for those wanting to do their own PCB. To begin construction, follow these steps:

1. Lay out and identify all the parts and pieces, and check them with the parts list. Note that some parts may sometimes vary in value. This

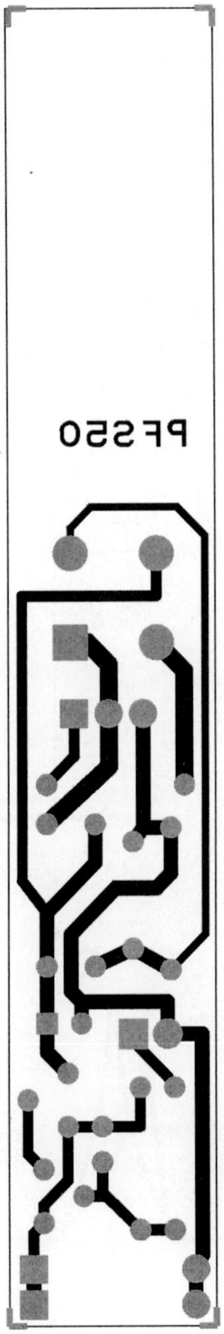

This view is helpful for those who want to design their own layout on a vector or perforated circuit board. Component leads may be used for most routing traces with heavier leads for the wider traces.

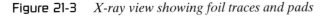

Figure 21-3   *X-ray view showing foil traces and pads*

is acceptable as all components are 10 to 20 percent tolerant unless otherwise noted.

2. Identify the pins on the base of transformer T1, as shown in Figure 21-4.

3. Insert components into PCB PCPF5 as shown in Figure 21-4. Note to leave at least $\frac{1}{8}$ to $\frac{1}{4}$ of an inch of lead between the actual compo-

nents and the surface of the board. Also notice the polarity on C2 and the proper position of transistors Q1 and Q3. Solder the connections and cut away the excess leads. Connect the T1 transformer using short pieces of bus wire and secure it to the board using some tape. Then attach the leads for B1 and B2. Note these leads are strain relieved by passing through

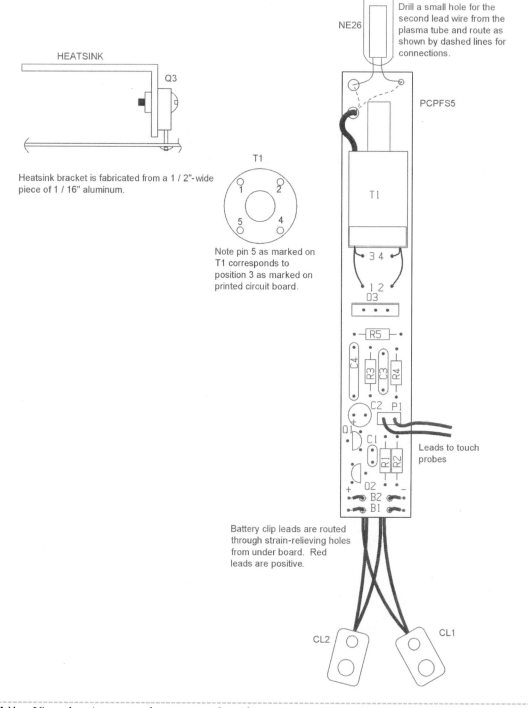

Figure 21-4   *View showing parts placement on board*

the holes on the foil side of the board. Solder 11-inch leads for contact pads located on the enclosure tube. These may be shortened later. Check for accuracy, the quality of solder joints, potential shorts, and so on.

4. Obtain a 12- to 36-inch neon tube, NE26, as shown in Figure 21-5. Note that the neon tube is shown utilizing only one internal electrode. An external electrode consisting of a piece of metallic tape wrapped around the tube end will also work. The internal electrode approach seems to work slightly better as the input impedance of the feedpoint is obviously nonreactive, now being only resistive.

   Note that the assembly of the tube is beyond most hobbyists and probably should be obtained as indicated on the parts list. Solder the tube to the assembly board as shown in Figure 21-5 and secure it with some *room temperature vulcanizing* (RTV) silicon rubber. Insert a piece of plastic vinyl tubing between the tube and the PCB to protect the tube from hitting the assembly board and possibly fracturing.

5. Connect two 9-volt batteries or a 9-volt power converter. Note that batteries are connected in a parallel way to supply more current and consequently last longer.

6. You may verify the circuitry by connecting a current meter in series with the batteries and note it reading zero. Turn the meter range down a step at a time to 50 micro-amps or the lowest range. The meter should still read zero. Note that any current flow in this test will wear down the battery over a period of time, indicating transistor leakage or a wiring error. The battery will drain down even if the saber is not in use.

7. Now set the meter range to read 300 to 400 milliamperes and reverse if necessary. Make contact between the pads' − and + leads. Note that the neon tube fully ignites and the current meter indicates around 300 milliamperes. Please refer to the test points in Figure 21-2 if you experience difficulty. These are explained in the supplementary test points section.

8. Attempt to make contact between these points using the resistance of your finger and verify the partial ignition of the neon tube. Dampen your finger if the skin is dry. This verifies the proper operation of the electronics. A

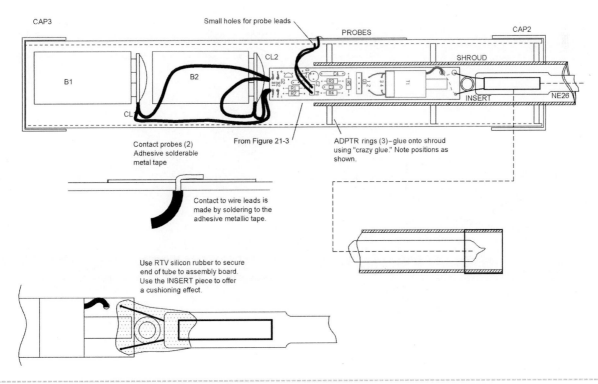

Figure 21-5   *X-ray view of handle innards*

dry hand may require a tighter grip, whereas a damp hand requires only a light touch to achieve full plasma ignition.

Mechanical Assembly Steps

To begin the project assembly, follow these steps:

1. Cut a 29 ½-inch length of 1-inch OD plastic tube for the shroud, as shown in Figure 21-6. Even up and remove sharp edges from the ends. A red transparent tube greatly enhances the display when using red neon gas.

2. Fabricate four flexible spacer rings from a sheet of clear, flexible ⅛-inch vinyl, as shown in Figure 21-6. These spacers position the neon tube, NE26, inside the plastic shroud and offer some shock protection in case the units

are mishandled. The center holes should be a snug fit to the neon with the outer diameter providing reasonable friction to the inner walls of the shroud tube. These spacer rings should be positioned on the NE26 display tube as shown. Note that other materials may be used for this part.

3. Fabricate three adapter rings as shown in Figure 21-6. The outside diameter must fit snugly into handle HA1. A 1-inch hole must be in the true center and fit snugly around the shroud. These are positioned and glued to the shroud tube, as shown in Figure 21-5.

4. Fabricate CAP2 with a 1-inch center hole and position it on the shroud assembly as shown in Figure 21-5. It should abut closely to the forward adapter ring.

5. Fabricate handle HA1 from a 10-inch-long piece of 1 ⅝- × 1/16-inch rigid, wall *polyvinyl chloride* (PVC) or an equivalent material. Place two, small 1/16-inch holes as shown in Figure 21-5 for contact probe leads. Note that

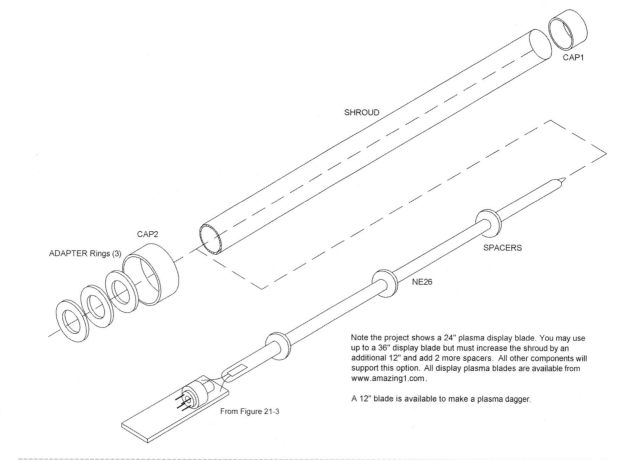

CAP1

SHROUD

CAP2

ADAPTER Rings (3)

SPACERS

NE26

Note the project shows a 24" plasma display blade. You may use up to a 36" display blade but must increase the shroud by an additional 12" and add 2 more spacers. All other components will support this option. All display plasma blades are available from www.amazing1.com.

A 12" blade is available to make a plasma dagger.

From Figure 21-3

Figure 21-6 *Assembly of display tube, shroud, and spacers*

the position is not critical and can be placed to suit your preference.

6. Insert the NE26 tube assembly (see Figure 21-6) with spacer rings along with the assembly board into the shroud as shown. It may help to moisten the inner walls of the shroud by breathing into one end and quickly inserting the neon tube assembly. You may also use a mist bottle. Work gloves should be used to avoid injury in case of breakage.

7. Insert the above assembly into handle HA1 with the forward adapter ring recessed into the handle approximately $\frac{1}{4}$ of an inch (see Figure 21-5). You may glue it in place or leave it as is for disassembly. Slide CAP2 into place as shown. Please note that it may take considerable force as the fit is tight and will not normally need gluing.

8. Using extreme patience and your own ingenuity, attempt to thread the probe wires through the holes in HA1, as shown in Figure 21-5. Sandwich the stripped ends to the handle with small pieces of metallic tape as shown. Note that electrical contact is made to the metallic probes by this sandwiching action. Cut the pads to shape for appearance using an x-acto knife.

9. Insert two fresh, standard 9-volt alkaline or lithium batteries into CL1 and two battery snaps into the handle. Obtain some foam rubber and use some pieces to hold the batteries in place. Secure the end with the CAP3 cap. Test the unit by touching the probes and verify that the correct operation occurs. Verify the heatsink is not too hot to touch.

10. Finally, assemble everything as shown (see Figure 21-7) by attaching the caps into place along with any labels or decals. Now go forth, have fun, and may the force be with you.

Test Points and Troubleshooting Suggestions

The following are some measurements for testing the device:

- Measure 9 VDC at the TPA point and at the COM line. If the previous tests cannot be obtained, it is suggested to double-check all

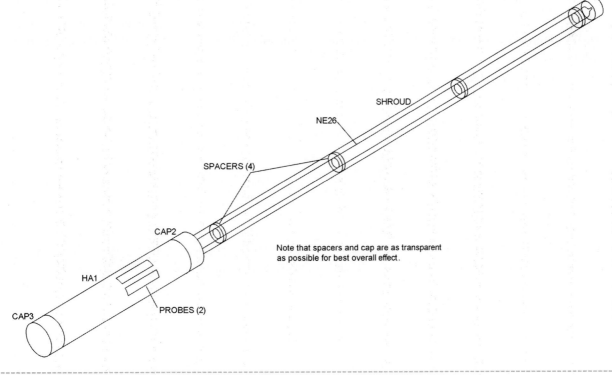

Note that spacers and cap are as transparent as possible for best overall effect.

Figure 21-7 *Plasma lightsaber final view*

wiring and components for correct placement. Pay attention to Q1 and Q2, as they are not the same.

- Measure 0 VDC at the TPB point and at the COM line.

- Short contact probes A and B together and now measure 8 to 9 VDC at TPB.

- If the display tube fails to ignite, double-check all the wiring and components for correct placement. Pay attention to Q1 and Q2, as they are not the same.

Special Notes

Gas tubes longer than 30 inches may require a 12-volt battery pack to activate to full length.

The electronics assembly and batteries will easily fit into the often preferred Graphlex and Heiland camera flash handles. Contact the factory for facsimiles of this hard-to-find part.

This chapter's plans show the unit built with a 26-inch gas display tube. The system is designed to use any length of tube up to 36 inches long. The only modifications will be the length of the outer shroud tube and adding or subtracting the spacer rings. Our standard lengths of tubes are 12, 26, and 36 inches.

The touch probes may be replaced by a 1 megohm adjustable pot and switch for varying the discharge length. R1 must be changed to a 47-kilo-ohm value.

Batteries

The unit will work with two alkaline batteries connected in parallel to double the current rating for about five hours of "off and on" operation. Two lithium batteries, while more expensive, will allow eight hours of operation. The suggested *nickel-cadmium* (NiCad) rechargeables are Varta #TR $^7/_8$ for 9.8 volts. The running time for two batteries will be almost an hour before recharging.

Thirty-six-inch red neon units may require 12 volts, which is obtainable by using 8 AA cells to fully ignite the display. You may also use two AAA cells connected in a series with the existing two 9-volt batteries. The wiring is shown in Figure 21-2. This approach allows fitting with the Graphlex and Heiland handles, as were the original props. Double-check the heatsink tab for Q3.

Applications

The property of the moving, ignited plasma your lightsaber produces can be utilized to provide an indicator of direction, such as turn signals for vehicles, semaphore signaling, pointing devices, or an excellent safety device for nighttime jogging or walking as a visible, piercing colored light is produced.

Table 21-1 Plasma lightsaber parts list

Ref. #	Qty.	Description	DB #
R1		5.6 meg, $^{1}/_{4}$-watt resistor (gr-bl-gr)	
R2		1K, $^{1}/_{4}$-watt resistor (br-blk-red)	
R3		2.2K, $^{1}/_{4}$-watt resistor (red-red-red)	
R4		4.7K, $^{1}/_{4}$-watt resistor (yel-pur-red)	
R5		330-ohm, $^{1}/_{4}$-watt resistor (or-or-br)	
C1		.01 mfd, 50-volt plastic capacitor	
C2		10 mfd, 25-volt electrolytic capacitor	
C3		.022 mfd, 250-volt plastic capacitor	
C4		.1 mfd, 400-volt metal polypropylene capacitor	
Q1		PN2907 PNP GP transistor	
Q2		PN2222 NPN GP transistor	
Q3		MJE182 power tab NPN transistor	
T1		Ferrite high-voltage transformer	#28K077
CL1, 2	2	Battery snap connectors	
PCPFS5		PCB	#PCPFS5
HS1/SW1		Heatsink bracket and #6 × 1/4-inch sheet metal screw	
BUSWIRE		Three-inch piece of bus wire for connecting T1 pins	
WR2	2	12-inch lengths of #24 vinyl hookup wire	
ADPAPTERINGS	3	1 $^{1}/_{2}$ OD × 1 hole × $^{1}/_{16}$ Lexan washer fabrication	
SHROUD		29 $^{1}/_{2}$ × 1 OD × $^{1}/_{16}$ wall Lexan or other clear or colored plastic tubing	
SPACER	4	$^{7}/_{8}$ × $^{3}/_{8}$ hole × $^{1}/_{8}$ flexible, clear vinyl washer	
CAP1		1-inch clear plastic cap	
CAP2		1 $^{5}/_{8}$-inch black plastic cap with 1-inch hole	
CAP3		1 $^{5}/_{8}$-inch black plastic cap	
PROBE	2	2- × $^{1}/_{4}$-inch strips of adhesive, solderable, metallic tape	
HA1		10 $^{1}/_{2}$- × 1 $^{5}/_{8}$-inch black plastic handle	
INSERT		$^{3}/_{8}$- × $^{3}/_{8}$-inch Tygon tubing	
NE26		26-inch × 10-millimeter special prepared plasma tube	#NE26 available in "phaser" green, "photon" blue, "starfire" purple or "neon" red

Part Five

Chapter Twenty-Two

Ion Ray and Charge Gun

This project shows how to build a device that produces a bizarre yet interesting electrical phenomena that possibly may provide the propulsive forces used for space travel in the near future. A high potential source is allowed to leak off charges by being terminated into a sharp and pointed object. A charge emission now occurs due to the repulsive force of like charges occuring at the pointed end, producing a high density of charged particles that possess mobility over a distance. The finished unit is intended for use

by the electrical physicist (see Figure 22-1) as well as for educational demonstrations, material testing, and many other applications as presented in this chapter.

Expect to spend $75 to $100 for this useful electrical device. Assembly will require only basic hand tools, wiring, and soldering. The finished project must be used with caution as moderate electrical charges can be generated at a distance. The parts list for this project is shown in Table 22-1.

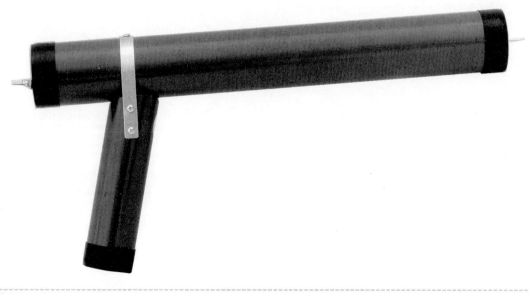

Figure 22-1   *Ion ray gun*

Circuit Operation

Your ion ray gun requires a high DC voltage at a very low current. The driver power supply, as shown in Figure 22-2, generates over 600 micro-coulombs (600 micro-amps per second). This amount is a large number of ions and is sufficient to induce shocks at a distance, charge objects, and perform a host of bizarre electrical experiments. Even though the current is low, improper contact can result in a harmless yet painful shock.

The output voltage of the driver is obtained using a Cockcroft Walton voltage multiplier with 4 to 10 stages of multiplication. This method of obtaining high voltages was used in the first atom smasher ushering in the nuclear age. The multiplier section requires a high-voltage/frequency source for input supplied by transformer T1, producing 6 to 8 Kv at approximately 30 kHz. You will note that this transformer is a proprietary design owned by Information Unlimited. The part is small and lightweight for the power produced.

The primary winding of T1 is current driven through inductor L1 and is switched at the desired frequency by *field-effect transistor* (FET) switch Q1. Capacitor C6 is resonated with the primary of T1 and zero-voltage switches when the frequency is properly adjusted. (This mode of operation is very similar to class E operation.) The timing of the drive pulses to Q1 is therefore critical to obtain optimum operation.

The drive pulses are generated by a 555 timer circuit (I1) connected as an astable multivibrator with a repetition rate determined by the setting of trimpot (R1) and a fixed-value timing capacitor (C2). I1 is now turned on and off by a second timer, I2. This timer operates at a fixed frequency of 100 Hz but has an adjustable "duty cycle" (ratio of on to off time) determined by the setting of control pot R10. I1 is now gated on and off with this controlled pulse, providing an adjustment of output power.

Even though the output is short circuit protected against a continuous overload, constant hard discharging of the output can cause damage and must be limited. A pulse current resistor, R7, helps to protect the circuit from these potential damaging current spikes.

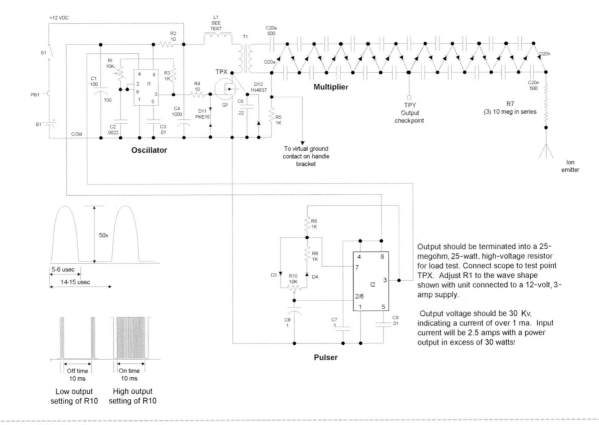

Figure 22-2 *Ion ray gun schematic*

Power input "enable" is controlled by switch S1 that is part of control pot R10. A trigger switch (PB1) allows instantaneous control. The actual power is a battery pack placed in the handle that consists of 8 AA cells in a suitable holder. A virtual ground is produced by user contact to the circuit return via a metallic probe built into the handle.

Construction Steps

Here you'll begin construction of the electronic ion-generating power supply. The ion generator is shown built using a perforated circuit board, as this is the preferred approach for science projects because the system looks more homemade.

The perforated board approach is more challenging, as the component leads must be routed and used as the conductive metal traces. We suggest that you closely follow the figures in this section and mark the actual holes with a pen before inserting the parts. Start from a corner, using it a reference, and proceed from left to right.

The *printed circuit board* (PCB) only requires that you identify the particular part and insert it into the respective marked holes. Soldering is now greatly simplified.

Board Assembly Steps

To assemble the perforated circuit board, follow these steps:

1. Lay out and identify all the parts and pieces. Verify them with the parts list, and separate the resistors as they have a color code to determine their value as noted.

2. Cut a piece of .1-inch grid perforated board to a size of 4.8 × 2.9 inches. Then cut a piece of 10- × 2.9- × .063-inch polycarbonate for the multiplier section. Locate and drill the holes as shown in Figure 22-3. An optional PCB is available from Information Unlimited.

3. Fabricate the metal heatsink for Q1 from a piece of .063 aluminum at 1.5 × .75 inches, as shown in Figure 22-4.

4. Assemble L1 as shown in Figure 22-4.

5. If you are building from a perforated board, insert components starting in the lower left-hand corner, as shown in Figure 22-5 and 22-6. Pay attention to the polarity of the capacitors with polarity signs and to all the semiconductors. Route the leads of the components as shown and solder as you go, cutting away unused wires. Attempt to use certain leads as

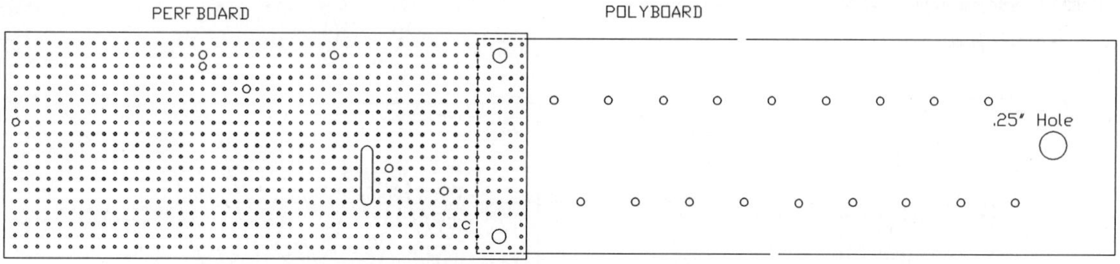

PERFBOARD   POLYBOARD

.25" Hole

The assembly board is in two sections attached together by two outer 6-32 nylon screws and nuts. The middle hole is used to fasten the entire assembly to the base of the enclosure.

The circuit section is 4.8" x 2.9" x .1" perforated board. The high voltage polycarbonate section as shown is 10" x 2.9" x .063" thickness. This is sufficient to accommodate 10 stages of multiplication.
Drill .063" holes in the perforated section and the polycarbonate section located as shown.

Drill the three .125" holes in both sections for attaching together.

Drill and drag the .125" slot as shown. This cutout and the enlarged holes are for mounting transformer T1.

Using the optionally available printed circuit board will still require fabrication of the Plexiglas board.

Hole diameters are not critical.

Always use the lower left-hand corner of the perf board for position reference.

Figure 22-3   *Driver and multiplier board fabrication*

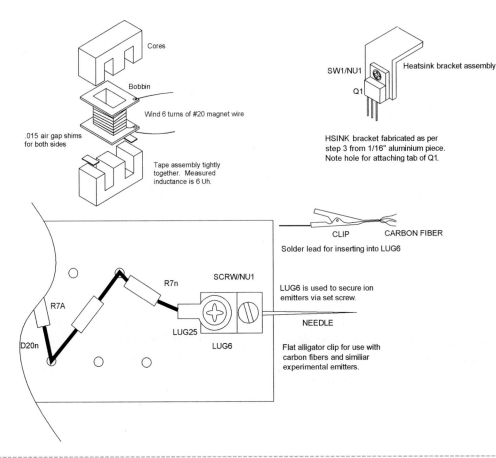

Figure 22-4   *L1 current feed inductor and heatsink bracket*

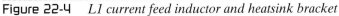

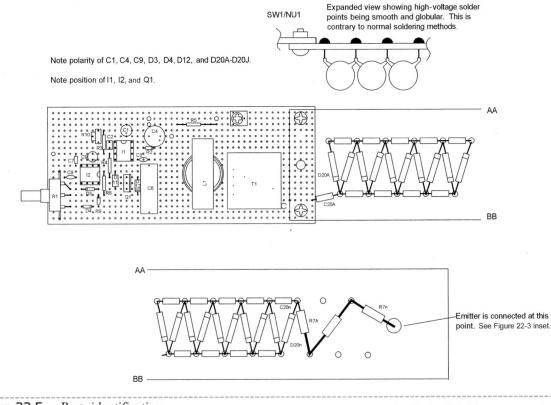

Figure 22-5   *Parts identification*

wire runs or use pieces of the included #22 bus wire. Follow the dashed lines on the assembly drawing as these indicate connection runs on the underside of the assembly board. The heavy dashed lines indicate use of thicker #20 bus wire, as this is a high-current discharge path and common ground connection.

6. Attach the external leads as shown in Figure 22-6. Figures 22-7a and 22-7b are enlarged views of the assembly board wiring.

7. Assemble the voltage multiplier as shown in Figure 22-5. The project shows 10 stages of voltage multiplication. Each stage consists of two capacitors (C20xx) and two diodes (D20xx). The stages can be reduced to a number of 10 where you will obtain 7 to 10 Kv of output as each stage contributes this amount of additional voltage. Additional stages over 10 will produce more ions but will only generate a higher potential when terminated into a smooth 4- to 5-inch terminal.

8. Double-check the accuracy of the wiring and the quality of the solder joints. Avoid wire

bridges, shorts, and close proximity to other circuit components. If a wire bridge is necessary, sleeve some insulation onto the lead to avoid any potential shorts. See the note in Figure 22-5 showing smooth, globular solder joints for all high-voltage points on the multiplier board.

Testing Steps

To run a test on your device, follow these steps:

1. Preset trimpot R1 to midrange and R10 to full *clockwise* (CW).

2. Obtain a 25-megohm, 20-watt high-voltage resistor. You can make this part by connecting 25 1-megohm, 1-watt resistors in a series and sleeving them into a plastic tube. Then seal the ends with silicon rubber

3. Obtain a 12-volt, DC, 3-amp power converter or a 12-volt battery. You may use the 8 AA cells in the specified holder.

See Figures 22-7a and 22-7b and B for enlarged views of this figure.

- - - - - Thick dashed lines are direct connection runs beneath board of #20 bus wire (WR20BUSS) and are extended for the spark switch electrodes.

Thinner dashed lines are #24 bus wire (WR24BUSS) and component leads wherever possible.

▷ Triangles are direct connection point junctions.

Solid black lines are external leads for input and output lines. Use red (WR20R) for +12 input.
Use green (WR20G) for lifter connection.
Use black (WR20B) for com -12 input.

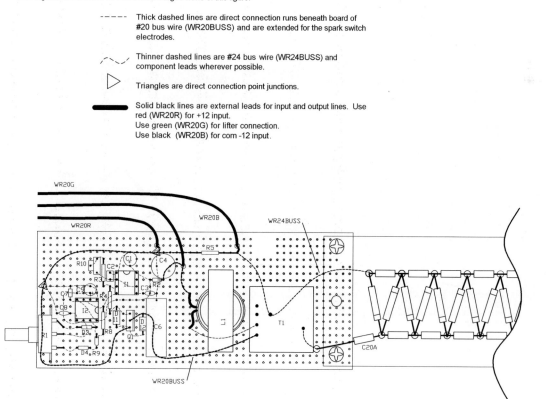

Figure 22-6 *Wiring connections and external leads*

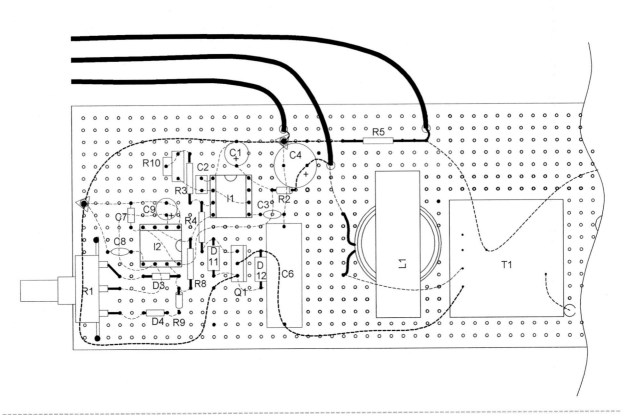

Figure 22-7A   *Enlarged view of the assembly board*

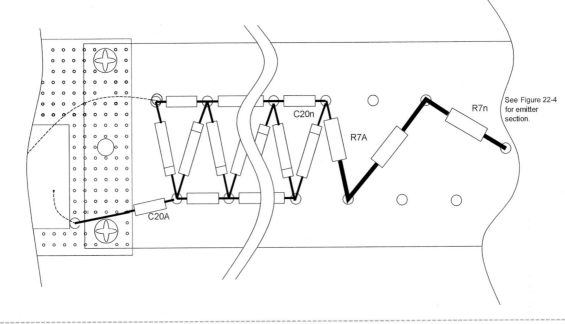

Figure 22-7B   *Enlarged view of the high-voltage section of the assembly board*

4. Connect the input to the power converter and the midsection of the multiplier section to the 25-megohm load resistor. Connect oscilloscope to drain pin of Q1 and set it to read 100 volts with a sweep time of 5 *microseconds* (usecs).

5. Apply power and quickly adjust R1 to the wave shape shown in Figure 22-2. The spark gap may fire intermittently and should be respaced just to the point of triggering. This is usually between 25 to 30 Kv.

6. Rotate R10 *counterclockwise* (CCW) and the input current smoothly drops almost to zero. This control varies the ratio of off to on time and nicely controls the system current to the ion emitters.

If you have access to a high-voltage probe meter, such as a B&K HV44, it will be possible to measure the direct output, noting 20 to 30 Kv across the 25-meg load resistor. This equates to over 30 watts! You will see a smooth change in output as R10 is varied.

Also note that the output voltage indicates only half the output value and is also heavily loaded by the load resistor along with the rest of the system. In actual usage in the intended ion and charge gun, the input current will be adjusted low by the setting of R10 as excessive power is not necessary.

Also, do not continually allow a hard spark discharge as circuit damage can occur.

Mechanical Assembly

To put the rest of the device together, follow these steps:

1. Create the main enclosure (EN1) from a 15-inch section of 2 3/8-inch schedule 40 PVC tubing, as shown in Figure 22-8. Note the 2 1/2-inch clearance holes accessing the retaining screw on LUG6 that are necessary to secure the emitter selection and passage of the leads to the HA1 handle section. You may want to use a piece of 2 1/4-inch OD clear plastic tubing with a 2-inch ID for this piece as it allows viewing of the circuit innards and can be impressive if your assembly is neat and orderly.

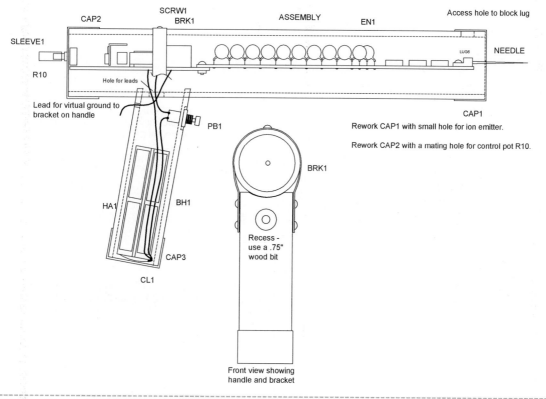

Figure 22-8 *X-ray view on circuit innards*

2. Build the handle section HA1 from a 6-inch piece of 1.9-inch schedule 40 PVC tubing. Note the contour that is filed out to fit the curvature of the enclosure section is at a slight angle, providing a gun-like look. You will need a small hole for the passage of the BRK1 grounding lead and a hole for the pushbutton switch PB1. The hole will require recessing using a 3/4-inch wood bit because the width of the handle tubing is too wide to allow adequate clearance for the securing nut.

3. Create the bracket, BRK1, from a 10-inch length of 1/2-inch .035 aluminum strip and shape as shown in Figure 22-8. Drill holes for the screws (SCRW1).

4. Assemble everything as shown in Figure 22-9, reading all the data in the figure.

Operation and Applications

The unit's output will be a soft, bluish flame forming at the emitter point. Ions are produced by charge concentration occurring at the end of the ion emitter. In order to be optimized, a return path to ground is necessary and is provided by a conductive hand grip connected to the common line of the circuit. The user now creates the ground return or electrical image necessary for enhancing the charge and ion mobility.

Control of the system is done via a pushbutton switch. This switch can easily be modified or changed to suit the user's needs. A low-current, spring-loaded push button is shown.

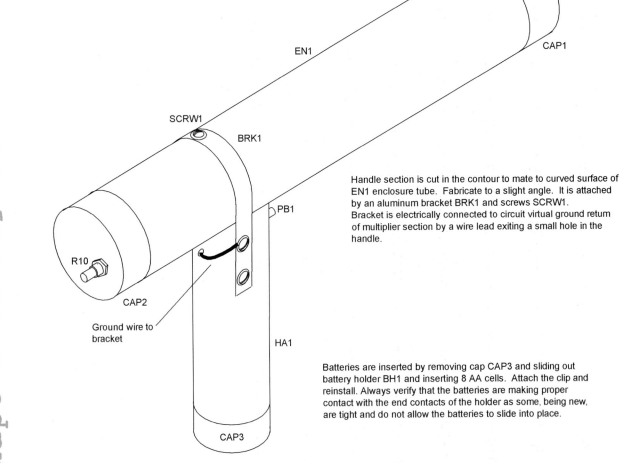

Handle section is cut in the contour to mate to curved surface of EN1 enclosure tube. Fabricate to a slight angle. It is attached by an aluminum bracket BRK1 and screws SCRW1. Bracket is electrically connected to circuit virtual ground return of multiplier section by a wire lead exiting a small hole in the handle.

Batteries are inserted by removing cap CAP3 and sliding out battery holder BH1 and inserting 8 AA cells. Attach the clip and reinstall. Always verify that the batteries are making proper contact with the end contacts of the holder as some, being new, are tight and do not allow the batteries to slide into place.

Figure 22-9   *Ion ray and charge gun*

Power to the unit is via an eight AA cell battery pack fitted in the handle. This approach provides a neat compact unit. Rechargeable batteries may be used, utilizing a built-in charging circuit for those applications where constant use is required.

Your ion ray gun demonstrates an interesting phenomenon involving the mobility of charged particles. It is capable of producing the following effects:

- Inducing electrical shocks in other people
- Causing lamps to flicker and ignite without contact
- Causing paper to stick to surfaces, playing cards, and so on
- Causing motion of objects/ion motors
- Charging of objects to a high potential without contact
- Static electricity experiments
- Kirlian photography and ozone production
- Strange and bizarre effects on certain materials

- Effects on painted and insulated surfaces that require darkness
- Effects on vapors, steam, and liquids
- Visual discharge of plasma force, corona, and so on
- Effect on electronics equipment, TV, and computers—use caution

The device accomplishes all the previous effects without any direct connections other than the traveling of ions through the air. In order to demonstrate this effect, it is necessary to produce voltages of magnitudes that may be at a hazardous shock potential but at a relatively small amperage. Even though the device is battery operated with low-input voltage, it must be treated with caution. Use discretion when using, as it is possible for a person wearing insulated shoes to accumulate enough of a charge to produce a moderately painful or irritating shock when he touches a grounded object. The effect could cause injury to a person in weak physical condition (note

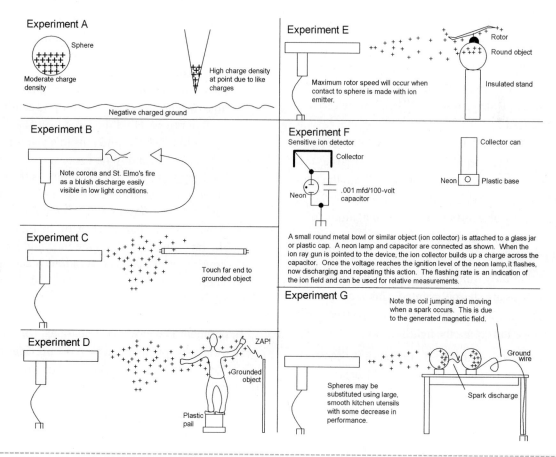

Figure 22-10   *Experiments*

---

the warnings). The effects depend on many parameters, including humidity, leakage amounts, types of objects, and proximity.

The device can be used in two ways. When the output is terminated into a large, smooth-surface collector such as a large metal sphere or oblate, it becomes a useful high-potential source capable of powering particle accelerators and other related devices. It may be built as a producer of negative or positive ions, demonstrating a phenomenon that is often regarded as a demerit when building and designing high-voltage power supplies.

The device is then terminated into a sharp point where the leakage of positive or negative ions can occur. This will result in corona and the formation of nitric acid via the production of the ozone produced when combining with nitrogen and forming nitrous oxide, which, with water, produces this strong acid. The production of ions as leakage also robs the available current from the supply.

Experiments

The following experiments are graphically shown in Figure 22-10. It is well known that high-voltage generators usually consist of large, smooth-surface collectors where leakage is minimized, allowing these collective terminals to accumulate high voltages with less current demand. Leakage of a high voltage point is the result of the repulsion of similar charges to the extent that these charges are forced out into the air as ions. The rate of ions produced is a result of the charge density at a certain point. The magnitude of this quantity is a function of voltage and the reciprocal of the angle of projection of the surface.

Note that Experiment A in the figure shows why lightning rods are sharply pointed. This causes the charges to leak off into the air before a voltage can be developed to create the lightning bolt.

It is now evident that to create ions it is necessary to have a high voltage applied to an object such as a needle or another sharp device used as an emitter. Once the ions leave the emitter, they possess a certain mobility that allows them to travel moderate distances, contacting and charging up other objects by accumulation and collision.

Experiment B shows St. Elmo's Fire. This glow discharge occurs during periods of high electrical activity. It is a corona discharge that is brush-like, luminous, and often may be audible when leaking from charged objects in the atmosphere. It occurs on ship masts, aircraft propellers, wings, and other projecting parts, as well as on objects projecting from high terrain when the atmosphere is charged and a sufficiently strong electrical potential is created between the object and the surrounding air. Aircraft most frequently experience St. Elmo's fire when flying in or near cumulonimbus clouds, in thunderstorms, in snow showers, and in dust storms.

Experiment C shows a flashing fluorescent or neon light. This experiment demonstrates the mobility of the ions and their ability to charge up the capacitance in a fluorescent light tube and discharge in the form of a flash. For this experiment, perform the following steps:

1. Have a friend carefully hold a 10- to 40-watt fluorescent light or neon-filled tube and turn off the lights. Allow your eyes to become accustomed to the total darkness.

2. Hold the end about 3 feet from the output of the ion ray gun and note the lamp flickering. Increase the distance and note the flicker rate decreasing. Under ideal conditions and total darkness, the lamp will flicker to a considerable distance from the source. Use caution in total darkness. Hold the lamp by a glass envelope and touch the end pins to a water pipe, metal objects, and so on for best results and the brightest flash. The flash time is the equivalent of the equation $T = CV/I$, where T is the time between flashes and V is the flash breakdown voltage characteristic of the tube. C is the inherent capacity in the tube and I is the equivalent of the amount of ions reaching the lamp and obviously decreases by the $5/2$ power of the distance.

Experiment D concerns ion charging. This demonstrates the same phenomena as in the previous experiment but in a different way. Perform the following steps:

1. Set up the unit as shown with a ground contact about $1/4$ of an inch from the charge sphere.

2. Note the spark occurring at the grounded contact as a result of the ion accumulation on the sphere. Increase the distance and note where the spark becomes indistinguishable.

3. Obtain a subject brave enough to stand a moderate electric shock (use caution as a person with a heart condition should not be near this experiment).

4. Have the subject stand on an insulating surface and then touch a grounded or large metal object. Shoes with rubber soles often work to an extent.

Experiment E shows an ion motor. This dramatically demonstrates Newton's law of action producing a reaction. Escaping ions at a high velocity produce a reactive force. This is a viable means of propulsion for a spacecraft where hyper-velocities may approach the speed of light in this frictionless environment. Here you will work with a rotor and pin attached to a sphere via a lump of clay or something similar. A piece of folded paper will sometimes work.

Form a piece of #18 wire as shown. For maximum results, carefully balance and provide minimum friction at the pivot point. There are many different methods of performing this experiment with far better results. We leave this to the experimenter, bearing in mind that a well-made, balanced rotor can achieve amazing rpms. Note that as the rotor spins, giving off ions, one's body hair will bristle, nearby objects will spark, and a cold feeling will persist.

Experiment F shows accumulated ions on the insulated spherical object, charging it theoretically to its open circuit potential (this in practice doesn't occur due to leakage). The object accumulates a voltage equal to $V = it/c$. Note that the unit is also directly grounded to increase this effect by producing the necessary electrical mirror image. The quantity Q (coulombs) of the charge is equal to CV where C equals the capacitance of the object and V equals the voltage charged. The energy W (joules) stored is equal to ½ *capacitance* × voltage squared (CVE2). The capacitance can be calculated by approximating the area of the object's shadow projected directly beneath it and calculating the mean separation distance. The capacitance is now approximately equal to .25 times the projected area in square inches, divided by the separation in inches.

Also regarding Experiment F, note that the IOD1 ion detector described in the Information Unlimited catalog is an excellent device and provides extraordinary sensitivity. The ESCOPE electroscope also is an excellent detector.

Experiment G demonstrates the transmission of energy via mobile ions. The objects used here are round spheres placed on glass bottles used as insulators. One object is grounded using thin, insulated wire. Another object discharges to the grounded object. Note that the grounding wire physically jumps at the time of discharge; these phenomena are the result of current producing a mechanical force. As the unit is brought closer, the length of the spark discharge and the discharge rate will increase. A discharge length of a half-inch may be obtainable with the unit 4 to 5 feet away. This demonstrates the potential effectiveness of the device.

Other experiments and uses would be materials and insulation dielectric breakdown testing, ozone production for odor control, X-ray power supplies, and capacitance charging using a Leyden jar. Other uses include the ignition of gas tubes and spark gaps, particle acceleration and atom smashers, Kirlian photography, electrostatics, and ion generation. Other related material may easily be obtained on these subjects.

For example, an experiment on charge attraction demonstrates the force between *unlike* charges. Place an 8- × 11-inch piece of paper on a wooden desk or tabletop. Scan the paper with the unit approximately 2 to 3 inches from its surface. Note the paper pressing to the surface and becoming strongly attracted as indicated when attempting to lift it up.

An experiment on change repulsion would demonstrate the force between *like* charges. For example, place a small paper cup on top of the output. Obtain some small pieces of Styrofoam and place them in the cup. Note that some of the pieces fly out of the cup. Bring a grounded lead near the cup and note the reaction.

The effects on many materials can supply hours of interesting experiments, producing sometimes weird and bizarre phenomena. If you discover or happen to come up with any new experiments or data, contact us at www.amazing1.com.

Negative Ion Information

In the last two decades, a medical controversy has evolved pertaining to the beneficial effects of these minute electrical particles. As with any device that appears to affect people in a beneficial sense, there are those who sensationalize and exaggerate these claims as a cure for all ailments and ills. Such people manufacture and market these devices under false pretenses and consequently give the products a bad name. The Food and Drug Administration now steps in on these claims and the product, along with any beneficial facets, goes down the tubes.

People are affected by negative ions from the property of these particles to increase the rate of activity by cilia (whose property is to keep the trachea clean from foreign objects), thus enhancing oxygen intake and increasing the flow of mucous. This property neutralizes the effects of cigarette smoking, which slows down this activity of the cilia. Hay fever and bronchial asthma victims are greatly relieved by these particles. Burns and surgery patients are relieved of pain and heal faster. Tiredness, lethargy, and a general feeling of fatigue are replaced by a sense of well-being and renewed energy. Negative ions destroy bacteria and purify the air with country air freshness. They cheer people up by decreasing the serotonin content of the blood. As can be seen in countless articles and technical writings, negative ions are a benefit to man and his environment.

Negative ions occur naturally from static electricity, certain kinds of wind, waterfalls, crashing surf, cosmic radiation, radioactivity, and ultraviolet radiation. Positive ions are also produced from some of the previous phenomena and usually neutralize each other out as a natural statistical occurrence. However, many man-made objects and devices have a tendency to neutralize the negative ions, thus leaving an abundance of positive ions, which create sluggishness and most of the opposite physiological effects of its negative counterpart.

One method of producing negative ions is obtaining a radioactive source rich in Beta radiations (electrons). Alpha and gamma emissions from this source produce positive ions that are neutralized electrically. The resulting negative ions are electrostatically directed to the output exit of the device and are fur-

ther dispersed by the action of a fan (this method has recently come under attack by the Bureau of Radiological Health and Welfare) for the use of tritium or other radioactive salts. This approach appears to be the most hazardous one according to the product consumer safety people.

A more accepted method is to place a small tuft of stainless steel wool as the ion emitter at the output terminal of a negative high-voltage DC power supply. The hair-like property of the stainless steel wool allows ions to be produced at a relatively low voltage yet with reduced ozone output. Ions are produced by the leakage of the particles charging air molecules in the immediate vicinity of the steel wool emitter. The unit should be operated below 15 Kv as too much voltage can produce substantial amounts of ozone that can mask the beneficial effects of the increased ions obtained.

Special Notes

The ion emitter plays an important function in the proper operation of this unit for the particular application. It is suggested that you use a small, sharp, stainless steel needle for starters.

The average current is about 500 microamps and can be considered constant for most loads. It charges objects of an electrical capacitance (C) to a voltage by the following formula: $V = it/c$, where t equals the time in seconds and c equals the capacitor of the object in farads. The average human body usually equates out to 10 to 20 *picofarads* (pfd). Objects of a larger capacity could mathematically be charged up to dangerous energy amounts if they were well insulated. Consequently, this must be taken into consideration.

Here's an example. An object of a capacity equal to .001 mfd, insulated up to 25,000 volts, would charge up to near this value in approximately $1/3$ of a second. This should equate out to .2 joules and can result in a painful electric shock, as anyone knows who has gotten across a charged capacitor.

An interesting phenomenon is that a human body on a dry day can accumulate a sufficient charge to cause a neon or fluorescent lamp to flash reasonably brilliantly when contact is made with a grounded object.

Table 22-1 Ion ray and charge gun project parts list

Ref. #	Qty.	Description	DB #
R1		10K trimpot vertical	
R2, 4	2	10-ohm, $\frac{1}{4}$-watt resistor (br-blk-blk)	
R3, 5, 8, 9	4	1K, $\frac{1}{4}$-watt resistor (br-blk-red)	
R7	3	10-megohm, 1-watt resistor (br-blk-bl)	
R10/S1		10K pot and switch	
C1		100 mfd/25-volt vertical electroradial leads	
C2		.0022 mfd/50-volt green plastic cap (222)	
C3, 8	2	.01 mfd/50-volt disk (103)	
C4		1000 mfd/25-volt vertical electrolytic capacitor	
C20a–n	20	500 pfd, 10 Kv ceramic disk cap	#500P/10KV
C6		.22 mfd/250-volt metallized polypropylene	
C9		1 mfd, 25-volt vertical electro cap.	
C7		.1 mfd, 50-volt cap	
D20a–n	20	16 Kv, 5-milliampere avalanche diodes	#VG16
D3, 4	2	IN914 silicon diodes	
D11	1	PKE15 15-volt transient suppressor	
D12		1N4937 fast-switching 1 Kv diode	
Q1		IRF540 *metal-oxide-semiconductor field effect transistor* (MOSFET) TO220	
I1, 2	2	555 DIP timer	
T1		Mini-switching transformer. 7 Kv, 10 milliampere	#IU28K089
L1	1	6 Uh inductor; see text on assembly	#IU6UH
CL1		Battery clip #22 with 12-inch leads	
BH1		Eight AA cell battery pack	
PB1		Pushbutton switch (normal open)	
PERFBOARD		5 × 2.9 × .1 grid perforated board; cut to size per Figure 22-3	
PCGRA		Optional PCB	#PCGRA
POLYBOARD		10 × 2.9 × .063 polycarbonate (Lexan) plastic	
WR20R	12 inches	#20 vinyl red wire for positive input	
WR20B	12 inches	#20 vinyl black wire for negative input	
WR20G	12 inches	#20 vinyl green wire for output ground to craft return	
WR20BUSS	12 inches	#20 bus wire for light leads	
WR24BUSS	12 inches	#24 bus wire for light leads	
SCRW1	5	#6 × $\frac{3}{8}$ sheet metal screws	
SW1/NU1	3	#6-32 × $\frac{1}{2}$-inch screws and nuts	
HSINK		1.5- × 1-inch .063 AL plate fabricated as per Figure 22-4	
LUG6		#6 aluminum block lug	
LUG25		$\frac{1}{4}$-inch ring lug	
CLIP		Small alligator clip, duckbill type	
EN1		15 × 2 $\frac{3}{8}$ OD × $\frac{1}{8}$ wall PVC or clear plastic tube	
HA1		6- × 1 $\frac{7}{8}$-inch OD PVC schedule 40 tube	
BRK1		10 × $\frac{1}{2}$ × .035 aluminum strip fabricated as shown	
CAP1, 2	2	2 $\frac{3}{8}$-inch plastic caps fabricated as shown with holes	
CAP3		1 $\frac{7}{8}$-inch plastic cap	
Optional items			
B1-8	8	AA alkaline cells, 1 $\frac{1}{2}$ volts	
NEEDLE		Stainless steel	
CARBONFIB		Carbon fiber hair for high-output emitters	#CARBFIB

Chapter Twenty-Three

See-in-the-Dark Project

This useful and interesting project shows how to build a device capable of seeing in total darkness. Unlike conventional devices requiring the minute light from the stars or other ambient background light, this system contains its own infrared source, allowing covert viewing of the desired subject (see Figure 23-1).

Assembly is shown in two parts, the high voltage power supply and the final enclosure with optics and an illuminator. Expect to spend $50 to $100 for this useful infrared imaging system with all the specialized parts available from www.amazing1.com.

General Description

This project shows how to construct a device capable of allowing one to see in total darkness. It can be used to view a subject for recognition or evidence-gathering reasons without any indication to the target subject that he or she is under surveillance. It is an invaluable device when used for detection, the alignment of infrared alarms, invisible-laser gun sights, and in communications systems. This technology can also be used to detect diseased vegetation in certain types of crops from the air, to serve as an aid to nighttime varmint hunting, and to view high-temperature thermographic scenes where heat is used to produce the image. This device is excellent for use with the infrared laser described in Chapter 9, "Handheld Burning Diode Laser Ray Gun," with a performance that is as good operationally as units that cost much more.

The unit is built using readily available parts for the enclosure and basic optics. The batteries are enclosed into the housing and do not require side packs, cables, and so on. The range and field of viewing are determined by the intensity of the integrated infrared source and the viewing angle of the optics. Readily available and low-cost optics are usable, but they may have spherical aberration and other adverse effects. This approach keeps the basic cost down for those not requiring actual viewing of detailed scenes. Improved optics will eliminate these effects and can be obtained at most video supply houses as an option.

Assembly focuses around common *polyvinyl chloride* (PVC) tubing as the main housing and a specially designed, patented, miniature power source for energizing the image tube. The tube is a readily available image converter being used by most manufacturers of similar devices. This tube establishes the limits of viewing resolution and is suitable for most applications but may be limited if one desires video perfection.

The viewing range is determined mainly by the intensity of the infrared source and can be controlled by varying this parameter. Our basic unit is shown

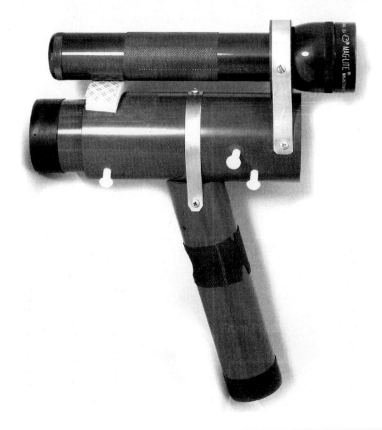

Figure 23-1 *See-in-the-dark viewer*

utilizing a 2-D cell flashlight with an integrated filter placed over the lens to prevent the subject from seeing the source. This provides a working range of up to 50 feet (reliably) and can be increased to several hundred using a more powerful source such as a 5 to 6 cell flashlight. Needless to say, the builder can choose his or her infrared source and adjust the optics to meet his or her needs. Infrared *light-emitting diodes* (LEDs) or lasers, as described in our catalog, are also good illumination sources. Long-range, quick viewing may utilize a small, two-cell light with eight *nickel-cadmium* (NiCad) AA cells to replace the normal two D cells, providing a significantly brighter infrared source yet lasting for less time than the normal D cell would.

The unit can also be operated using external sources such as super-intense Q-beam handheld lamps with an added filter extending the range out to 400 to 500 feet, providing a wide field of illumination. Note the viewing of active infrared sources such as lasers does not require the internal infrared source.

An optional long-range, infrared illuminator for viewing up to 300 feet is available. See #HLR10 at www.amazing1.com and optional equipment can be found on the parts list at the site.

Basic Theory

A subminiature high-voltage power supply produces approximately 15 Kv at several hundred microamperes from a 7- to 9-volt rechargeable *nickle cadmium* (NiCad) or alkaline battery. This voltage is applied to the tube (IR16) with the "plus" going to the viewing end and the "negative" to the objective end. A focus voltage is taken from a tap in the multiplier circuit and is approximately $1/6$ of the total potential.

An objective lens (LENS1) with an adjustable focal length gathers the reflected image, illuminated by the infrared lens, and focuses this image at the objective end of the tube. Image conversion now takes place inside the tube and is displayed on the viewing screen of the tube in a greenish tinge. The viewing resolution is usually adequate to provide subject identification at a distance of 50 feet or more depending on the intensity of the infrared source and the quality of the optics.

Circuit Description

Transistor Q1 is connected as a free-running resonant oscillator with a frequency determined by the combination resonance of capacitor C3 and the primary winding of the stepup transformer, T1. This oscillating voltage is stepped up to several thousand in the secondary winding of T1. Capacitors C4 through C15, along with diodes D1 through D12, form a full-wave voltage multiplier where the output is multiplied by six and is converted to *direct current* (DC). Output is taken between C5 and C15, as shown, and may be either positive or negative depending on the direction of the diodes. Different values of voltage may be obtained at various taps of the capacitors. Figure 23-2 shows the connections for the taps to the image tube.

The base of Q1 is connected to a feedback winding of T1 where the oscillator voltage is at the proper value to sustain oscillation. Resistor R2 biases the base into conduction for the initial activation. Resistor R1 limits the base current, wheras capacitor C2 speeds up the deactivation of Q1 by supplying a negative bias and capacitor C1 bypasses any high-frequency energy. The input power is supplied through switch S1 via a "snap-in" battery clip.

Circuit Assembly

To put the circuitry together, follow these steps:

1. Lay out and identify all the parts and pieces, checking them with the parts list. Note that some parts may sometimes vary in value. This is acceptable as all components are 10 to 20 percent tolerant unless otherwise noted. A length of bus wire is used for long circuit runs.

2. Create the PB1 perforated circuit board as shown in Figure 23-3. Enlarge the holes as follows:

    Thirteen $1/16$-inch holes for the junctions of the diodes and the capacitors in the multiplier

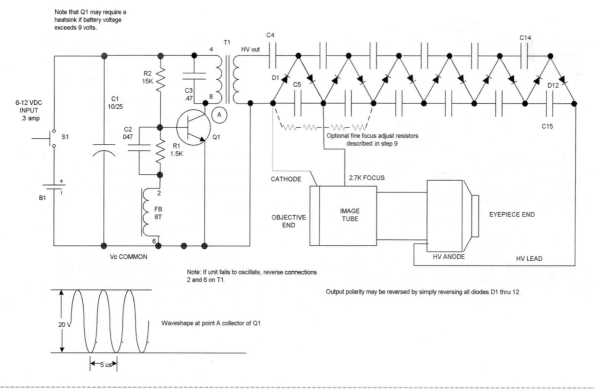

Figure 23-2 *Power supply schematic*

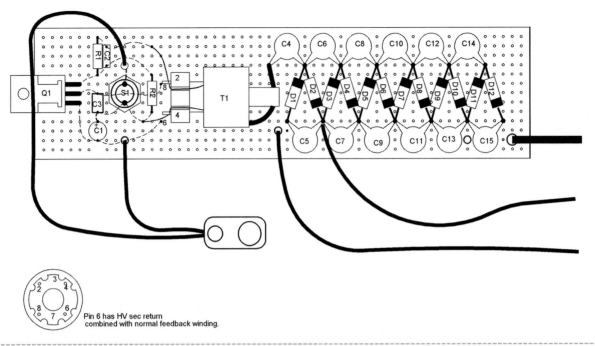

Figure 23-3 *Assembly board*

Seven $\frac{1}{8}$-inch holes for mounting the switch (S1) and the external connection leads

The switch is shown attached to the assembly board but may be remotely located using interconnecting leads.

3. Assemble the board as shown in Figure 23-3. Start to insert components into the board holes as shown. Note to start and proceed from right to left, attempting to obtain the layout as shown.

Certain leads of the actual components will be used for connecting points and circuit runs. Do not cut or trim at this time. It is best to temporarily fold the leads over to secure the individual parts from falling out of the board holes for now.

Note that the solder joints in the multiplier section, consisting of C4 through C15 and D1 through D12, should be globular shaped and smooth to prevent high-voltage leakage and corona. The solder globe size is that of a BB. Run your fingers over the joints and verify the absence of sharp points and protrusions.

Also note that T1 is laying on its side and uses short pieces of bus wire soldered to its pins as extensions for connections to the circuit board.

Circuit Board Testing

To test the circuit board, follow these steps:

1. Separate high-voltage output leads approximately 1 inch from one another.

2. Connect 9 volts to the input and note a current draw of approximately 150 to 200 milliamperes when S1 is pressed.

3. Decrease the separation of the high-voltage leads until a thin, bluish discharge occurs, usually between $\frac{1}{2}$ to $\frac{3}{4}$ of an inch. Note the current input increasing. The increased value depends on the length of the spark, corona, and so on, but should not exceed 300 milliamperes.

4. Check the collector tab of Q1 and add a small heatsink if too hot to touch. A heatsink tab is shown in Figure 22-4 (Chapter 22).

For those with a scope, it may be interesting to note the wave shape at the collector tab, as shown in Figure 23-2. Note this is without any sparking occurring.

Note the takeoff point for the focus lead. This point is approximately at $\frac{1}{6}$ the output voltage. The unit may be powered up to 12 *volts-direct current*

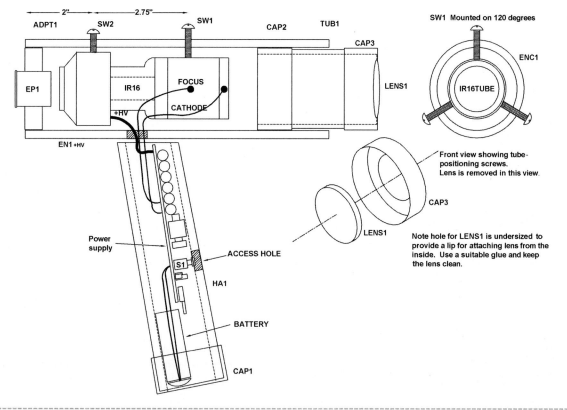

Figure 23-4   *X-ray view showing innards*

(VDC) but will positively require a heatsink on the tab of Q1.

This unit is capable of producing 10 to 20 Kv from a small, standard 9-volt battery. It is built on a *printed circuit board* (PCB) or a small piece of perforated circuit board and can easily be housed or enclosed, as the application requires. Applications include powering image converter tubes for night vision devices, ignition circuits for flame-throwing or -producing units, capacitor charging for energy storage, shocking electric fences, insect eradication, Kirlian photography, ion propulsion electric field generators, ozone producing, and more.

Fabrication and Mechanical Assembly

To begin the assembly of the device's machinery, follow these steps:

1. It is assumed the power board as outlined is properly operating. Check for the absence of corona in the high-voltage section. Corona dope is a coating that reduces electrical leakage. Remove all sharp points and insulate with corona dope and so on.

2. Take a window screen and place it flush against the objective end of the image tube, TUB1, with a piece of clear scotch tape. Secure the tube on the bench via modeling clay and temporarily connect it to the leads from the power board, as shown in Figure 23-2. Observe the proper clearance of the leads and components. Darken the room and place a source of infrared filter light pointing toward the tube. (Use a flashlight preferably with an IR filter.) Note the tube glowing greenish and an image of the screen appearing either sharp or blurred. If the image is good and sharp, you are in luck. You may further improve the focusing by adding the 22-megohm resistors as shown in Figure 23-2. This is usually not necessary.

3. Fabricate EN1 from a 7-inch length of 2 3/8-inch ID schedule 40 PVC tubing. Note the hole adjacent to the HA1 handle for feeding high-voltage wires to the tube from the power

board and ¼-20 threaded holes are dimensioned in Figure 23-4 for securing and centering the image tube. These holes are located on a 120-degree radius.

4. Fabricate the HA1 handle from an 8-inch length of 1½-inch ID schedule 40 PVC tubing. The tube must be shaped and fitted where it abuts to the EN1 main enclosure.

5. Fabricate the BRK 1 and 2 brackets from a half-inch-wide strip of 22-gauge aluminum as shown. Note the holes for #6 × ¼ sheet metal screws for securing the assembly together.

6. Fabricate TUB1 from a 3 ½-inch length of 2-inch ID schedule 40 PVC tubing for the objective lens. Note this is only 2 inches long when using the optional optics and "C or T" mount adapter fitting.

7. In order for TUB1 to telescope into the main enclosure EN1, suitable cylindrical shims, CAP2 and CAP3, must be fabricated. These are the 2 ⅜-inch plastic caps. CAP2 has its end removed by cutting out the center using the wall of tubing as a guide for the knife. CAP3 has a smaller section cut out for LENS1. This method is cheap and works reasonably well. You obviously could substitute the pieces with properly fitted parts fabricated from aluminum or plastic if you desire. This approach is more professional looking but can be much more costly.

8. The lens shown is a simple, uncorrected convex that is adequate for most infrared source viewing. It is not a quality viewing lens such as the optional 50 mm wide-angle or 75 mm telephoto with the C mount threads. When using this lens, you should either create or purchase an adapter ring that will adapt to the lens threads and fit snugly into the enclosure. See CMT1.

9. The IR16 image tube has preconnected leads. The negative short lead attached to the objective end must have a 10-inch lead spliced to it. Insert the tube partway into the enclosure and snake the leads through the access hole. Position the tube and gently screw in the retaining screws by hand to secure and center it.

10. Connect the leads from the tube to the power board as shown.

11. Insert the power board into the HA1 handle. You will have to determine the access hole and drill for the switch S1 once the board is secured in its final position. Wires should be long enough for the complete removal of the assembly when the handle is secured in place via the BRKI bracket. This allows any preliminary adjustment or service. Leads may be shortened once proper operation is verified. Connect the battery to the power board and energize switch SI. If you did your homework, you will not have to readjust the focus taps or divider values. Once the operation is verified, check for any excessive corona and eliminate it. Position the board to switch S1 adjacent to the access hole in the handle. It may be necessary to further secure the board in place via foam rubber pieces, a *room temperature vulcanizing* (RTV) adhesive, and so on. Slide a flexible rubber membrane over the access hole and insert the battery and cap CAPI.

12. Finally, assemble everything as shown in Figure 23-5 and mount the infrared filtered flashlight. You will have to seal any light leaks using plumbers' "monkey dung" or coax seal.

13. Adjust the objective and then the eyepiece for the clearest image.

Special Notes

The unit is shown with a built-in infrared source consisting of a common, everyday two-cell flashlight fitted with a special infrared filter. Any visible light leaks must be sealed with electrician's gunk, coax seal, or black liquid rubber.

This approach allows total flexibility in viewing sources not requiring infrared illumination as the light need not be energized or may even be removed. The light source may also be intensified by replacing the two D cells with an eight AA cell NiCad pack providing approximately 9 volts. A suitable lamp may be substituted, providing several times more illumination. The lamp and battery life will be greatly

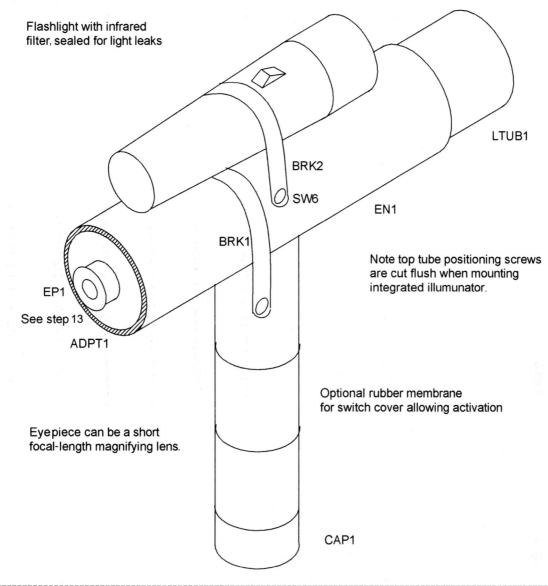

Flashlight with infrared filter, sealed for light leaks

LTUB1

BRK2

SW6

EN1

BRK1

Note top tube positioning screws are cut flush when mounting integrated illumunator.

EP1

See step 13

ADPT1

Optional rubber membrane for switch cover allowing activation

Eyepiece can be a short focal-length magnifying lens.

CAP1

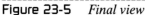

Figure 23-5   *Final view*

reduced as this approach is only intended for intermittent use. Note the now available halogen lamps are far more intense and make excellent infrared sources.

Longer-range viewing may be accomplished by using other, more intense sources such as higher-powered lights, auto headlamps, and so on. These must be fitted with the proper filters to be usable. A range of several hundred meters may be possible with these higher-powered sources. A source capable of allowing viewing from up to 500 feet is referenced in the project parts list.

Obtaining maximum performance and range from the system may require the optional lens system specified. The viewing of externally illuminated infrared sources will not require the integral infrared source.

You will note that this device is excellent for viewing the output of most solid-state, gallium arsenide laser systems, LEDs, or any other source of infrared energy in the 9000 A spectrum. No internal infrared source is necessary when viewing these actual sources.

Table 23-1 See-in-the-dark project

Ref. #	Qty.	Description	DB #
R1		1.5K, $\frac{1}{4}$-watt resistor (br-gr-red)	
R2		15K, $\frac{1}{4}$-watt resistor (br-gr-or)	
C1		10 m/25-volt electrical vertical capacitor (blue or green can)	
C2		.047 m/50-volt plastic capacitor (473)	
C3		.47 m/100-volt plastic capacitor (474)	
C4–15	12	270 pfd/3 Kv plastic disk capacitors	
D1–12	12	6 Kv, 100-nanosecond high-voltage avalanche diodes	
Q1		MJE3055 NPN TO 220 case transistor	
T1		Special transformer info #28K077	#IU28K077
S1		Pushbutton switch	
PB1		5 $\frac{1}{2}$- × 1 $\frac{1}{2}$-inch perforated board with .1 × .1 grid	
CL1		Snap battery clip	
WR22	2	24-inch length of #22 vinyl hookup wire	
WRHV20		12-inch, 20 Kv silicon wire	
IR16		Image converter tube	#IUIR16
EN1		8- × 2 $\frac{3}{8}$-inch schedule 40 gray PVC tube, created as shown	
TUB1		3 $\frac{1}{2}$-inch length × 2-inch ID schedule 40 gray PVC	
BRK1, 2	2	9- × $\frac{1}{2}$-inch-thin aluminum strip as shown	
CAP1		2-inch plastic cap for handle	
CAP2, 3	2	2 $\frac{3}{8}$-inch plastic cap as shown	
LENS1		45 × 63 double convex glass lens	
SW1, 2	6	$\frac{1}{4}$-20 × 1-inch nylon screws	
SW6	6	#6 × $\frac{1}{4}$-inch sheet metal screws	
Optional parts			
PCPBK		PCB	#PCPBK
CMT1		Prefabricated C mount adapter for EN1 enclosure	
EP1		Small eyepiece	
FIL6		6-inch glass infrared filter, 99.99 percent dark, for Q beam light	
HLR10		200,000-candle-power infrared illuminator invisible to the naked eye, at 12 VDC	#HLR10

Chapter Twenty-Four

Fish Stunning and Wormer Project

This useful project is intended used for driving worms out of the ground or stunning fish while in their normal environment. This stupefying effect allows tagging and relocating as the fish float to the surface. It is best to check before trying this in public waters, as many states do not allow "electric fishing," except by those qualified.

The device (see Figure 24-1) must be used with caution as improper use can cause electric shock.

The finished project should have a danger label affixed to it.

Expect to spend $50 to $100 for this outdoor project. Most parts are readily available, and any specialized parts are obtainable through www.amazing1. com. The parts for this project are listed in Table 24-1.

General Description

As shown in Figure 24-2, this project generates an adjustable 1 to 2 joules of 600-volt pulses at a 30-reps-per-second rate. The output of the circuit is electrically floating (no ground reference) to minimize but not totally eliminate a dangerous shock potential. The output leads are intended to be connected to probes or drag chains, as shown in the operating instructions.

The device is housed in a 3-inch *polyvinyl chloride* (PVC) tube with splash-proof plastic end caps. The front cap retains the control panel with power switch (S1), pulse energy pot (R5), *light-emitting diode* (LED) power on the indicator, and the input power leads. The rear cap has a passage hole for exiting the output leads. This arrangement helps protect the device from moisture but in no way makes the unit submersible.

Figure 24-1 *Fish shocker*

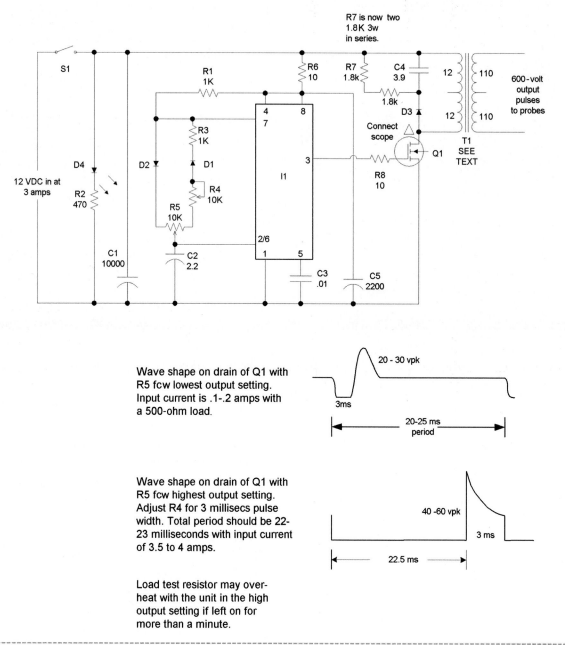

Wave shape on drain of Q1 with R5 fcw lowest output setting. Input current is .1-.2 amps with a 500-ohm load.

Wave shape on drain of Q1 with R5 fcw highest output setting. Adjust R4 for 3 millisecs pulse width. Total period should be 22-23 milliseconds with input current of 3.5 to 4 amps.

Load test resistor may overheat with the unit in the high output setting if left on for more than a minute.

Figure 24-2 *Circuit schematic*

Circuit Description

The circuit utilizes inductive charging similar to that used in the ignition systems of older automobiles. The primary winding of the transformer TI current charges through *metal-oxide-semiconductor field effect transistor* (MOSFET) switch Q1. The current ramps up to a value determined by i = Et/L, where E equals the 12 *volts-direct current* (VDC) input, L the

inductance of the T1 primary, and t the "on" time of switch Q1. The pulse energy is now equal to $Li^2/2$ because it is being controlled by the "on" time as determined by pot R5.

Transformer T1 requires cutting in an air gap necessary to store the inductive energy because the core itself would saturate, making T1 useless.

Timer I1 is wired as a stable pulse generator with a fixed frequency determined by the total value of pot R5 and trimpot R4, along with timing capacitor C2.

Trimpot R4 is used to set the maximum core charging time range of R5. Resistor R6 and capacitor C5 decouple the operating voltage (Vc) to timer I1 from the main 12 VDC. LED D4, along with the current-limit resistor, indicate when power switch S1 is on.

Construction Steps

To begin assembly, follow these steps:

1. Lay out and identify all parts and pieces, checking them with the parts list. Note that certain parts may sometimes vary in value. This is acceptable as all components are 10 to 20 percent tolerant unless otherwise noted.

2. Create the PB1 assembly board at $2^{1}/_{4} \times 5$ inches from a piece of .1 × .1 vector board. Note that it is a good idea to duplicate the placement of the perforated holes as shown in Figures 24-3 and 24-4. This makes placement of the components identical to what is shown.

3. Assemble the board as shown, inserting the components into the board holes. Proceed from right to left, attempting to obtain the layout as shown. Dashed lines indicate connection runs on the underside of the board.

Note that certain leads of the actual components will be used for connecting points and circuit runs. Do not cut or trim them at this time. It is best to temporarily fold the leads over to secure the individual parts from falling out of the board holes for now.

4. Rework the T1 transformer as shown in Figure 24-4.

5. Put the frame assembly (FRAME) together, as shown in Figure 24-5. You may want to trial-fit the components before actually fabricating.

6. Mount the components to the frame as shown in Figure 24-6, using TYE1 wraps to secure T1. Note the mounting of Q1, using a thermo pad and nylon screw. The large capacitor C1 is mounted behind the assembly board with the ground side attached to the common grounding lug LUG1.

7. Preconnect all the leads as shown in Figure 24-4. Note that the wires intended for input and output leads are 3 to 4 feet in length. They may be longer or shorter.

8. Check the wiring for any potential shorts, wire bridge shorts, poor solder joints, the correctness of the components, and the position and orientation of semiconductors and capacitors.

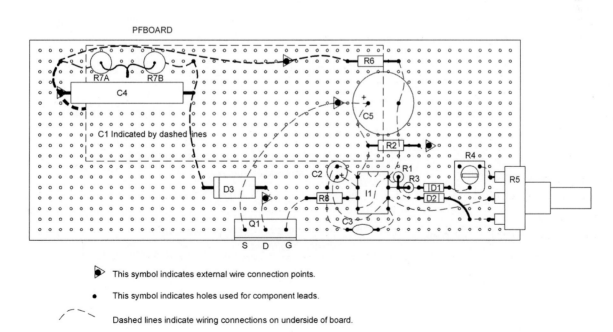

Figure 24-3   *Parts layout and wiring of assembly board*

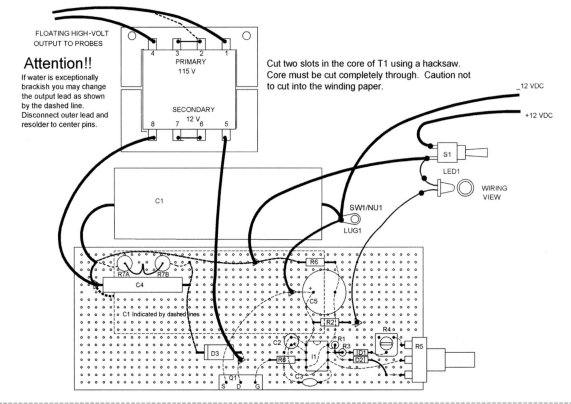

FLOATING HIGH-VOLT
OUTPUT TO PROBES

4   3   2   1
PRIMARY
115 V

SECONDARY
12 V
8   7   6   5

Attention!!

If water is exceptionally
brackish you may change
the output lead as shown
by the dashed line.
Disconnect outer lead and
resolder to center pins.

Cut two slots in the core of T1 using a hacksaw.
Core must be cut completely through. Caution not
to cut into the winding paper.

_12 VDC

+12 VDC

S1

LED1

WIRING
VIEW

C1

SW1/NU1

LUG1

R7A   R7B

C4

C1 Indicated by dashed lines

R6

C5
+

R2

R4

C2
R1   R3
R8   I1   D1
D2
R5

D3

C3

Q1
S   D   G

Figure 24-4   *Final wiring showing transformer rework*

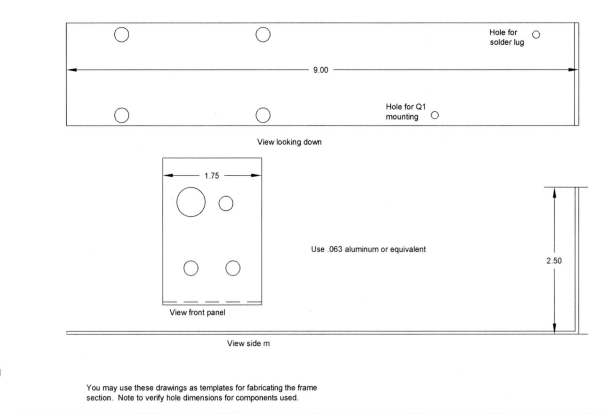

Hole for
solder lug

9.00

Hole for Q1
mounting

View looking down

1.75

Use .063 aluminum or equivalent

2.50

View front panel

View side m

You may use these drawings as templates for fabricating the frame
section. Note to verify hole dimensions for components used.

Figure 24-5   *Construction of the frame assembly template*

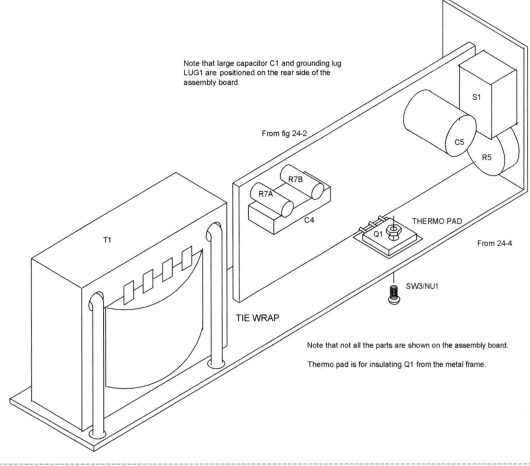

Note that large capacitor C1 and grounding lug LUG1 are positioned on the rear side of the assembly board.

From fig 24-2

THERMO PAD

From 24-4

SW3/NU1

TIE WRAP

Note that not all the parts are shown on the assembly board.

Thermo pad is for insulating Q1 from the metal frame.

Figure 24-6   *Isometric overall view*

Subassembly Pretest

To run a test on the device, follow these steps:

1. Turn pots R5 to full *counterclockwise* (CCW), R4 trimpot to midrange, and S1 to off (down position).

2. Connect a 10- to 25-watt, 500-ohm testing power resistor to the output leads. If you have a scope, it is suggested that you connect it to the drain of Q1.

3. Apply a 12-volt DC to the input leads, and turn S1 on (to the up position). Adjust R5 full *clockwise* (CW) and note that the input current does not exceed 2.5 amps. Adjust trimpot R4 to limit this maximum value with R5 at full CW. Check for wave shapes as shown with R5 at full CW and CCW. Do not operate into the

test resistor for too long as the resistor may overheat.

4. Verify D4 LED lighting when S1 is energized.

Final Assembly

To complete the assembly, follow these steps:

1. Cut a 10-inch piece of 3¹/₂-inch OD schedule 40 PVC tubing for the enclosure (EN1). Rework the 3¹/₂-inch plastic end caps by simply poking two small holes for the output leads (CAP1). Cut out the center of cap CAP2 by placing it onto a suitable form and remove the center with an x-acto knife using the wall of the form tube as a guide.

Fish Stunning and Wormer Project

2. Retest using a load resistor and verify all the controls before actually using the device. It is a good idea to attach suitable connection clips to the input leads for battery terminals.

3. Connect the output leads to a suitable probe scheme and study the instructions section of these plans.

4. Place a high-voltage warning label on the enclosure (see Figure 24-7).

Operation

Your fish stunner is intended for survey tagging and population evaluation of certain species. The system is designed to operate from a 12-volt battery and draws 3 amps at maximum output. The unit is shock circuit protected and utilizes our highly efficient induction charging and switching to obtain the high-peak currents necessary for the high conductivity often found in brackish waters. A power adjuster controls the duration of the pulse, therefore controlling the current flowing in the water. The pulse repetition rate is factory set at 25 pulses per second. The ratio of the pulse's "on" to "off" time is controlled by the power adjuster control. The output voltage with a load of 500 ohms is over 300 volts at its peak at a corresponding current of over half an amp. No load voltage rises to a high value and open-circuit operation must be avoided. Five hundred ohms correspond to a water resistance representing a typical freshwater pond found in southern New Hampshire.

At these parameters, the power dissipated into the water is around 25 watts and is effective up to 10 feet from the boat.

The effectiveness of the system is dependent on the following:

• Is the target fish within the area?

• Are the fish bottom dwellers?

• The fish size. Larger fish are easier to stun than the smaller ones.

• The water temperature may be too cold. This is important for proper operation.

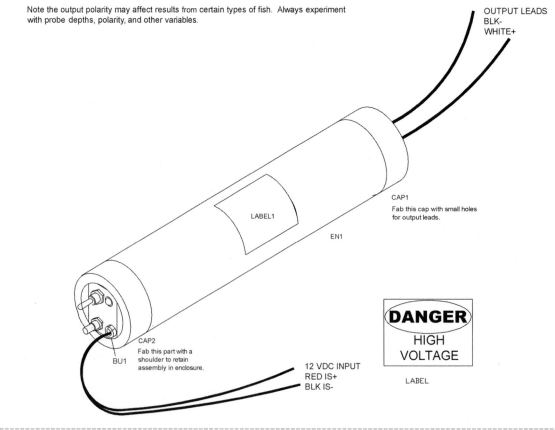

Note the output polarity may affect results from certain types of fish. Always experiment with probe depths, polarity, and other variables.

OUTPUT LEADS
BLK-
WHITE+

CAP1
Fab this cap with small holes
for output leads.

LABEL1

EN1

CAP2
Fab this part with a
shoulder to retain
assembly in enclosure.

BU1

12 VDC INPUT
RED IS+
BLK IS-

DANGER
HIGH
VOLTAGE

LABEL

Figure 24-7 Fully assembled unit

- Use of the correct dragline, chains, wire mesh, and weights for the particular fish. Several examples are shown in Figure 24-8.

Please note that the system utilizes a floating output, which means you do not have to use the electri-cal system of the boat or the boat itself as an electrode. If it is metal, you may choose to do this for certain applications. This is simply done by connecting the negative output lead to the -12 VDC input or craft common, forcing a grounded return.

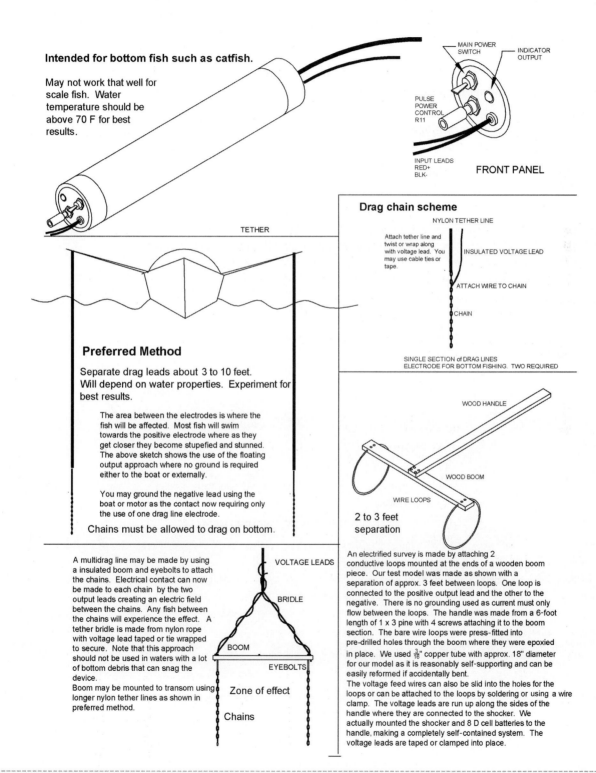

Intended for bottom fish such as catfish.

May not work that well for scale fish. Water temperature should be above 70 F for best results.

MAIN POWER SWITCH
INDICATOR OUTPUT
PULSE POWER CONTROL R11
INPUT LEADS RED+ BLK-
FRONT PANEL

TETHER

Drag chain scheme

NYLON TETHER LINE

Attach tether line and twist or wrap along with voltage lead. You may use cable ties or tape.

INSULATED VOLTAGE LEAD

ATTACH WIRE TO CHAIN

CHAIN

SINGLE SECTION of DRAG LINES ELECTRODE FOR BOTTOM FISHING. TWO REQUIRED

Preferred Method

Separate drag leads about 3 to 10 feet. Will depend on water properties. Experiment for best results.

The area between the electrodes is where the fish will be affected. Most fish will swim towards the positive electrode where as they get closer they become stupefied and stunned. The above sketch shows the use of the floating output approach where no ground is required either to the boat or externally.

You may ground the negative lead using the boat or motor as the contact now requiring only the use of one drag line electrode.

Chains must be allowed to drag on bottom.

WOOD HANDLE

WOOD BOOM

WIRE LOOPS

2 to 3 feet separation

A multidrag line may be made by using a insulated boom and eyebolts to attach the chains. Electrical contact can now be made to each chain by the two output leads creating an electric field between the chains. Any fish between the chains will experience the effect. A tether bridle is made from nylon rope with voltage lead taped or tie wrapped to secure. Note that this approach should not be used in waters with a lot of bottom debris that can snag the device.
Boom may be mounted to transom using longer nylon tether lines as shown in preferred method.

VOLTAGE LEADS

BRIDLE

BOOM

EYEBOLTS

Zone of effect

Chains

An electrified survey is made by attaching 2 conductive loops mounted at the ends of a wooden boom piece. Our test model was made as shown with a separation of approx. 3 feet between loops. One loop is connected to the positive output lead and the other to the negative. There is no grounding used as current must only flow between the loops. The handle was made from a 6-foot length of 1 x 3 pine with 4 screws attaching it to the boom section. The bare wire loops were press-fitted into pre-drilled holes through the boom where they were epoxied in place. We used $\frac{3}{16}$" copper tube with approx. 18" diameter for our model as it is reasonably self-supporting and can be easily reformed if accidentally bent.
The voltage feed wires can also be slid into the holes for the loops or can be attached to the loops by soldering or using a wire clamp. The voltage leads are run up along the sides of the handle where they are connected to the shocker. We actually mounted the shocker and 8 D cell batteries to the handle, making a completely self-contained system. The voltage leads are taped or clamped into place.

Figure 24-8 *Electric fishing probes*

A floating output provides a certain degree of safety for the operator because he or she must make actual contact with both output leads simultaneously to get shocked. If you now ground one of the output leads, it is possible to get shocked just by being in water that may be inside the boat and making accidental contact with only one of the output leads.

Verification of Operation

To confirm that the device works properly, follow these steps:

1. Connect the output to a household 15-watt, 115 VAC fluorescent lamp.

Table 24-1 Fish stunner project parts

Ref. #	Qty.	Description	DB #
R1, 3	2	1K, 1/4-watt resistor (br-blk-red)	
R2		470-ohm, 1/4-watt resistor (yel-pur-br)	
R4		10K trimpot (103)	
R5		10K control pot linear	
R6, 8	2	10-ohms, 1/4-watt (br-blk-blk)	
R7	2	1.8K, 3-watt *metal-oxide-semiconductor* (MOX) resistors, two in a series for 3600 ohms	
C1		10,000 mfd/16-volt electrolytic capacitor axial leads	
C2		2.2 mfd/50-volt nonpolarized electrolytic capacitor	
C3		.01 mfd/50-volt disk (103)	
C5		2,200 mfd/25-volt vertical electrolytic capacitor	
C4		3.9 mfd, 350-volt plastic capacitor	#3.9M
D1, 2	2	IN914 silicon diodes	
D3		IN5408 1 Kv, 3-amp rectifier	
D4		Bright-green LED	
Q1		IRF450 MOSFET transistor	
LAB1		DANGER HIGH VOLTAGE label	
I1		555 DIP timer IC	
T1		24-volt 4 A with 240-volt primary 60 Hz, reworked per text	#IUTR2412R
S1		SPST 3-amp toggle switch	
PBOARD		2 1/4 × 5 × .1 grid perforated board, cut to size per Figure 24-3	
WR20R	6 feet	#20 vinyl wires, red	
WR20B	6 feet	#20 vinyl wires, black	
WRBUSS	24 inches	#20-inch bus wire	
THERMO		Thermo pad for mounting under Q1	
LUG1		6-32 solder lug	
TYEWRAP	2	12-inch heavy-duty tie wraps for holding T1 to frame	
FRAME		11.5 × 1.75 .063 Al plate fabricated per Figure 24-5	
BU1		3/8" plastic clamp bushing	
SW1	1	6-32 1/2 Phillips screw	
SW3	1	6-32 × 1/2-inch nylon screw	
NUT1	2	6-32 kep nut	
CAP1, 2	2	3 1/2-inch plastic caps fabricated per Figure 24-7	
EN1		10- × 3 1/2-inch OD schedule 40 PVC tubing	

2. Connect the input to a 12 VDC source or a battery capable of supplying 2 amps.

3. Turn on the power and rotate the control, noting that the bulb lights and gets brighter as the control is turned clockwise. Also note the output indicator lighting.

4. Connect to the required electrodes and test it out on a target fish. Use as suggested in Figure 24-8.

Chapter Twenty-Five

Electromagnetic Pulse (EMP) Generator

This advanced project shows how to produce a multimegawatt pulse of electromagnetic energy that can cause irreversible damage to computerized and sensitive communication equipment. A nuclear detonation causes such a pulse, which must be countermeasured to protect electronic devices. This project requires lethal amounts of electrical energy storage and must not be attempted unless in a qualified laboratory environment. Such a device can be used to deactivate the computer systems in automobiles, avoiding dangerous high-speed chases. Sensitive electronic equipment can be tested for susceptibility to lightning and potential nuclear detonations.

The project is semidetailed with references made only to the major components. A low-cost, open-air spark switch is shown but will provide only limited results. A gas-filled or isotope doped switch is required for optimum results (see Figure 25-1).

Figure 25-1   *Electromagnetic pulser*

Basic Description

Shockwave generators are capable of producing focused acoustic or electromagnetic energy that can break up objects such as kidney stones and other similar materials. *Electromagnetic pulse* (EMP) generators can produce pulses of electromagnetic energy that can destroy the sensitive electronics in computers and microprocessors. Destabilized inductive and capacitive (LC) circuits can produce multigigawatt pulses by using an explosive wire disruption switch. These high-power pulses can be coupled into parallel plate transmission lines for EMP hardness testing, parabolic and elliptical antennas, horns, and so on for directional far-field effects.

For example, research is currently being undertaken to develop a system that would disable a car during a dangerous high-speed chase. The trick is generating a high enough power pulse to fry the electronic control processor modules. This would be a lot simpler if the vehicle were covered in plastic or fiberglass rather than metal. The shielding of the metal body offers a challenge to the researcher developing a practical system. A device could be built to do this, but it would be costly and could produce collateral damage to friendly targets.

Project Objective

The objective here is to generate a high-peak power pulse of electromagnetic energy to test the hardness of sensitive electronic equipment. Specifically, this project explores the use of such a device for disabling vehicles by jamming or destroying computerized control chips. We'll experiment with disruptive LCR circuits with focused shockwave capabilities.

Hazards

The project uses deadly electrical energy that can kill a person instantly if improperly contacted. The high-energy system that will be assembled uses exploding wires that can create dangerous shrapnel-like effects. A discharge of the system can severely damage nearby computers and other related equipment.

Theory

A capacitor (C) is charged from a current source to an energy source over a period of time. Once it reaches a certain voltage corresponding to a certain energy level, it is allowed to discharge quickly into a resonant circuit. A wire now is made to explode, disrupting this high-peak current through the circuit inductance. A powerful, undamped wave is now generated at the natural frequency and at the associated harmonics of this resonant circuit. The inductance (L) of the resonant circuit may consist of a coil and associated lead inductance, along with the intrinsic inductance of the capacitor, which is around 20 nanohenries. The capacitor of the circuit determines the energy storage and also has an effect on the resonant frequency of the system.

Radiation of the energy pulse can be made via a conductive conic section or a metal, horn-like structure. Some experimenters have used lumped, half-wave elements center-fed by a coil coupled to the coil of the resonant circuit. This half-wave antenna consists of two quarter-wave sections tuned to the resonant circuit frequency. These are in the form of coils wound with an approximate length of wire equal to a quarter-wavelength. The antenna has two radiation lobes parallel to its length or broadside. Minimum radiation occurs at points axially located or at its ends, but we have not validated this approach. For example, a gas discharge lamp, such as a household fluorescent lamp, will flash brightly at a distance from the source, indicating a powerful directional pulse of electromagnetic energy.

Our test pulse system produces conservative, multimegawatt electromagnetic pulses (1 megawatt of broadband energy) and is radiated preferably via a conical section antenna consisting of a parabolic reflector of 100 to 300 millimeters in diameter. A 25- × 25-centimeter-square metallic horn, flaring out to 100 centimeters square, will also provide a degree of performance. A .5-microfarad, special, low-inductance capacitor charges up in about 20 seconds with the ion charger described in Chapter 1, "Antigravity Project," and is modified as shown. Faster charging rates can be obtained by a higher-current system available on special request for more serious research from www.amazing1.com.

A high-power radio frequency pulse can be generated where the output of the pulser may also be coupled to a full-size, center-fed, half-wave antenna tuned between 1 and 1.5 MHz. The actual length at 1 MHz is over 150 meters (492 feet) and may be too large for many experiments. However, it is normalized for a radiation coefficient of 1, with all other schemes being less. The actual elements may be reduced in length by using tuned quarterwave sections consisting of a 75-meter (246-foot) length of wire spaced and wound on 2- to 3-meter pieces of *polyvinyl chloride* (PVC) tubing. This scheme produces a pulse of low-frequency energy.

Please note, as stated, that the pulse output of this system will cause damage to computers and any devices using microprocessors or similar circuitry up to a considerable distance. Always use caution when testing and using this system—just being close can damage sensitive electronic equipment. Figure 25-2 provides a description of the strategic parts used in our lab-assembled system.

Capacitor

The capacitor (C) used for this type of application must have very low inherent inductance and discharge resistance. At the same time, the part must have the energy storage sufficient to produce the necessary high-powered pulse at the target frequency. Unfortunately, these two requirements do not go hand in hand. Higher-energy capacitors always will have more inductance than lower-energy units. Another important point is the use of relatively high-discharge voltages (V) to generate high-discharge currents. These values are required to overcome the inherent complex loss impedance of the series inductance and resistance of the discharge path.

The capacitor used in our system is .5 mfd at 50,000 volts with a .03-microhenry series inductance. Our tar-

get fundamental frequency for the low-power nondisruptive circuit is 1 MHz. The system energy is 400 joules, as determined by $E = \frac{1}{2} CV^2$, with E at 40 Kv.

Inductor

The inductor can be easily made for a low-frequency radio pulse. The inductance shown as L1 is a lumping of all stray connecting leads, the spark switch, the exploding wire disrupter, and the inherent inductance of the capacitor. This inductance resonates at a wide band of frequencies and must be able to handle the high-discharge current pulse (I). The value of the lumped value is around .05 to .1 Uh. The conductor sizes must take into effect the high pulse current, ideally equal to $V \times (C/L)^{1/2}$. This fast current transition wants to flow on the conductor surface due to the high-frequency skin effect.

You may use an inductor of several turns for experimenting at the lower frequencies along with a coupled antenna. Dimensions are determined by the air inductance formula: $L = (10 \times D^2 \times N^2)/l$, where D is the diameter in centimeters, l is the length in centimeters, and N is the number of turns. A coil from 3 turns of 10 millimeters (.375 inches) of copper tubing on a 7.5-centimeter (3-inch) diameter spread out to 15 centimeters (6 inches) will have a calculated inductance of .3 Uh.

Spark Switch

The spark gap switches the energy from the capacitor into the inductor where a resonant tank is momentarily set up (see Figure 25-2). The current rise time occurs over the period $pi/2 \times (LC)^{-5}$. The gap separation distance is set to fire at the desired breakdown voltage. The impedance of the spark switch is determined by the equation $Z_{SP} = (k \times l)/Q$, where k equals $.8 \times 10^{-3}$, l equals the spark gap distance in centimeters, and Q equals the amps per second (coulombs) of discharge. The gap is self-firing and requires no external triggering. The gap assembly is an integral part of the discharge path and must be constructed to minimize inductance and resistance.

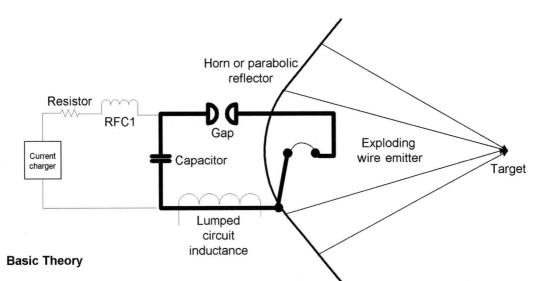

Basic Theory

A resonant LCR circuit consisting of components as in the above figure. Capacitor C1 is charged up from constant current charger at Ic. The voltage V across C1 is now related to V = It/C. The spark switch (GAP) is set to fire just before V reaches 50,000 volts. Once fired, a peak current rise of di/dt=V/L occurs. The period of circuit response is functional of .16 x (LC)$^{.5}$. The capacitor now discharges into the circuit inductance in 1/4t with the peak current now causing the wire to explode and interrupting this current just before it peaks. The inductive energy (LI^2) is released in an explosive burst of broadband electromagnetic radiation. The peak power is derived via the following and is in excess of many megawatts!!!!!

1. Charging Cycle: dv=Idt/C (Expresses the voltage charging on the capacitor as a f(t) with I constant current)

2. Storage energy in C as a f(v):   E=.5Cv^2 (Expresses energy in joules as the voltage increases)

3. Response time 1/4 cycle current peak: 1.57(LC)$^{.5}$. (Expresses the time for the first resonant current peaking when the spark switch fires)

4. Peak current in 1/4 cycle: V (C/L)$^{.5}$ (Expresses the peak current)

5. Initial response as a (f)t: Ldi/dt+iR+1/C+1/C$_{int}$ idt = 0 [Expresses voltages as a f(t)]

6. Energy joules in inductor: E= .5Li^2

7. Response when circuit is disrupted at max current through L:
   Ld^2 i/dt^2 + Rdi/dt + it/C = dv/dt. One now sees the explosive effects of the first term of this simple equation as the energy in the inductor must go somewhere in a very short time, resulting in an explosive E X B field energy release.

An appreciable pulse of many megawatts in the upper RF energy spectrum can be obtained by destabilizing the LCR circuit as shown above. The only limiting factor is the intrinsic real resistance that is always present in several forms, such as leads, skin effect, dielectric and switching losses, etc. These losses must be minimized for optimum results. The RF output can be coupled to a parabolic microwave dish or tuned horn. The Q of the output will depend to an extent on the geometry of the wire switch. Longer lengths will produce more "B" field characteristics while short more "E" field. These parameters will enter into the coupling equations regarding the radiation efficiency of the antenna. Experimenting is the best approach using your math skills only for approximating key parameters. Damage to circuitry usuallly is the result of very high di/dt (B field) pulse properties. This is a point of discussion!!

Figure 25-2   *EMP pulser schematic*

The fabrication of our lab test unit is shown and you may deviate with your own ideas, but the objective must be minimal circuit impedance.

You will note that the bottom ball of the spark gap switch is at a high potential and is made adjustable by a threaded rod and locking jam nut scheme. The top ball also uses a threaded rod that fits into the 3/4-inch PVC tubing used for the structural support of the wire disruption scheme.

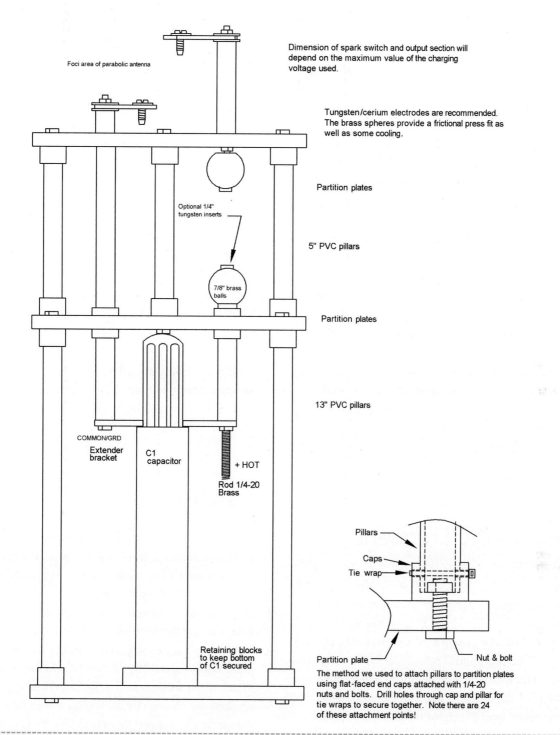

Foci area of parabolic antenna

Dimension of spark switch and output section will depend on the maximum value of the charging voltage used.

Tungsten/cerium electrodes are recommended. The brass spheres provide a frictional press fit as well as some cooling.

Partition plates

Optional 1/4" tungsten inserts

5" PVC pillars

7/8" brass balls

Partition plates

13" PVC pillars

COMMON/GRD
Extender bracket

C1 capacitor

+ HOT
Rod 1/4-20 Brass

Retaining blocks to keep bottom of C1 secured

Pillars

Caps

Tie wrap

Partition plate

Nut & bolt

The method we used to attach pillars to partition plates using flat-faced end caps attached with 1/4-20 nuts and bolts. Drill holes through cap and pillar for tie wraps to secure together. Note there are 24 of these attachment points!

Figure 25-3 *Front view of pulser showing spark switch*

Low-inductance extension pieces are used to lengthen the capacitor terminals and are fabricated from ¼-inch brass plates with mating holes to the existing block terminals of the capacitor. The edges are rounded and smoothed to prevent corona.

The open-air, self-triggering spark switch is intended as a low-cost approach to this very strategic part. A system of this type requires the fastest switching times possible. Gas-filled, triggered gaps will switch faster than open air, and isotope doped electrodes will further enhance performance.

Exploding Wire Disruption Switch (High-Power, High-Frequency Pulse)

This is where the stored energy in the circuit inductance is released as an explosion of electromagnetic energy at broadband proportions. The released energy is a function of LI^2 where I is the current rise in the spark gap switch at the moment the wire explodes and L is the inductance. The actual power lost in the spark switch is but a fraction of that emitted in the explosion of the wire.

Selection of the wire size must take into consideration the electrical circuit parameters for the proper timing of the optimum release of energy. We experimented with a .1-millimeter (.004 inch) to .3-millimeter (.012 inch) diameter of brass and aluminum wire about 50 centimeters in length. The wire is attached by a sandwiching action between two flat brass washers as shown. Note that a longer wire will tend to produce more of a magnetic pulse, whereas a shorter will produce more of an electric pulse.

RFC1 Radio Frequency Choke

This component is necessary to keep the fast current pulse rise isolated from the charger multiplier diodes that could be avalanched by the rapid dv/dt. The suggested value is around .2 mh with .3 Uh tertiary coils. The assembly can be a single-layer close wound winding of 150 turns for the .2 mh section. Wind three turns spaced over 1 inch for the .3 Uh section. Use #28 magnet wire on a 1 1/2- × 12-inch PVC plastic tube.

R1 Resistor

Intended as a safety precaution, this resistor provides a high impedance should a short occur in the output stage of the current driver. Use approximately 50 to 100K with at least 100 watts.

Charger

The charger for the system can be any current-limited source with an open-circuit voltage in excess of 50 Kv. The charging current rate will determine the amount of time necessary to reach a firing level and need not be that fast for this experimental system. A single charging cycle produces approximately 500 joules per shot and requires reloading of the wire for an exploding disrupter switch. A 2-milliampere current source will charge the .5 mfd capacitor to 50 Kv in approximately 5 seconds. This is shown mathematically by t = cv/i(.5)(10e − 6)(5 × 10,000/.002). This rate is more than ample and there is no advantage to a higher-current system unless you are planning to do a multiple discharge system using a spark gap-driven radiator or wire-dispensing scheme.

The ion supply described in Chapter 1 can supply an open-circuit voltage of 50 Kv by easily adding four more capacitors and high-voltage diodes to the multiplier section. Everything else can remain the same.

Assembly

Our lab pulser is shown constructed using materials and parts available from any hardware store. The structure uses a combination of 3/4-inch schedule 40 PVC tubing for the pillars and flat-faced end caps for the retainers (see Figure 25-4). Partitions are made from nonconductive material of structural integrity for the application. We used 3/8-inch, clear, acrylic plate stock. Figure 25-3 shows the scheme we used to attach these sections to the flat-faced end caps. The sections are secured by drilling clearance holes and plastic tie wraps keep them together. PVC cement is obviously stronger but prevents disassembly unless you destroy the support structure.

The pillar and cap assemblies are attached to the partition plates using 1-inch × 1/4-20 bolts and nuts. A cradle assembly fabricated from wood or plastic secures the capacitor C1 to the bottom partition. This scheme stabilizes the bottom of the capacitor.

The metal plate sections extend the terminal connections of the capacitor. These attach to the terminals connecting to the exploding wire cavity section

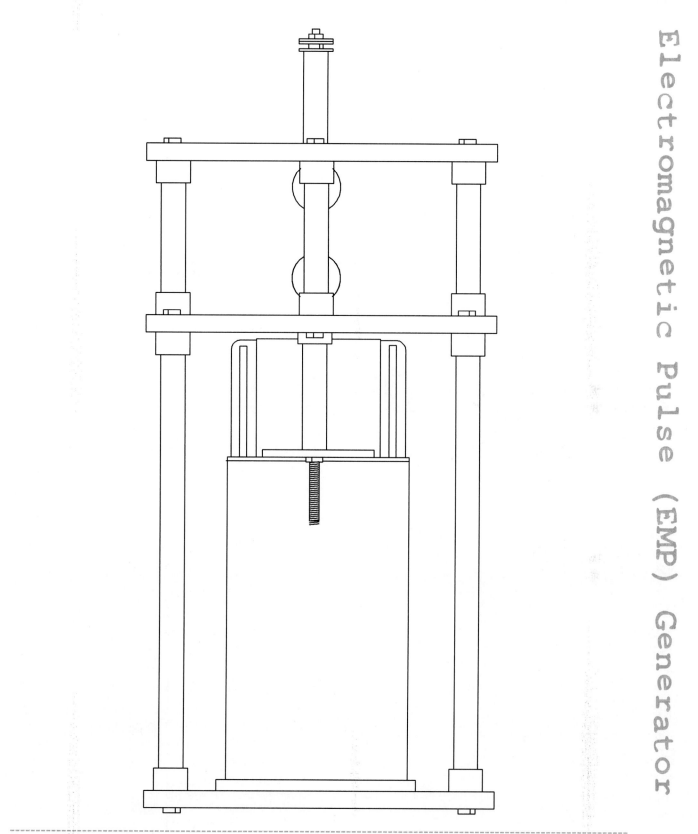

Figure 25-4   *Side view of EMP pulser*

via brass-threaded rods sleeved into pieces of ³/₄-inch PVC pillar tubing. The bottom spark gap electrode is

made adjustable by adjusting the bottom nut on the extended rod.

You will note the four longer pillars are positioned at the corners of the bottom and middle partition plates. The shorter pillars, however, are positioned at the midsections of the middle and top partition plates. This layout is shown in Figure 25-5. Figure 25-6 illustrates the final view of the pulser and Figure 25-7 shows the spark switch setup for coupling to the antenna.

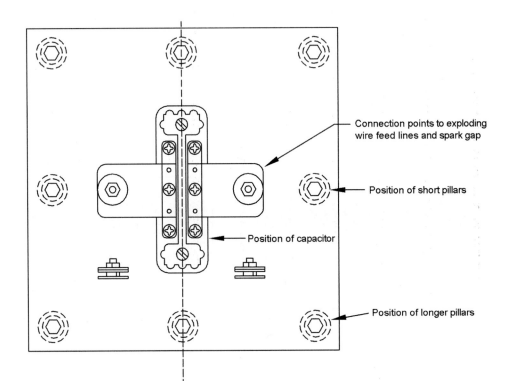

Connection points to exploding wire feed lines and spark gap

Position of short pillars

Position of capacitor

Position of longer pillars

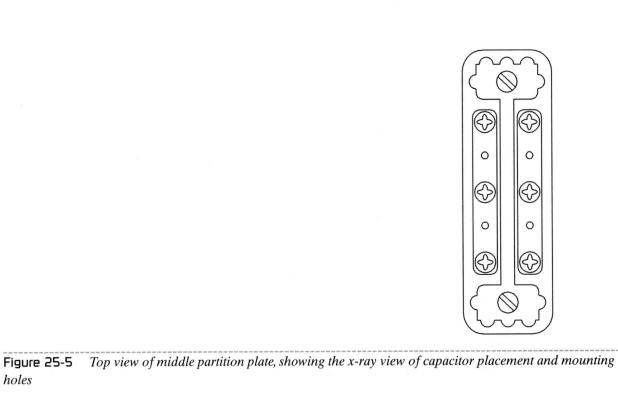

Figure 25-5 *Top view of middle partition plate, showing the x-ray view of capacitor placement and mounting holes*

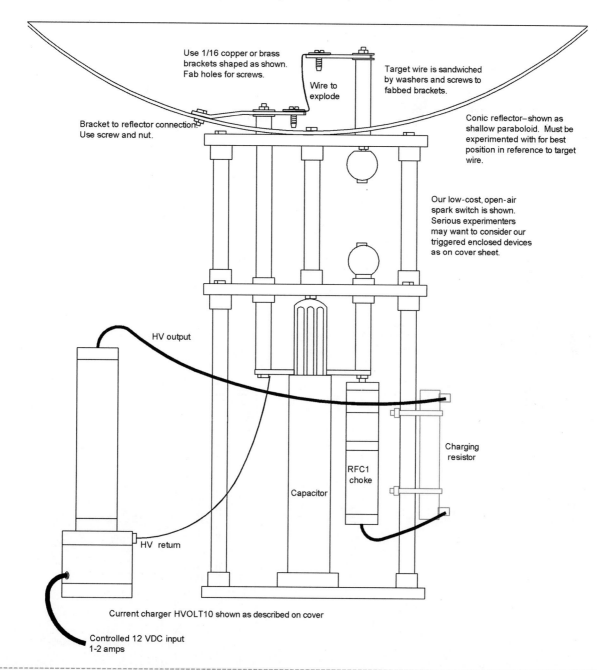

Event action occurs along the wire that explodes. The foci point of the dish is not exact for all points of the explosion event. Experiment for best results.

Use 1/16 copper or brass brackets shaped as shown. Fab holes for screws.

Wire to explode

Target wire is sandwiched by washers and screws to fabbed brackets.

Bracket to reflector connection. Use screw and nut.

Conic reflector–shown as shallow paraboloid. Must be experimented with for best position in reference to target wire.

Our low-cost, open-air spark switch is shown. Serious experimenters may want to consider our triggered enclosed devices as on cover sheet.

HV output

Charging resistor

RFC1 choke

Capacitor

HV return

Current charger HVOLT10 shown as described on cover

Controlled 12 VDC input
1-2 amps

Figure 25-6   *Final view of pulser, showing conic antenna*

Application

This system is intended for research into the suscepti-bility of sensitive electronic equipment to an EMP. The system can be scaled down for portable field use and operate on rechargeable batteries. It can be scaled up to produce kilojoule pulses at the user's own risk. No attempt to construct or use this device should be considered unless thoroughly experienced in the use of high-pulse energy systems.

The electromagnetic energy pulse can be focused or made parallel by use of a parabolic reflector. Experimental targets can be any sensitive electronic

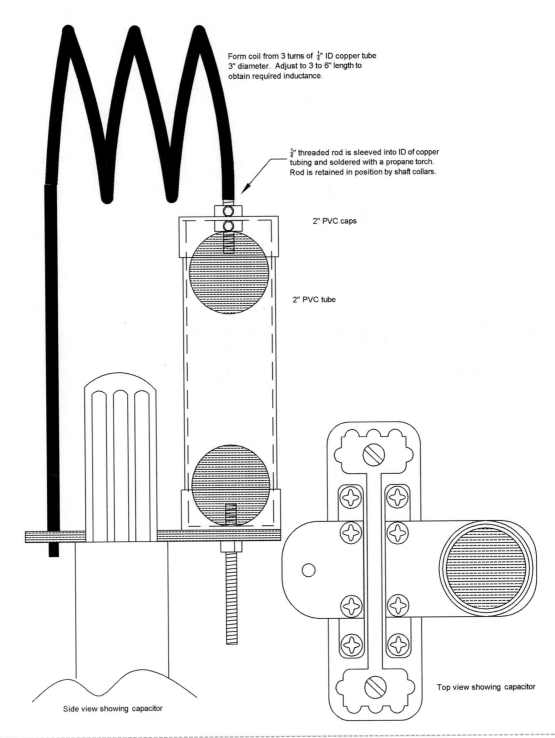

Form coil from 3 turns of $\frac{1}{4}$" ID copper tube 3" diameter. Adjust to 3 to 6" length to obtain required inductance.

$\frac{1}{4}$" threaded rod is sleeved into ID of copper tubing and soldered with a propane torch. Rod is retained in position by shaft collars.

2" PVC caps

2" PVC tube

Side view showing capacitor

Top view showing capacitor

Figure 25-7   *Spark switch setup and layout for low-frequency coupling to antenna*

equipment or even a gas-discharge lamp. The acoustical spark energy can produce a sonic shockwave of high sound pressure at the focal length of the parabolic antenna.

Sources

High-voltage chargers, transformers, capacitors, gas-filled and isotope doped spark switches, MARX impulse generators up to 2 megavolts, and EMP generators are all available at www.amazing1.com.

Part Six

Chapter Twenty-Six

Ultrasonic Microphone

This interesting electronic project enables you to clearly hear a world of sounds beyond that of human ability. The possible applications of the ultrasonic microphone you will create (see Figure 26-1) range from the detection of leaking gases, liquids, the mechanical wear of bearings, rotational and reciprocating devices, and electrical leakage on power-line insulators. A whole world of sounds coming from living creatures is also audible. Simple events like a cat walking across wet grass, the rattling of key chains, and even a collapsing plastic bag all are clearly heard. On a warm summer night, the sounds that can be heard are remarkable, as bats to small insects all perform a cacophony of nature's own orchestra at its best.

This handheld and directional microphone easily detects and locates these high-frequency sounds. The addition of a parabolic reflector further enhances the performance of this project. Expect to spend $30 to $50 for this rewarding effort.

Figure 26-1   *Ultrasonic microphone with parabolic reflector*

This project enables you to listen to a world of sound that few people even know exists. The unit is built in a gun configuration with the barrel housing the electronics. A rear panel contains the on/off volume, the tuning control pot, and the headphone jack. The front of the unit contains the directional receiving transducer. The handle houses the batteries.

The addition of an optional parabolic reflector greatly enhances the device's performance, providing super-high gain and directivity.

Applications

One of the most interesting sources of high-frequency sound is the many species of insects emitting their mating and warning calls. On a typical summer night, one can spend hours listening to bats and other strange insect noises. A whole new world of natural sound awaits the user. Many man-made sounds also generate high-frequency sounds easily detected by the device. Several examples are as follows, yet these only represent a small fraction of the potential sources of high-frequency sounds:

- Leaking gases and rushing air.

- Water from sprinklers or leaks.

- Corona leakage, sparking devices, or lightning.

- Fires and chemical reactions.

- Animals walking in wet grass or in the brush can be heard over a considerable distance. This is an excellent aid for hunters or trackers, or for just finding pets at night.

- Computer monitors, TV sets, high-frequency oscillators, mechanical wear bearings, rattle detection often developing in automobiles, plastic bags, and rattling change.

An excellent demonstration of this ultrasonic microphone would involve Doppler shifts where motion toward the source produces a rise in frequency and motion away from it does the opposite.

Doppler shift is when an observer moving toward a source of sound experiences an increasing frequency. This is easy to visualize when one realizes that sound propagates as a longitudinal wave at a rel-

atively constant velocity. As the observer moves toward the direction of the sound source, he intercepts more waves in a shorter period of time, thus hearing a sound that seems to be shorter in wavelength or higher in frequency.

A fun game both for children and adults is to hide a small test oscillator and attempt to have your opponent locate it in a minimal amount of time.

Circuit Description

An ultrasonic transducer mike (TD1) picks up high-frequency sounds and converts them to electrical signals via the piezoelectric effect (see Figure 26-2). Inductor L1 tunes the inherent capacity of the transducer to a window frequency centering around 25 kHz. This parallel, equivalent resonant circuit produces a high-impedance signal source that is coupled to *field-effect transistor* (FET) Q1 amplifier through capacitor C2. Resistor R1 and capacitor C1 decouple the bias voltage to the drain. Layout and input lead shielding is important, as this section is prone to noise, feedback, and extraneous signal sensitivity.

The output of Q1 is taken across the drain resistor R2 and is capacitance coupled to amplifier I1A. Gain is set to ×50 by the ratio of resistors (R6/R4).

The output of I1A is AC coupled to the combination mixer/amplifier I1B through capacitor C4. The output of oscillator I1C is coupled into the circuit by a "gimmick" capacitor, CM. This is a short lead from pin 8 of IC1 and is twisted with a similar lead from pin 2 of I1B. (It is suggested to check performance without this gimmick.) The oscillator now generates a frequency that is mixed with the picked-up signals. The resultant is two signals, one being the sum and the other the difference.

Capacitor C7 and resistor R17 form a filter network attenuating the higher-frequency component of the mixed frequencies while allowing the lower frequency to pass by a factor of 20 *decibels* (db). The lower-frequency results are the difference between the oscillator frequency and the actual signal frequency. This is similar to a superheterodyne effect. The high frequency is obviously bypassed by filter C7 and R17. The filtered signal, being the composite dif-

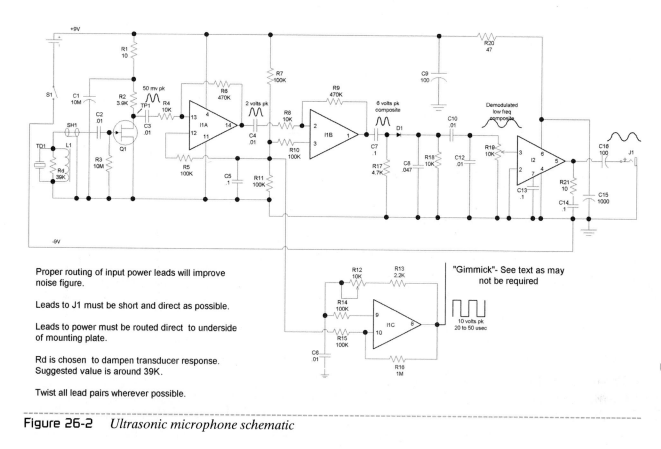

Figure 26-2 *Ultrasonic microphone schematic*

Proper routing of input power leads will improve noise figure.

Leads to J1 must be short and direct as possible.

Leads to power must be routed direct to underside of mounting plate.

Rd is chosen to dampen transducer response. Suggested value is around 39K.

Twist all lead pairs wherever possible.

ference, is rectified by diode D1 and integrated with capacitor C8. This signal is in the audio frequency range and is what you actually hear. It is tuned by control pot R12 in the oscillator section and enables the selective tuning of certain target frequencies within an acceptance window of the transducer TD1. These resulting audio frequencies are coupled to volume control R19 through DC blocking capacitor C10. Capacitor C12 bypasses any higher frequencies that may leak through. The arm of R19 feeds the audio signal into headphone amplifier I2. The output is 8 ohms and is capacitance-coupled to the headset jack J1. You may use a small speaker in a quiet location for group listening. Network R21/C4 further attenuates any further higher frequencies.

Power to I2 is decoupled through resistor R20 and capacitor C15. This provides circuit stability, preventing feedback oscillations and other undesirable effects.

The operating points of I1A, I1B, and I1C are set at the supply voltage midpoint selected by divider resistors R7 and R11. Resistors R5, R10, and R15 compensate for offset currents.

Construction Steps

To begin constructing the device, follow these steps:

1. Identify all the parts and pieces and verify them with the bill of materials.

2. Insert the components, starting from one end of the perforated circuit board, and follow the locations shown in Figure 26-3, using the individual holes as guides. The board is cut 2.25 × 2.25 × .1. A *printed circuit board* (PCB) is also available from www.amazing1.com.

   Use the leads of the actual components as the connection runs, which are indicated by the dashed lines. It is a good idea to trial-fit the larger parts before actually starting to solder.

   Always avoid bare wire bridges, messy solder joints, and potential solder shorts. Also check for cold or loose solder joints.

   Pay attention to the polarity of the capacitors with polarity signs and to all the semiconductors. The positioning of the control pots must

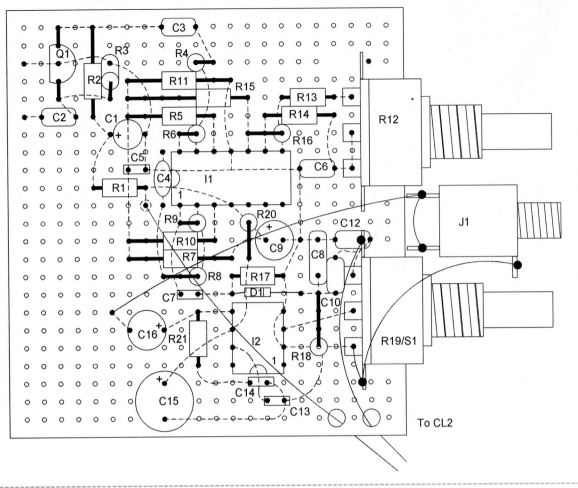

Figure 26-3 *Assembly board*

allow physical alignment with the mounting holes in RP1.

3. Cut, strip, and tin the wire leads for connecting to J1 and solder them. These leads should be 2 inches long and twisted.

4. Fabricate the CHAS1 chassis, the RP1 front panel section, the EN1 enclosure, and the HAND1 handle, as shown in Figure 26-4.

5. Prepare the SH1 shielded cable at both ends as follows. If the optional reflector is used, you will need a length of 18 inches; if not, you will need 6 inches. This is shown in Figure 26-5.

    a. Carefully remove ³/₄ of an inch of the outer insulation, being careful not to nick the shielded braid.

    b. Shred the shielding braid using a pointed object, such as a pin, and twist it into a lead. Carefully tin only the ends to hold the wires together.

    c. Carefully strip ¹/₄ of an inch of the insulation from the center conductor and tin.

    d. Check the finished cable for shorts or leakage with a high-resistance meter.

6. Solder the inductor L1 and the damping resistor Rd, as shown in Figure 26-5. Solder one end of the SH1 cable, being careful not to overheat the transducer pins or the insulation of the center conductor. Overheating these pins will ruin this part. If in doubt, you must perform a simple test of measuring a short circuit to the metal case of the part to the shorter pin. If this resistance is above 1 ohm, you have trashed this part and need a replacement. The inset in Figure 26-5 suggests using mechanical connections such as crimpling, wire nuts, and so on.

7. Assemble as shown. Figure 26-6 shows an assembly using the parabolic reflector, whereas Figure 26-7 shows one without.

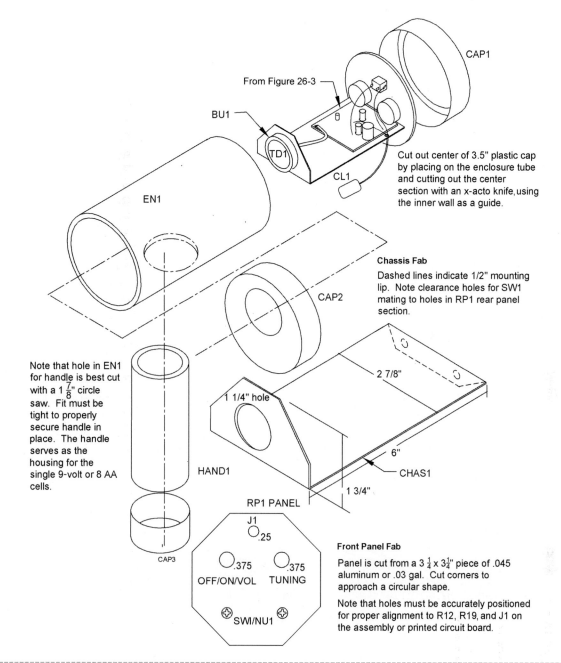

CAP1

From Figure 26-3

BU1

TD1

EN1

CL1

Cut out center of 3.5" plastic cap by placing on the enclosure tube and cutting out the center section with an x-acto knife, using the inner wall as a guide.

Chassis Fab

Dashed lines indicate 1/2" mounting lip. Note clearance holes for SW1 mating to holes in RP1 rear panel section.

CAP2

2 7/8"

1 1/4" hole

Note that hole in EN1 for handle is best cut with a $1\frac{7}{8}$" circle saw. Fit must be tight to properly secure handle in place. The handle serves as the housing for the single 9-volt or 8 AA cells.

HAND1

6"

CHAS1

1 3/4"

RP1 PANEL

CAP3

J1

.25

OFF/ON/VOL .375   .375 TUNING

SWI/NU1

Front Panel Fab

Panel is cut from a $3\frac{1}{4}$ x $3\frac{1}{4}$" piece of .045 aluminum or .03 gal. Cut corners to approach a circular shape.

Note that holes must be accurately positioned for proper alignment to R12, R19, and J1 on the assembly or printed circuit board.

Figure 26-4 *Final blowup and fabrication*

Electrical Pretest

To run a test on the system, follow these steps:

1. Turn the controls to "off," plug in the HS30 headset, and insert a 9-volt battery. Connect a meter set to read 100 milliamperes across the switch pins of R19 and quickly read a current of around 20 milliamperes. Remove the meter and turn control pot R19 to the midway point.

Note a smooth, rushing sound in the headsets. Then turn on a computer or TV monitor, and adjust the R12 control until you hear a clear tone. Turn off the sound source and gently rub two fingers together, noting a distinct sound. Check the range of the controls for unwanted feedback or spurious signals.

The unit is now ready for final assembly. Note the test points and wave shapes shown in Figure 26-2.

It cannot be stressed the importance of proper heat sinking of the pins to the transducer TD1 before attempting soldering. If you are in doubt, use a wire nut or "slip on" pin connections. Note the shorter pin is internally connected to the transducer case and connects to the ground side of the circuit.
You should measure a short circuit from this pin to the aluminum shell or the transducer is no good!

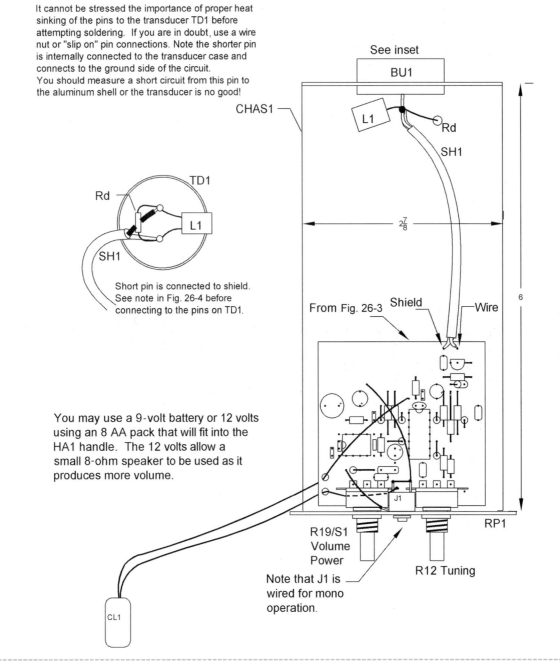

Short pin is connected to shield. See note in Fig. 26-4 before connecting to the pins on TD1.

You may use a 9-volt battery or 12 volts using an 8 AA pack that will fit into the HA1 handle. The 12 volts allow a small 8-ohm speaker to be used as it produces more volume.

Figure 26-5 *Chassis assembly with board connections*

2. Complete the final assembly, adding parabolic reflector PARA12 for a greatly enhanced performance, as shown in Figure 26-6.

Note that your unit may pick up strong magnetic fields, as it is not shielded for such. Performing the Doppler shift test, as noted earlier in these plans, easily differentiates these fields.

Special Note: Utilizing a Standing Wave

It is possible to produce a standing wave at the face of the transducer TD1 and improve the system sensitivity. Point the device at a steady, low-intensity source of high-frequency energy and carefully adjust

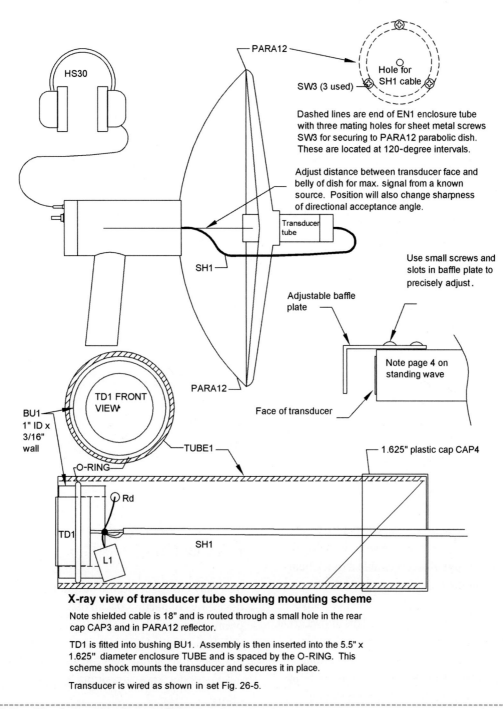

PARA12

Hole for SH1 cable

SW3 (3 used)

Dashed lines are end of EN1 enclosure tube with three mating holes for sheet metal screws SW3 for securing to PARA12 parabolic dish. These are located at 120-degree intervals.

Adjust distance between transducer face and belly of dish for max. signal from a known source. Position will also change sharpness of directional acceptance angle.

HS30

Transducer tube

SH1

Use small screws and slots in baffle plate to precisely adjust.

Adjustable baffle plate

Note page 4 on standing wave

PARA12

Face of transducer

TD1 FRONT VIEW

BU1 1" ID x 3/16" wall

TUBE1

1.625" plastic cap CAP4

O-RING

Rd

TD1

SH1

L1

X-ray view of transducer tube showing mounting scheme

Note shielded cable is 18" and is routed through a small hole in the rear cap CAP3 and in PARA12 reflector.

TD1 is fitted into bushing BU1. Assembly is then inserted into the 5.5" x 1.625" diameter enclosure TUBE and is spaced by the O-RING. This scheme shock mounts the transducer and secures it in place.

Transducer is wired as shown in set Fig. 26-5.

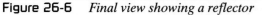

Figure 26-6   *Final view showing a reflector*

the distance of the 1 × 1" metal flat plate relative to the transducer face, noting an increase in the signal. This effect will occur at half-wave multiples with the most pronounced being closest to the face. Use your own ingenuity in retrofitting this simple step.

Supplementary Application Note

This device can provide hours of entertainment for adults and children alike. For example, a small box containing an oscillator at a frequency of 25 kHz

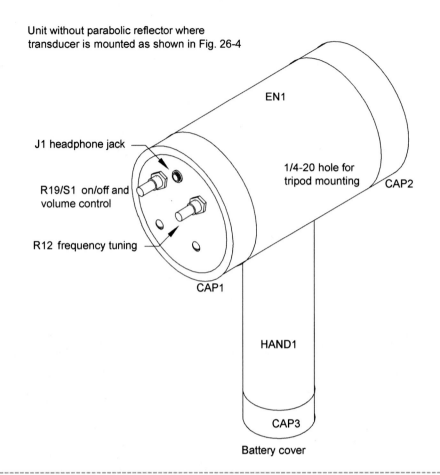

Unit without parabolic reflector where
transducer is mounted as shown in Fig. 26-4

EN1

J1 headphone jack

1/4-20 hole for
tripod mounting

CAP2

R19/S1 on/off and
volume control

R12 frequency tuning

CAP1

HAND1

CAP3

Battery cover

Figure 26-7 *Final assembly without a reflector*

could be well hidden somewhere. The highly directional characteristics of the device with its ability to respond to various signal levels enable a quick bearing to be made on the hidden source and the fun begins. It should be noted that the detection range could be in excess of a quarter-mile! This allows great flexibility in hiding the oscillator, using only your imagination for making detection difficult. Along with my children and friends, I have had many enjoyable hours and good laughs playing with this equipment.

Recording

You may easily record the output to a recording device by connecting the auxiliary input to the headphone output jack. A "Y" adapter can be used for simultaneously plugging in the headset to monitor the recorded sounds. Two sets of headphones can be used with the adapter, and the output impedance is 8 ohms.

Table 26-1 Ultrasonic microphone parts

Ref. #	Qty.	Description	DB#
R1, 21	2	10 ohm, $\frac{1}{4}$ watt (br-blk-blk)	
R2		3.9K, $\frac{1}{4}$ watt (or-wh-red)	
R3		10 megs, $\frac{1}{4}$ watt (br-blk-blue)	
R4, 8, 18	3	10K, $\frac{1}{4}$ watt (br-blk-or)	
R12		10K, 17-millimeter pot	
R5, 7, 10, 11, 14, 15	6	100K, $\frac{1}{4}$ watt (br-blk-yel)	
R6, 9	2	470K, $\frac{1}{4}$ watt (yel-pur-yel)	
R13		2.2K, $\frac{1}{4}$ watt (red-red-red)	
R16		1 meg, $\frac{1}{4}$ watt (br-blk-grn)	
R17		4.7K, $\frac{1}{4}$ watt (yel-pur-red)	
R19/S1		10K, 17-millimeter linear pot and switch	
R20		47 ohm, $\frac{1}{4}$ watt (yel-pur-blk)	
Rd		Select for dampening transducer circuit—10 to 47K, $\frac{1}{4}$ watt	
C1		10 mfd, 25-volt vertical electrolytic capacitor	
C2, 3, 4, 6, 10, 12	6	.01 mfd, 25-volt disk or plastic capacitor	
D1		IN914 small-signal diode	
C8		.047 mfd, 50-volt plastic capacitor	
C5, 7, 13, 14	4	.1 mfd, 25-volt disk or plastic capacitor	
C9, 16	2	100 mfd, 25-volt vertical electrolytic capacitor	
C15		1,000 mfd, 25-volt vertical electrolytic capacitor	
L1		27 millihenry inductor	#IU27H
Q1		J202 n-channel FET	
I1		LM074 4-second OP amp DIP	
I2		LM386 audio amplifier DIP	
J1		3.5-millimeter stereo audio jack wired for mono	
TD1		25 kHz acoustical receiving transducer	#IUTR8925
SH1	18 inches	Shielded mike cable	
PB1		PC or .1 grid perforated vector board, 2.25- $\times$ 2.25-inch	
BAF1		Thin piece of 2.25- $\times$ 2.25-inch plastic for insulating	
BU1		$1 \times \frac{1}{2} \times \frac{3}{16}$ wall neoprene bushing	
CL1		Battery clip with 12-inch leads	
PARA12		Optional parabolic reflector	#IUPARA12
PCHT9		Optional PCB	#IUPCHT9

Chapter Twenty-Seven

Phaser Pain Field Gun

This project shows how to build a directional, hand-held device capable of driving off both wild and domestic animal pests (see Figure 27-1). This device produces high-pressure-level sound at complex frequencies mostly above that of human hearing. This sound is painful to many animals as their hearing is much more acute than humans. The unit is tunable and can also be adjusted to be very annoying to people when exposed at close range.

The circuitry is housed inside an easily obtainable piece of *polyvinyl chloride* (PVC) tubing and resembles a large ray gun. The output controls are on the rear panel with the batteries in the handle. Expect to spend around $25 to assemble this sound gun. All the parts are readily available, with the hard-to-find items available at www.amazing1.com.

Figure 27-1   *Phaser pain field gun*

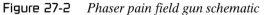

Caution: Exposure below 20 kHz at a sound-pressure level of 105 decibels (db) in excess of 1 hour may cause a hearing impairment.

This project shows how to construct a moderately powered, directional source of continuous, adjustable, high-frequency, time-variant acoustical shockwaves. The output energy of the device is conservatively rated at 125 db. The frequency is programmable or is constantly sweeping at an internal preset rate and is variable approximately from an audible 5 to 20 kHz. The unit is intended for use as a research tool in the study of animal behavior, acoustical experimentation, or as a source of intense, directional, acoustical, high-frequency sound for other scientific and laboratory applications. The device is excellent for use in agricultural applications such as flushing out rats and other rodents from granaries, silage bins, chicken coups, and so on.

It is to be used with discretion and not treated as a toy. Caution must be used as exposure to most people causes pain, headaches, nausea, and extreme irritability (younger women are especially affected). Do not, under any circumstances, point the unit at a person's ears or head at close range, as severe discomfort and possible ear damage may result. Usage on dogs and other animals must be done with discretion to avoid excessive discomfort.

Circuit Theory

A timer (1C2) is connected as a stable, free-running multivibrator whose frequency is externally controlled by pot R9 (see Figure 27-2). Resistor R10 selects the range limit of R9. Capacitor C5, along with the resistors, determines the frequency range of the device. The square wave output of 1C2 is via pin 3 and is resistively coupled to power amplifier Q2. The drain of Q2 is DC biased through resonator choke L1.

The square wave output signal is now fed into transducer TD1 in a series with resonating coils L2 and L3 in a parallel combination. The resonant action among the inherent capacities of TD1, the tuning capacitor C7, and the inductance now produces a sinusoidal-shaped wave peaking around 25 kHz or the upper limit of the tuning range. This signal waveform now has a peak-to-peak voltage several times that of the original square wave. Transducer TD1 now can take advantage of these peak voltages to produce the high-pressure-level sounds necessary without exceeding the high average ratings of an equivalent voltage-level square wave.

Timer IC1 is similarly connected as a stable running multivibrator and is used to produce the sweeping voltage necessary for modulating the frequency of

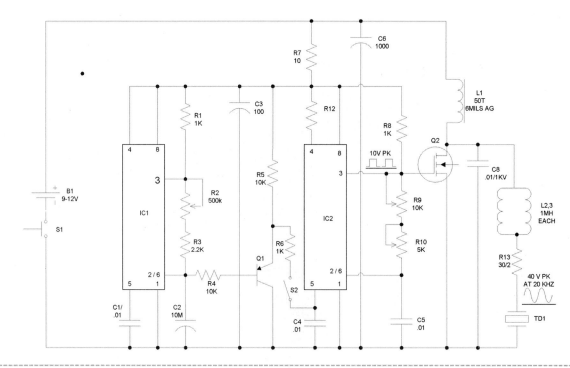

Figure 27-2 *Phaser pain field gun schematic*

1C2. This sweep repetition rate is controlled by pot R2, and resistor R3 limits the range of this repetition time. Resistor R1 selects the duty cycle of the pulse, whereas capacitor C2 sets the sweep time range. The output for Il is via pins 6 and 2 where the signal ramp function voltage is resistively coupled to inverter transistor Q1 via resistor R4. The output of Q1 is fed to pin 5 of 12 and provides the output modulation voltage necessary to vary the frequency as required. Note that the modulation signal is easily disabled via R2/S2.

Power to the system comes from battery B1 and pushbutton S1. Capacitor C6 guarantees an AC return path for the output signal. Power to the driver circuits IC1 and IC2 come through a decoupling network consisting of resistor R7 and capacitor C3.

Construction Steps

To begin the assembly of the gun, follow these steps:

1. Lay out and identify all the parts and pieces, checking them with the parts list.

2. Cut the HS1 heatsink bracket from a .75 × 2 × .065 aluminum piece. Bend it 90 degrees at its midsection and drill a hole for the SW1/NU1 screw and nut. Then attach it to Q2 as shown later in Figure 27-4.

3. Figure 27-3 shows the assembly using a *printed circuit board* (PCB) available through www.amazing1.com. Experienced builders may use a piece of .1-inch grid perforated vector board. Use the drawing for parts location and the schematic for the connections. Certain leads of the actual components will be used for connecting points and circuit runs. Do not cut or trim them at this time. It is best to temporarily fold the leads over to secure the individual parts from falling out of the board holes. If you obtain a PCB, you may omit this step.

4. Insert and then solder in the designated components:

   a. Insert the ¼-watt resistors: R1, R3, R4, R6, R7, R8, and R12.

   b. Insert R10 and R13.

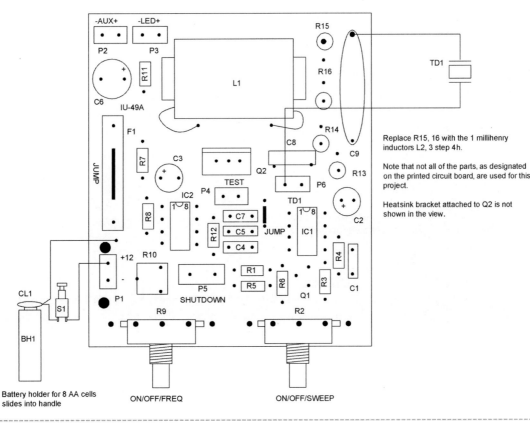

Figure 27-3 *PCB wiring*

c. Solder in the control pots R2/S2 and R9. Use short pieces of bare-wire leads from the component pins to the respective holes in the PCB. Note these controls must be as close to the board surface as possible.

Note that R11 and R14 are not used and the positions for R15 and R16 are used for components designated as L2 and L3.

d. Insert capacitors C1, C4, C5, and C8.

e. Insert electrolytic capacitors C2, C3, and C6, noting the polarity. Note that C7 is not used in this circuit.

f. Insert semiconductors Q1, Q2, IC1, and IC2. Note that Q2 is attached to heatsink HS1 via SW1/NU1.

g. Insert two wire jumps.

h. Insert the two 1 millihenry inductors in place of R15 and R16.

i. Insert the red lead of CL1 and the 10-inch lead for S1 as shown.

j. Insert two 10-inch leads for transducer TD1.

5. Assemble the Ll choke coil by wrapping 50 turns of #24 magnet wire on the nylon bobbin as evenly as possible. Leave two inches of leads for connection to the circuitry. Assemble the E core, as shown in Figure 27-4. Shim each side with pieces of yellow cardboard strips of

.003 inches each for a total of .006 inches. If you have an inductance capacitance bridge, measure 1.5 millihenrys.

6. Wire in the inductor and secure it to the board with *room temperature vulcanizing* (RTV) silicon rubber or another suitable adhesive.

7. Preconnect all the leads to TD1, S1, and CL1.

8. Check the complete wiring for potential shorts, wire bridge shorts, poor solder joints, the correctness of the components, and the position and orientation of Q1, Q2, C3, C6, IC1, and IC2.

Subassembly Pretest

To conduct a test on the device, follow these steps:

1. Turn pots counterclockwise and install eight fresh AA batteries into the BH1 holder, as shown in Figure 27-5.

Note that a variable bench supply capable of 12 *volts-direct current* (vdc) at 1 amp with a volt and current meter can be a great convenience for the remaining steps and for the testing of other similar circuits.

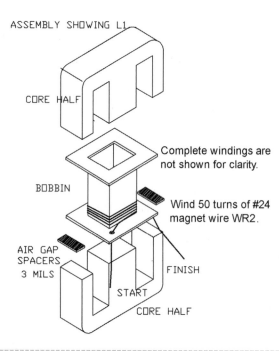

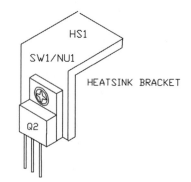

Figure 27-4 *Assembly of inductor L1 and heatsink HS1*

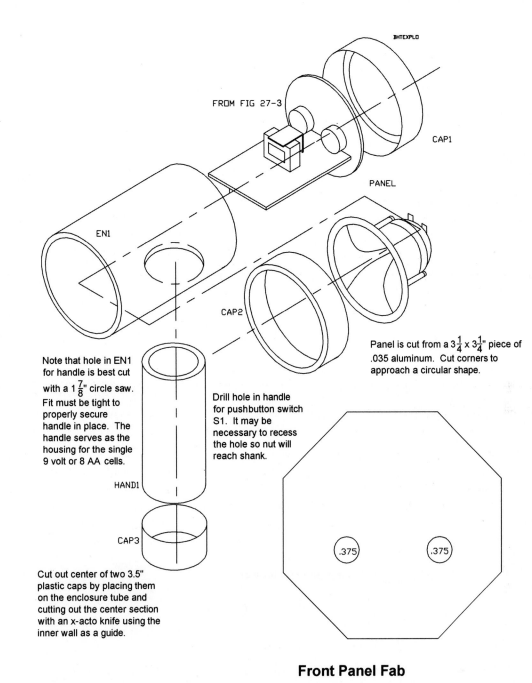

BHTEXPLD

FROM FIG 27-3

CAP1

PANEL

EN1

CAP2

Panel is cut from a $3\frac{1}{4}$ x $3\frac{1}{4}$" piece of .035 aluminum. Cut corners to approach a circular shape.

Note that hole in EN1 for handle is best cut with a $1\frac{7}{8}$" circle saw. Fit must be tight to properly secure handle in place. The handle serves as the housing for the single 9 volt or 8 AA cells.

Drill hole in handle for pushbutton switch S1. It may be necessary to recess the hole so nut will reach shank.

HAND1

CAP3

.375   .375

Cut out center of two 3.5" plastic caps by placing them on the enclosure tube and cutting out the center section with an x-acto knife using the inner wall as a guide.

Front Panel Fab

Figure 27-5   *Final blowup*

2. Push button S1 and note a loud, piercing wave coming from TD1. Measure a current of 300 to 400 milliamperes in a series with a battery.

3. Rotate R9 and note that the frequency increases to above the audible range. Then measure a current of 100 to 200 milliamperes.

4. Note the wave shapes shown in Figure 27-2 for those who have a scope.

5. Turn on sweep control R2/S2 and note the frequency being modulated by a changing rate as this control is adjusted. Use caution as certain sweep rates may cause epileptic fits and other undesirable effects. Sweep rates between 7 and 20 per second should be used with caution.

Final Assembly

The final steps are as follows:

1. Cut the rear panel from a $3^{1}/_{4}$- × $3^{1}/_{4}$-inch piece of aluminum. Add holes for the controls as shown in Figure 27-5.

2. Create the CAP2 and CAP3 retaining caps from a $3^{1}/_{2}$-inch plastic cap. This is easily done by placing a cap over the end of the main enclosure and carefully removing the center section with a sharp knife, using the wall thickness as a guide. These reworked caps now retain the transducer and the rear panel to the main enclosure.

3. Create EN1 and HA1 from the material indicated on the parts list. Add a hole in the main enclosure for the insertion of the HA1 handle section. The hole must provide a tight fit. Glue the parts together.

4. Create the final assembly as shown.

Note that the frequency range with the values shown is 4 to 20 kHz. The frequency range can be changed by increasing the value of C5 to lower it or by decreasing the range to raise it. Experiment for a desired result. The sweep range with the value shown is 4 to 50 Hz. C2 can be changed in a similar function.

Sound pressure measurements will vary from 130 to 100 db at 18 inches depending on the frequency. Certain sweep rates between 5 to 20 per second may cause dizziness or epileptic fits.

Table 27-1 Phaser pain field gun project parts

Ref. #	Qty.	Description	DB#
R1, 6, 8, 12	4	1K, $^{1}/_{4}$-watt resistor (br-blk-red)	
R2/S2		500K to 1 meg pot/switch, 17 millimeters	
R3		2.2K, $^{1}/_{4}$-watt resistor (red-red-red)	
R4, 5	2	10K, $^{1}/_{4}$-watt resistor (br-blk-or)	
R7		10-ohm, $^{1}/_{4}$-watt resistor (br-blk-blk)	
R9		10K pot, 17 millimeters	
R10		5K trimpot, horizontal	
R13	1	30-ohm, 3-watt resistors (or-blk-blk)	
C1, 4	2	.01 mfd/50-volt disk capacitor (103)	
C3		100 mfd/25-volt vertical electrolytic capacitor	
C2		10 mfd/25-volt vertical electrolytic capacitor	
C5		.01 mfd/50-volt polyester capacitor	
C6		1000 mfd/25-volt vertical electrolytic capacitor	
C8		.01 mfd/2 Kv disk capacitor	
C9			
L2, 3	2	1 millihenry inductors in place of R15 and R16, as marked on PCB	IU1MH
L1		Inductor—2 Hitachi 30.48 E cores and mating bobbin, as shown in Figure 27-4	IUPPP1L1
Q1		PN2907 PNP GP transistor	
Q2		IRF530 or 540 N channel *metal-oxide-semiconductor field effect transistors* (MOSFETs)	

Table 27-1 Continued

Ref. #	Qty.	Description	DB#
IC1, 2	2	555 DIP timer	
HS1		Heatsink bracket shown in Figure 27-4	
SW1/NUT		6-32 × ¼-inch screw and nut	
CL1	1	9-volt battery clip	
BH1		Eight AA cell holders for 1.5 volt batteries (B1)	
S1		SPST pushbutton switch	
PCPPP1 IUPCSONIC		PCB or use perforated vector board	
TD1 IUMOTRAN		Polarized 130 db piezo transducer	
WR1	3 feet	#24 vinyl hookup wire, red and black piece	
WR2	5 feet	#24 magnet wire to wind L1	
CAP1		1 ⅞-inch plastic cap (A1 ⅞)	
CAP2, 3	2	3 ½-inch plastic cap (A3 ½), as shown in Figure 27-5	
RP1		3 ⅛-inch square #22-24 aluminum, as shown in Figure 27-5	
EN1		3 ½- OD × 7-inch PVC tube, as shown in Figure 27-5	
HAND1		1 ⅞- OD × 6-inch PVC tube, as shown in Figure 27-5	

Chapter Twenty-Eight

Pain Field Property-Protection Guard

This chapter offers an excellent home- and property-protection project when properly built that provides a low-liability deterrent to unauthorized intrusion from both two- and four-legged threats. Once activated, strategically placed transducers project an uncomfortable and disorientating field of complex acoustical sound and shockwaves. It is a harmless effect yet discourages most intrusions in a defined area.

A reasonably simple electronic system can power up to eight individual transducer emitters positioned in the target area (see Figure 28-1). Activation can be an open or closed fault switch or a voltage level such as that produced by our laser property guard described in Chapter 12, "Laser Property-Protection Fence." Together these projects can be interfaced to produce an effective intrusion detection and deterrent system. Expect to spend $50 to $100 with hard-to-find parts available at www.amazing1.com. The complete parts list is outlined in Table 28-1.

Warning: Do not operate this system at continuous, high output at frequencies below 20 kHz. Daily sound pressure exposures in excess of 1 hour at 105 *decibels* (db) may lead to hearing impairment. When properly used, this device provides a limited liability deterrent. It should not cause permanent damage or trauma.

Figure 28-1   *Phaser property protection guard*

Project Description

The following product is intended to be a property- or home-protection device. It consists of a field of acoustical, ultrasonic, high sound pressure energy that is triggered when unauthorized intrusion is detected.

Detection consists of the following functions:

- A "trip wire" or closed system such as taped glass windows and doors where a break or open triggers the unit. *This function is via J3.*

- A "switch" input where a closure to ground such as a pressure switch, door, or entrance switch triggers the unit. *This function is via J2.*

- A "+ level" input where a voltage pulse or level from other detection equipment such as an infrared intrusion, motion, or sound detection system triggers the unit. *This function is via J1.*

Test and Reset buttons enable total system control.

Upon activation via the above, a moderately powered source of acoustical, ultrasonic energy is produced, causing certain adverse effects to the intruder. These may be paranoia, severe headaches, disorientation, nausea, cranial pain, an upset stomach, or just plain irritating discomfort. Most people are affected in one way or the other, with young women unfortunately being the most sensitive. External adjustments enable the user to select clearly audible sounds that serve as an alarm, high-frequency energy that produces the physiological effects, or a combination of both.

The sound pressure level is less than 130 db and will not produce permanent damage if exposure is kept to a minimum. Obvious prolonged exposure is not encouraged for these reasons. A rule of thumb is to keep exposure to less than 1 hour with a frequency less than 20 kHz at a sound pressure of 105 db or over.

The system consists of the central power and control unit that powers up to six remotely located transducers. These are now positioned to take advantage of potential entrance and intrusion areas, considering

that each transducer can produce up to 118 db measured at 1 meter. Since the sound pressure level is logarithmic, an attenuation factor of -3 db must be factored in every time the distance is doubled from one of the transducer stations.

Driver Circuit Description

A timer (IC2) is connected as a stable, free-running multivibrator whose frequency is externally controlled by pot R9. The trimmer resistor (R10) selects the range limit of R9. Capacitor C5, along with the resistors, determines the frequency range of the device (see Figure 28-2).

The square wave output of 1C2 is via pin 3 and is connected to *metal-oxide-semiconductor field effect transistor* (MOSFET) Q2. The drain of Q2 is DC biased through L1. The amplified square waves are fed to the transducer via resonating coil L1 and capacitor C8, along with Q spoiling resistors, R13 through R16. Resonating coils (L2A and L2B) are selected to tune out the inherent capacity of the transducer at their upper-frequency limit, usually around 25 kHz. A sinusoidal wave is generated and allows the transducers to operate at a higher peak power level than the equivalent voltage square wave would. Resonant peaking of the voltage is also obtained. These transducers, unlike their electromagnetic counterparts, have a tendency to draw high current at higher frequencies. This effect is compensated to an extent by power resistors R17A and R17B. Note the wave shapes shown are at a fixed frequency of 20 kHz.

Timer IC1 is similarly connected as a stable, running multivibrator and is used to produce the sweeping voltage necessary for modulating the frequency of 1C2. The switch section of R2/S2 activates it, and this sweep repetition rate is controlled by the pot section R2. Resistor R3 limits the lower range of this repetition time. Capacitor C2 sets the sweep time range. Output from IC1 is via pins 6 and 2 where the signal ramp function voltage is resistively coupled to inverter transistor Q1 via resistor R4. The output of Q1 is fed to pin 5 of 1C2 and provides the modulation voltage necessary to generate the sweeping frequency action required. Note that this signal is easily

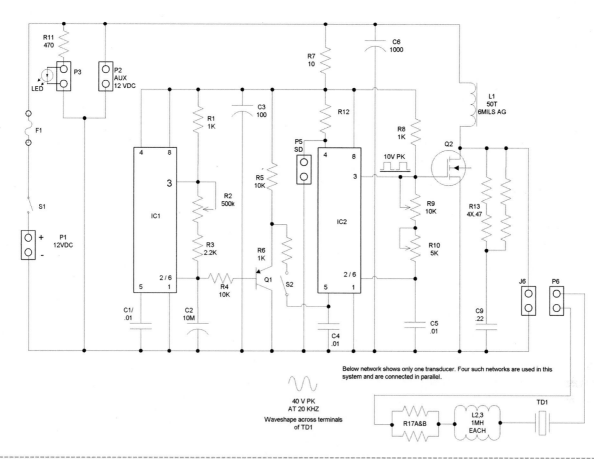

Figure 28-2 *Phaser pain field property guard schematic*

disabled via the switch section of R2/S2. This is a convenience when initially setting or checking the range of 1C2 as it eliminates the constant varying frequency.

Power is supplied to the system via a conventional step-down converter wall transformer, T1, which connects to the system via the DC JACK chassis mount connector. Power is controlled by switch S1 that is part of the frequency control pot R9. A *light-emitting diode* (LED) indicator lamp and an associated current-limit resistor (R11) tell when the system is energized.

Detection Circuit Description

The detection circuits consist of inputs J1, J2, and J3, which sense an intrusion and energize P5, enabling timer IC2. A reset switch (S4) enables the circuit to be reset. A test switch (S5) enables the verification of

system operation. Detection circuit functioning, along with the appropriate jack identification, is shown in Figure 28-3.

Construction

To begin assembling the project, follow these steps:

1. Lay out and identify all the parts and pieces to the assembly board (see Figure 28-4), sense board (refer to Figure 28-3), and the remaining parts for the final assembly.

2. Assemble inductor L1 as shown later in Figure 28-8. Drill a small hole in the bobbin for the start of the winding and wrap 50 turns of #24 magnet-enameled wire tightly and evenly along the bobbin length. Tape the winding in place and allow 2 inches for both the "start" and "finish" leads. Assemble everything as shown, inserting core halves into the bobbin, and place the air gap spacers between them.

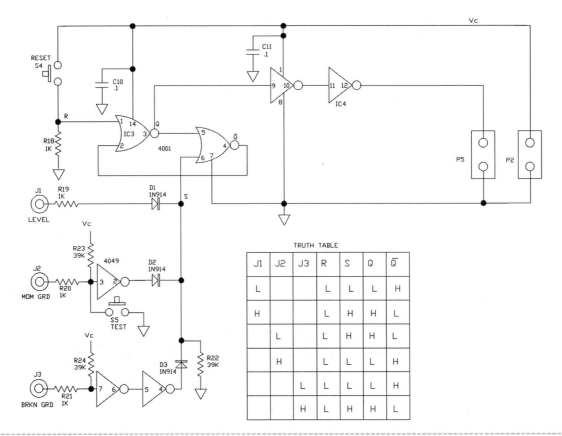

Figure 28-3 *Detection circuit schematic*

TRUTH TABLE

J1	J2	J3	R	S	Q	Q̄
L			L	L	L	H
H			L	H	H	L
	L		L	H	H	L
	H		L	L	L	H
		L	L	L	L	H
		H	L	H	H	L

These spacers can be from a cut-up business card. You should attempt to find a thickness of 3 mils (.003) and confirm with vernier calipers. Tape the assembly together or use elastics, O-rings, and so on. The finished inductor should read approximately 1 millihenry when measured on an inductance capacitance bridge.

3. Create the heatsink bracket HS1 from a .75 × 2 × .0625 aluminum piece. Bend it 90 degrees in the middle and drill a hole for the SW1/NU1 screw and nut, as shown later in Figure 28-8.

4. Assemble the PCB as shown in Figure 28-4. Note the polarity of all the diodes, transistors, integrated circuit, and electrolytic capacitors. Note the two wire jumps. Be careful of the solder bridges on the PCB foil as these can cause damage to circuit components. Attach approximately 6 inches of wire leads to the front and rear panel-mounted components per Figures 28-4 and 28-5. Note that most resistors are vertically mounted. Always leave

at least a $^1/_{16}$-inch lead between the body of the component and the board.

Experienced builders may use a piece of .1-inch grid perforated vector board. Use Figure 28-4 for parts location and the schematic for the connections. Certain leads of the actual components will be used for connecting points and circuit runs. Do not cut or trim them at this time. It is best to temporarily fold the leads over to secure the individual parts from falling out of the board holes. If you obtain a PCB, you may omit this step. Figure 28-6 shows the foil layout.

5. Assemble a sense board from a piece of 2 $^1/_4$- × 1 $^1/_2$-inch of .1-inch grid perforated board per Figure 28-7.

Insert the components into the board holes as shown. Certain leads are used for connecting points and circuit runs. Do not cut anything at this time. Fold over the leads to secure the parts in place.

Please read the following before doing any soldering:

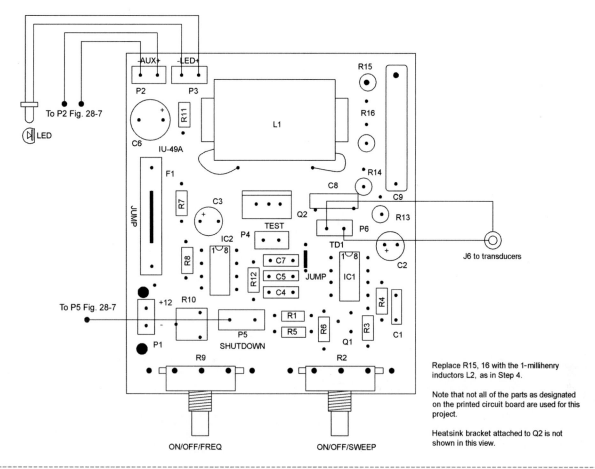

Figure 28-4 *Printed circuit board (PCB) wiring first level*

a. Components are mounted both horizontally and vertically. Leave at least $^1/_{16}$ of an inch between the part and the board surface.

b. Verify the polarity of IC3, IC4, D1, D2, and D3.

Use a good, pencil-type soldering iron, keeping the tip tinned and cleaned. Use rosin core solder. Do not overheat the solder joint as you may damage a component. All the connections should be smooth and shiny. Avoid excess solder.

Wire as shown, using the dashed lines as the circuit runs.

6. Attach the wire leads to the related front and rear panel components as shown.

Connect the interconnecting leads with the circuit board (see Figure 28-4) for P2, AUX12V, and P5, SHUTDOWN.

Note that most of these interconnecting leads are approximately 6 inches in length.

7. Wire L2A, L2B, R17A, and R17B to the rear of transducers TD1 through 4, as shown in Figure 28-8. Note securing the two-conductor speaker leads is, done via a small nylon clamp.

8. Check the wiring for accuracy, correct components, the quality of the solder joints, short circuits, foil shorts on the circuit board, pinched wires, and debris.

Final Assembly

To complete the assembly, follow these steps:

1. Fabricate the chassis from a piece of 6- × 7- × .063-inch aluminum, as shown in Figure 28-8. Bend up $1^1/_2$-inch sections for the front and rear panels. Then bend up $^1/_2$-inch flanges to mount the cover via four sheet metal screws (SW2). The location of most of the mounting holes is not critical and may be "eyeballed."

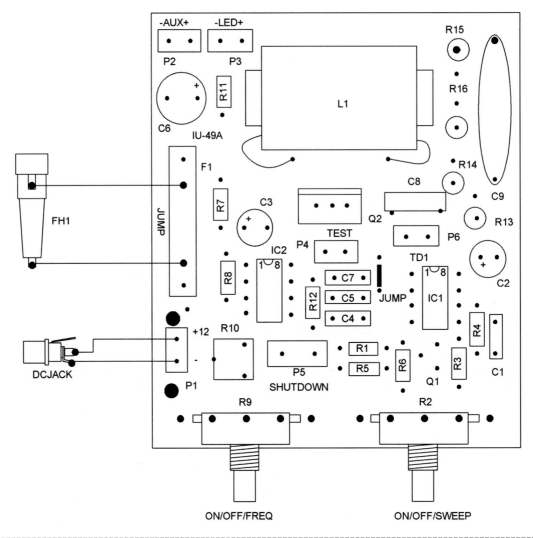

Figure 28-5 *PCB wiring second level*

However, the holes for R2 and R9 should be measured to match the PC1 PCB pads.

2. Fabricate the cover from a piece of $8^{1}/_2$ - × 5- × $^{1}/_{16}$-inch-thick plastic or aluminum. Form it to fit over the chassis and secure it via mating holes for the SW2 screws.

3. Fabricate a $4^{1}/_4$- × $3^{1}/_2$-inch piece of plastic or cardboard insulating material for placement between the circuit boards and the metal chassis.

4. Finally, assemble the circuit boards and the panel-mounted components to the chassis as shown in Figures 28-4 and 28-5. Twist the leads and dress as shown. Use a small piece of sticky tape to hold them in place.

Test Procedure

Testing the device can be done by following these steps:

1. Insert a 1-amp fuse into F1, connect a shorting plug into J3, as shown in Figure 28-9, preset controls to full *counterclockwise* (CCW), and click "off."

2. Apply 12 volts to the DC jack from a bench supply. You may use T1 or a 1-amp wall adapter if a supply is not available.

3. Connect a scope to pin 3 of IC2. Turn on R9 and note a square wave, as shown in Figure 28-2. Rotate R9 a full *clockwise* (CW) turn

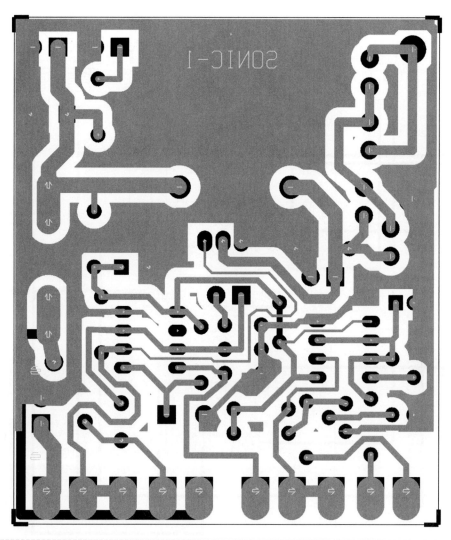

Figure 28-6 *View of foil routing and pad location*

and adjust R10 for a period of 50 *microseconds* (usec). This sets the high-frequency range to 20 kHz.

Obtain hearing protection for the next steps.

4. Temporarily connect the four-transducer assemblies in parallel and connect to J6. Turn R10 a full CCW turn and note a piercing, shrill, and uncomfortable shriek. Adjust R10 CW and note the pitch increasing to an upper-frequency limit.

5. Turn on R2 and note the frequency of the shriek periodically changing. Adjust and note a sweeping action varying from very slow all the way to a chirping sound.

Testing Detector-Sensing Circuit

To test the detector circuit, follow these steps:

1. Momentarily press S4 (RESET) and the unit shuts down. Press S5 (TEST) and the unit turns on. Repeat this action several times to verify proper operation.

2. With the unit in the reset mode, temporarily remove the shorting plug from J3 and note the unit turning on.

3. In the reset mode, temporarily short J2 to ground and note the unit turning on.

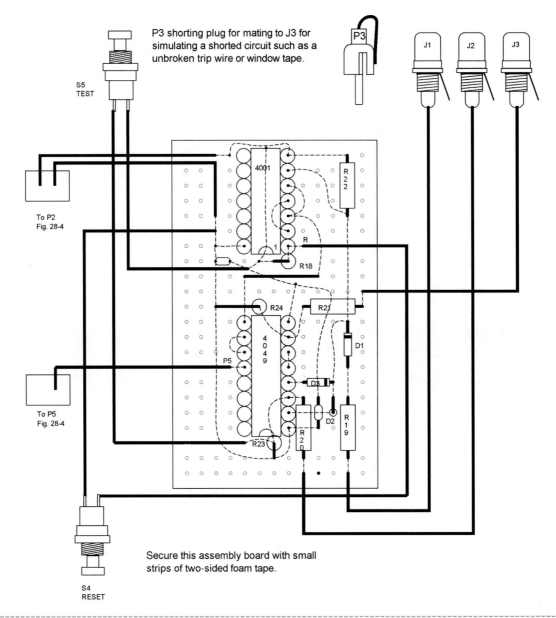

P3 shorting plug for mating to J3 for simulating a shorted circuit such as a unbroken trip wire or window tape.

S5 TEST

To P2 Fig. 28-4

To P5 Fig. 28-4

P3

J1 J2 J3

4001

R22

R

R18

R24

R21

P5

D1

D3

D2

R19

R23

R20

S4 RESET

Secure this assembly board with small strips of two-sided foam tape.

Figure 28-7   *Sense board layout and wiring*

4. Still in the reset mode, temporarily connect a 5-volt level to J1 and note the unit turning on.

for a particular range of frequencies, beam spread, and placement.

Special Notes

The transducers used in this system are piezoelectric and consequently are many times more efficient than the electromagnetic type. Several operational curves and charts are available and intended for those who may want to modify or optimize the existing circuitry

Application and Setup

Your phaser property guard system is capable of operating in two modes. Mode 1 is at a frequency known to produce paranoia, nausea, disorientation, and many other physiological effects. Mode 2 enables you to use the system as an audible alarm to frighten off intruders or warn the user. Both modes can be

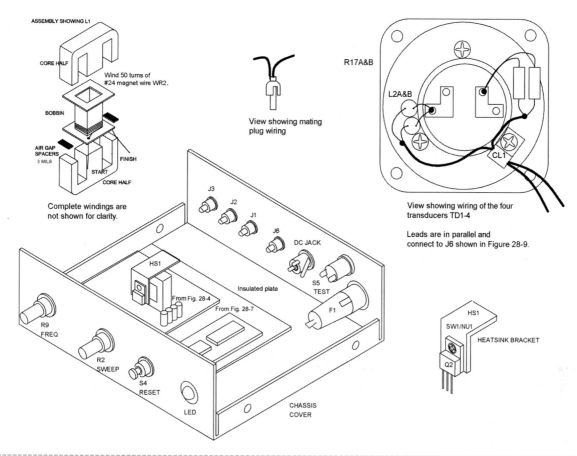

Figure 28-8   *Final assembly*

used in combination and are easily front-panel-controlled by the user. Three separate jacks enable the detection of a broken trip wire or contact foil, a pressure or actuating switch, and a positive voltage-level pulse obtained from other detection equipment such as those listed in the Information Unlimited catalog.

The position of the transducers should be set to direct as much energy as possible to the points of intrusion or access. They can be directed to any target area or be individually placed for multiple effects.

The power unit should be placed where the user can reset the device and preset the controls for maximum effect. Figure 28-9 shows the connections to the rear panel.

A Word of Caution

Ultrasonic is a gray area in many respects when an application involves the control of animals or even a

deterrent to unauthorized intrusion. It is always best to consult with local, municipal, and state laws before using this device to protect your home or property. Remember that many state laws lean more toward the right of the criminal rather than the victim.

General Information on Ultrasonics

Numerous requests have been made for information on the effect of these devices on people.

None of these devices have the ability to stop a person with the same effect as a gun, club, or more conventional weapon. They will, however, produce an extremely uncomfortable, irritating, and sometimes painful effect in most people. Not everyone will experience this effect to the same degree. As stated, young women are much more affected than older men due to being more acoustically sensitive. The range of the

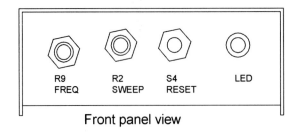

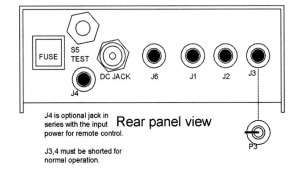

J4 is optional jack in series with the input power for remote control.

J3,4 must be shorted for normal operation.

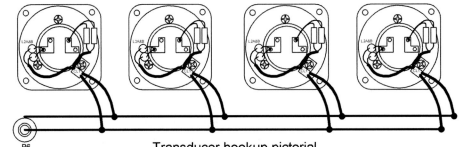

Figure 28-9   *Controls and hookup instructions*

devices depends on many variables and is normally somewhere between 10 and 100 feet from the transducers.

One possible use of the device (that deserves careful consideration) is for all the transducers to protect an area from unauthorized intrusion. This approach is excellent for protecting target areas such as jewelry boxes, gun cabinets, and safes and vaults. Normal use is to place each transducer to cover a given area.

Table 28-1 Phaser pain field property guard parts

Ref. #	Qty.	Description	DB#
R1, 6, 8, 18, 19, 20, 21	7	1K, $\frac{1}{4}$-watt resistor (br-blk-red)	
R2/S2		500K pot and switch	
R3		2.2K, $\frac{1}{4}$-watt resistor (red-red-red)	
R4, 5	2	10K, $\frac{1}{4}$-watt resistor (br-blk-or)	
R7		10-ohm, $\frac{1}{4}$-watt resistor (br-blk-blk)	
R9/S1		10K pot and switch	
R10		5K trimmer resistor	
R11		470-ohm, $\frac{1}{4}$-watt resistor (yel-pur-br)	Ohms
R13, 14, 15, 16	4	.47-ohm, 3-watt resistor (yel-pur-sil)	
R17A and B		120-ohm, 1-watt resistor (br-red-br)	
R22, 23, 24	3	39K, $\frac{1}{4}$-watt resistor (or-wh-or)	
C1, 4	2	.01 mfd, 50-volt disc capacitor	
C2		10 mfd, 25-volt electrolytic capacitor	
C3		100 mfd, 25-volt electrolytic capacitor	
C5		.033 mfd, 50-volt polyester capacitor	
C6		1000 mfd, 25-volt electrolytic capacitor	
C9		.22 mfd, 250-volt polypropylene capacitor	
C10, 11	2	.1 mfd, 50-volt polyester capacitor	
L1A and B	2	1-millihenry .25-amp inductor	IU1MH
L2		1-millihenry choke (see Figure 28-8)	IU1MHPPC
IC1, 2	2	555 DIP timer IC	
IC3		4001 Norgate *complementary metal oxide semiconductor* (CMOS) DIP IC	
IC4		4049 inverter/buffer CMOS dual in-line package integrated circuit	
Q1		PN2907 GP PNP transistor	
Q2		IRF540 power MOSFET TO220	
D1, 2, 3	3	IN914 silicon diodes	
LED		Light-emitting diode	
TD1, 2, 3, 4	4	Polarized high-output ceramic transducers	IUMOTRAN
S4, 5	2	Momentary pushbutton switches	
J1, 2, 3, 6	4	RCA chassis mount phono jacks	
P1, 2, 3, 6	4	RCA phono mating phono plugs	
DCJACK		2.1-millimeter DC jack	
F1		Fuse holder panel mount	
LEDRET		LED retainer bushing	

Table 28-1 Continued

Ref. #	Qty.	Description	DB#
PCPPP1		PCB	IUPCSONIC
PB1		Alternate 2- × 3-inch .1-inch grid perforated vector board	
HS1		Heatsink bracket fabricated as shown	
SW1/NU1		6-32 × $^3/_8$-inch screw and nut for HS1 attaching to Q2	
SW2	4	#6 × $^3/_8$-inch sheet metal screw for cover	
FEET	4	$^1/_2$-inch stick-on rubber feet	
CHASSIS		Metal chassis fabricated as shown	
COVER		Plastic cover to fit chassis	

Index

Index

Index

R

S

T

Index